Jacques Louis Lions (Ed.)

Numerical Analysis of Partial Differential Equations

Lectures given at a Summer School of the
Centro Internazionale Matematico Estivo (C.I.M.E.),
held in Ispra (Varese), Italy,
July 3-11, 1967

Springer

FONDAZIONE
CIME
ROBERTO CONTI

C.I.M.E. Foundation
c/o Dipartimento di Matematica "U. Dini"
Viale margagni n. 67/a
50134 Firenze
Italy
cime@math.unifi.it

ISBN 978-3-642-1105-6 ISBN 978-3-642-11057-3 (eBook)
DOI 10.1007/978-3-642-11057-3
Springer Heidelberg Dordrecht London New York

Reprint of the 1st Ed. C.I.M.E., Ed. Cremonese, Roma 1968
With kind permission of C.I.M.E.

Printed on acid-free paper

Springer.com

CENTRO INTERNAZIONALE MATEMATICO ESTIVO

(C.I.M.E.)

2° Ciclo - Ispra (Varese) dal 3-11 Luglio 1967

NUMERICAL ANALYSIS OF PARTIAL EQUATIONS

Coordinatore : J.L. LIONS

CENTRO INTERNAZIONALE MATEMATICO ESTIVO

- S. ALBERTONI -

" ALCUNI METODI DI CALCOLO NELLA TEORIA DELLA DIFFUSIONE DEI NEUTRONI"

Corso tenuto ad Ispra dal 3-11 Luglio 1967

ALCUNI METODI DI CALCOLO NELLA TEORIA A MULTIGRUPPI DELLA DIFFUSIONE DEI NEUTRONI

- S. Albertoni -

PARTE 1[a] - Teoria stazionaria a multigruppi, (1,2)

§ 1 - Notazioni e Problemi

Sia $\Omega \subset R_n$ un aperto limitato e $x \equiv (x_1, x_2, \ldots x_n) \in \Omega$. Nel seguito considereremo gli spazi $L^2(\Omega)$, $H^1(\Omega)$, $H^1_0(\Omega)$, e se g è intero >o, $(L^2)^g, (H^1)^g, (H^1_0)^g$ saranno i g - prodotti diretti di L^2, H^1, H^1_0.

Volendo considerare principalmente problemi di "trasmissione" supporremo $\Omega = \bigcup_1^r \Omega_i$, e siano $\Gamma = \partial\Omega$, $\Gamma_i = \partial\Omega_i$ rispettivamente le frontiere di Ω, Ω_i: $\gamma_{rs} = \partial\Omega_r \cap \partial\Omega_s$ sono le "interfacce", e cioè le parti di frontiera comuni a $\bar{\Omega}_r, \bar{\Omega}_s$.

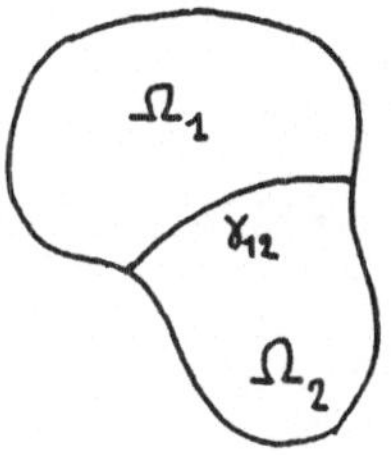

Infine ν_r sia la normale, diretta verso l'esterno, di $\partial\Omega_r$. Premesso ciò supponiamo assegnate le funzioni reali: $D^i_{pq}(x)$, $A^i(x)$, $C^i(x)$ $(c^1 \equiv o)$ $B^i(x), f^i(x)$. $(p,q = 1,2,\ldots n;\ i=1,2,\ldots g)$.

I problemi da risolvere sono due, e cioè:

Problema A - Trovare la soluzione $\boldsymbol{u}(x)$ del sistema:

$$
(1)\qquad
\begin{aligned}
&- \sum_1^n{}_{pq} \frac{\partial}{\partial x_p} \left(D^i_{pq} \frac{\partial \boldsymbol{u}^i}{\partial x_q} \right) + A^i \boldsymbol{u}^i = c^i \boldsymbol{u}^{i-1} + \frac{1}{\lambda} \omega^i(x) + f^i \\
&\omega^i(x) = \chi_i \sum_1^g{}_{j} \nu B^j(x) \boldsymbol{u}^j \ ; \ (i=1,2,\ldots g),
\end{aligned}
$$

soddisfacente a $u|_\Gamma = 0$, e alle condizioni di "trasmissione"

$$(1)' \qquad u^r = u^s \qquad \frac{du^r}{d\nu_r} = -\frac{du^s}{d\nu_s}$$

valide per $x \in \gamma_{rs}$, essendo $\frac{d}{d\nu_r}$ la derivata conormale a $\partial\Omega_r$ e u^r, u^s le restrizioni di u a Ω_r e Ω_s rispettivamente.

Osservazione I - Dal punto di vista fisico le (1) forniscono la distribuzione stazionaria del flusso neutronico entro un reattore Ω (composto da regioni di materiali diversi Ω_i) ripartito in g gruppi di energia decrescente (u^i è il flusso dei neutroni di i^ma^ energia) ove f^i sono le sorgenti neutroniche esterne e:

D^i un "coefficiente" di diffusione

A^i un "coefficiente" di assorbimento

C^i un "coefficiente" di rimozione dal gruppo i-1 al gruppo i (Ecco perchè $C^1 \equiv 0$)

B^i un "coefficiente" di fissione.

Osservazione II - E' utile per il seguito dare più esplicitamente la struttura formale del sistema (1):

$$(2) \qquad \begin{aligned} &-\sum_1^n{}_{pq} \frac{\partial}{\partial x_p}\left(D^1_{pq}\frac{\partial u^1}{\partial x_q}\right) + A^1 u^1 = \frac{1}{\lambda}\,\omega^1 + f^1 \\ &-\sum_1^n{}_{pq} \frac{\partial}{\partial x_p}\left(D^2_{pq}\frac{\partial u^2}{\partial x_q}\right) + A^2 u^2 = C^2 u^1 + \frac{1}{\lambda}\,\omega^2 + f^2 \\ &- \\ &-\sum_1^n{}_{pq} \frac{\partial}{\partial x_p}\left(D^g_{pq}\frac{\partial u^g}{\partial x_q}\right) + A^g u^g = C^g u^{g-1} + \frac{1}{\lambda}\,\omega^g + f^g \end{aligned}$$

Ipotesi sul problema A:

1) Positività dei coefficienti D^i, A^i, B^i, C^i (supposti misurabili a limitati in Ω) e cioè: $A^i(x)\geq a>o$, $B^i(x)\geq b>o$, $C^i(x)\geq c>o$ $(i=2,3\ldots)$, $D^i_{pq}(x)\geq d>o$; χ_i, ν sono coefficienti numerici $>o$. Inoltre si suppone $C^1(x) \equiv 0$.

2) Ellitticità uniforme e cioè: $\forall \{\lambda_p;\ p=1,2\ldots n,\ \lambda_p \in R_1, x\in\bar{\Omega}\}$ esistono m, $M>o$ tali che:

$$m \sum_1^n{}_p \lambda_p^2 \leq \sum_1^n{}_{pq} D^i_{pq} \lambda_p \lambda_q \leq M \sum_1^n{}_p \lambda_p^2$$

per ogni $i=1,2\ldots g$. Inoltre Ω sarà abbastanza regolare da garantire che tutte le operazioni di traccia abbiano un senso.

Problema B - Trovare il massimo autovalore $>o$, λ_o, e la corrispondente autosoluzione positiva $u_o(x)>o$, per il Problema A omogeneo, e cioè per $f^i=o$ e $u|_\Gamma=o$; tal caso fisicamente corrisponde ad una ripartizione di neutroni autosostenentesi (reattore critico).

§ 2 - Soluzione del problema A

Cercheremo u in $H^1_o(\Omega)(u|_\Gamma=o)$: le condizioni di trasmissione appariranno come "condizioni naturali", automaticamente soddisfatte nella formulazione variazionale, come è ben noto nel caso d'una equazione sola $(g=1)$. Per $g>1$ il Pb non è autoaggiunto. Si è allora preferito seguire un ragionamento elementare basato sul fatto che $c^1(x)\equiv o$. Introdotte le forme lineari:

$$a(u,v)=\sum_{1}^{g}{}_i \int_\Omega (\sum_{1}^{n}{}_{pq} D^i_{pq} \frac{\partial u^i}{\partial x_p} \frac{\partial v^i}{\partial x_q} + A^i u^i v^i) d\Omega$$

(3) $$b(u,v) = \sum_{1}^{g}{}_i \int_\Omega \omega^i(x) v^i d\Omega \qquad ; \qquad \forall\, u,v \in (H^1_0)^g$$

$$c(u,v) = \sum_{1}^{g}{}_i \int_\Omega c^i(x) u^{i-1} v^i d\Omega$$

siano A, B, C i corrispondenti operatori (matriciali) generati (nel senso delle distribuzioni). Per le ipotesi fatte sui coefficienti essi risultano definiti su tutto $(H^1_0)^g$ con codominio $\mathbf{C}(H^{-1})^g$. Gli elementi di matrice corrispondenti sono:

$$A_{ij} = \delta_{ij}\, A^i(x)$$

(4) $$B_{ij} = \chi_{i}\nu\, B^j(x) \qquad ; \qquad (i,j=1,2,\ldots g)$$

$$C_{ij} = \delta_{i,j+1}\, C^j(x)$$

ove:

(4)' $$A^i\cdot \equiv -\sum_{1}^{n}{}_{pq} \frac{\partial}{\partial x_p}(D^i_{pq} \frac{\partial}{\partial x_q}\cdot) + A^i\cdot \quad ; \quad (i=1,2,\ldots g)$$

Introdotti A, B, C il Problema A si riconduce alla risoluzione dell'equazione in u:

(5) $$A_\mu u \equiv (A-\mu B-C)u=f \quad ; \quad u \in (H^1_0)^g \quad ; \quad \mu = \frac{1}{\lambda}$$

Il procedimento esistenziale è allora il seguente:
Se $\mu=o$ allora la:

(6) $A_0 u=(A-C)u=f$ è costituita (si osservi la struttura delle (2)) da g equazioni disaccoppiate ($C^1\equiv 0$) del tipo:

(7) $$A^i u^i \equiv -\sum_{1}^{n}{}_{pq} \frac{\partial}{\partial x_p}(D^i_{pq} \frac{\partial u^i}{\partial x_q}) + A^i u^i = C^i u^{i-1} + f^i$$

Assumendo $f \in (H^{-1})^g$ (o più semplicemente $(L^2)^g$) per le ipotesi

1) e 2)) esiste allora un operatore G_o^1 di Green(°) che ci fornisce $u^1 = G_o^1 f^1$ (G_o^1 è un isomorfismo di $H^{-1}(\Omega)$ su $H_o^1(\Omega)$). Trovato u^1, u^2 è fornita da: $u^2 = G_o^2(C^2u^1+f^2) = G_o^2(C^2G_o^1f^1+f^2)$, e in generale si ha: $u^i = G_o^i(C^iG_o^{i-2}f^{i-1}+f^i) \in H_o^1(\Omega)$.

Ne consegue l'esistenza (μ=o) di un operatore matriciale di Green, G_o, per il nostro problema, che realizza un isomorfismo di $(H^{-1})^g$ su $(H_o^1)^g$, e $u=G_o f \in (H_o^1)^g$ soddisfa "naturalmente" alle condizioni di trasmissione.

Il caso $\mu \neq$ o si tratta riducendo la (5) nella forma:

(9) $$u = G_o f + G_o(Bu) = g+\mu Tu \quad ; \quad (g = G_o f \quad ; \quad G_o B = T)$$

Essendo G_o un operatore limitato da $(L^2)^g \to (H_o^1)^g$, B_{ij} moltiplicatori in $L^2(\Omega)$, e l'immersione di $(H_o^1)^g \to (L^2)^g$ compatta, allora T è compatto da $(H_o^1)^g \to (H_o^1)^g$, e pertanto il Problema A è ricondotto ad un classico Problema di Riesz-Fredholm. Lo spettro puntuale $\{\mu_j\}$ ammette l'unico punto di accumulazione all'infinito, e i $\lambda_i = \frac{1}{\mu_i}$ si accumulano solo verso zero.

§ 3 - Soluzione del Problema B

a) In tal caso si ha:

(10) $$(A-C)u \equiv Lu = \mu Bu \quad ; \quad \lambda Lu = Bu$$

(°) $G_o = (A-C)^{-1} = A_o^{-1}$)

Per risolvere la (10) si è trovata una rappresentazione a nucleo (funzione di Green) dell'operatore L^{-1}. Se $g=1$ la funzione di Green $G^1(x,y)$ è stata trovata, nel caso dei coefficienti discontinui, e per la 1^a volta, da Stampacchia(°) come soluzione di:

$$(11)\qquad A^1G^1(x,y) = \delta_{x,y} \qquad (\delta_{x,y} \text{ misura di Dirac in } (x\equiv y))$$

Allora

$$u = \int_\Omega G^1(x,y)\, f(y)dy \text{ è soluzione di } A^1u=f \text{ e si ha}$$

$$\int_\Omega G^1(x,y)A^1u(y)dy\ u(x) \quad ; \quad \forall\, u\ \ H_0^1(\Omega)$$

$$(10)\qquad G^1(x,y)>0 \;,\; G^1(x,y) = G^1(y,x) \;;$$

$$G^1(x,y) = \frac{M^1(x,y)}{(x-y)^{n-2}} \quad ; \quad \forall\, (x,y)\ \ \Omega' \text{ compatto} \subset \Omega \;.$$

<u>Osservazione</u> - Questo fatto di $\Omega'\subset\Omega$ non ci permette (benchè plausibile) di concludere che $G^1 \in L^2(\Omega\times\Omega)$ (e i suoi iterati).

Allora sfruttando il fatto che per l'operatore L si possono (vedi (6)) fare dei ragionamenti per singole equazioni (essendo queste disaccopiabili) si può costruire subito una matrice "formale" di Green di elementi ℓ_{iK}^{-1}:

$$\ell_{ik}^{-1}\cdot = \begin{cases} 0 & ; \quad i<K \\ \int_\Omega dy\ G^i(x,y)\,\cdot & ; \quad i=k \\ \ell_{ii}^{-1}\ C^i(x)\ \ell_{i-1,k}^{-1} & ; \quad i>k \end{cases}$$

(°) Si osservi che Habetler e Martino [8] già nel 1958 avevano considerato il Problema B assumendo però formalmente l'esistenza della funzione di Green.

Ad esempio:

$$(11) \qquad \ell^{-1}_{21}(\cdot)=\int_\Omega dy\, G^2(x,y)\, C^2(y)\int_\Omega dz\, G^1(x,y)(\cdot)(=\int_\Omega dy G^{21}(x,y)(\cdot)$$

se i G^i fossero $\in L^2(\Omega x\Omega)$.

Si verifica poi subito che $L^{-1}\equiv\{\ell^{-1}_{ik}\}$ è tale che $L^{-1}Lu\equiv u, u\in (H^1_0)^g$, e pertanto $L^{-1}\equiv\{\ell^{-1}_{ik}\}$ è la matrice di Green del problema $Lu = \phi$.

Osservazione - Per le ipotesi di positività fatte sulle C^i l'operatore L^{-1} lascia invariato il cono, in $(L^2)^g$, dei vettori >o, come pure, per le ipotesi sulle B^i, l'operatore $L^{-1}B$. (compatto in $(L^2)^g$).

Ne consegue subito, per noti risultati di Krein-Rutman, (1), che esiste un autovalore massimo (dominante e semplice) λ_o del problema, $\lambda_o u_o = L^{-1}Bu_o$, positivo e maggiore del valore assoluto di ogni altro autovalore, al quale corrisponde un'autosoluzione u_o pure >o in Ω.

b) - Determinazione iterativa di λ_o.

Richiamiamo il Teorema di I. Marek. (6)
Se H,K sono due operatori lineari con D_H, $D_K \subset K$, spazio di Hilbert, e se K è limitato e H^{-1} esiste limitato da $X\to D_H$, e $H^{-1}K$ è limitato con autovalore dominante λ_0 (e autosoluzione x_0) allora il processo iterativo di Rayleigh-Kellog descritto dalle:

$$u^{(o)} = x^{(o)} \qquad \text{(approssimazione zero)}$$

$$(12) \qquad v^{(m)} = K\, u^{(m)} \quad ; \quad H\, u^{(m+1)} = v^{(m)} \quad ; \quad u_{(m)} = \frac{u^{(m)}}{(u^{(m)}, u^{(m)})}$$

$$\lambda_m = \frac{(u^{(m+1)}, u^{(m+1)})}{(u^{(m)}, u^{(m)})} \quad ,$$

converge:

$$\lambda_m \to \lambda_o \quad , \quad u_{(m)} \to x_o$$

Per avere una determinazione iterativa di λ_0 basta applicare tale Teorema nel nostro caso K=B, H=L, $X\equiv(L^2)^g$.

Si osservi ora che il calcolo dell'iterazione m+1 si deve risolvere un'equazione del tipo:

$$(13)\quad -\sum_{1}^{n}{}_{pq}\frac{\partial}{\partial x_p}\left(D^i_{pq}\frac{\partial u^{(m+1),i}}{\partial x_q}\right)+A^i\,u^{(m+1),i}-C^i\,u^{(m+1),i-1}=v^{(m)}$$

e le (13), come al solito disaccoppiandosi, permettono di trovare le $u^{(m+1),i}$ una per volta (i =1,2,...g) risolvendo problemi del tipo:

$$-\sum_{1}^{n}{}_{pq}\frac{\partial}{\partial x_p}\left(D^1_{pq}\frac{\partial u^{(m+1),1}}{\partial x_q}\right)+A^1u^{(m+1),1}=v^{(m)}=\tilde{f}_1^{(m)}\quad\text{(nota)}$$

$$(14)\quad -\sum_{1}^{n}{}_{pq}\frac{\partial}{\partial x_p}\left(D^2_{pq}\frac{\partial u^{(m+1),2}}{\partial x_q}\right)+A^2u^{(m+2),2}=C^2u^{(m+1),1}+v^{(m)}=\tilde{f}_2^{(m)}\quad\text{(nota)}$$

- -

A ciascuna iterazione si devono perciò risolvere problemi di tipo ben noto. Questo consente perciò di applicare tecniche svariate studiate per problemi del tipo $A^iu=f$.

Ad esempio la $u^{(m+1),i}$ può essere trovata, [9], minimizzando il funzionale:

$$(15)\quad I^i(\phi)=\tfrac{1}{2}\int_\Omega\left(\sum_{1}^{n}{}_{pq}\,D^i_{pq}\,\frac{\partial\phi^i}{\partial x_p}\,\frac{\partial\phi^i}{\partial x_q}+A^i(\phi^i)^2-2\,\tilde{f}_i\phi\right)d\Omega$$

essendo $\tilde{f}_i$ nota dalla precedente iterazione.
Ne consegue che adattando questo metodo si ha un ciclo doppiamente iterativo-variazionale per trovare u_0, λ_0.

Evidentemente la $u^{(m+1),i}$ può essere trovata anche col metodo delle differenze finite, e appunto si sono fatte delle esperienze numeriche comparative al riguardo.

§ 4 - Esperienze numeriche

a) Risoluzione del Problema (15) attraverso il metodo di Riesz.

Se si assume una"base" finita $F_\nu(x)$; $\nu=1,2,\ldots N$, e se si rappresentano le ϕ^i come $\phi^i = \sum_1^N \alpha_{i,\nu} F_\nu$, le condizioni di minimo per il funzionale (15) diventano:

$$\frac{\partial I}{\partial \alpha_{i,\nu}} = 0 \quad ; \quad \begin{array}{l} i = 1,2,\ldots g \\ \nu = 1,2,\ldots N \end{array}$$

Queste ci danno un sistema algebrico lineare del tipo (per ogni i):

$$R^i \alpha^{(m+1),i} = \beta^{(m),i} \quad ; \quad \alpha = (\alpha_1^{(m+1),i} \ldots \alpha_N^{(m+1),i}) \quad ; \quad (i=1,2,\ldots g)$$

$$\beta^{(m),i} = \frac{\chi_i}{\lambda^{(m)}} \sum_1^g{}_j S^j \alpha^{(m),i} + Q^i \alpha^{(m+1),i-1} \quad ,$$

essendo R,S,Q matrici NXN di elementi noti.
Ad esempio:

$$S^i_{\mu\nu} = \sum_1^r{}_\ell \int_{\Omega_\ell} B^i F_\mu F_\nu \, d\Omega_\ell$$

L'inversione delle matrici R è stata fatta con l'algoritmo di Gauss.

b) Confronti tra il metodo iterativo variazionale descritto e quello delle differenze finite (assunto come elemento di confronto).

Caso monodimensionale: g=2, r=5, D,A,B,C funzioni costanti a tratti; N=20.

λ_o (con differenze finite) = λ_{df}

F_ν sono le autosoluzioni dell'operatore di Laplace monodimensionale

λ_o (con il nostro metodo) = λ_N

Ecco una tabella riferentesi ai vari casi:

λ_N	λ_{df}	e%
1,082447	1,082678	0,00637
1,226778	1,226704	0,00603
1,000041	1,000011	0,00299
0,439297	0,438788	0,116
1,141368	1,141318	0,00438

L'errore % è al più attorno allo 0,1%. Circa l'andamento delle soluzioni u^1, u^2 nei casi sperimentati si ha un accordo del nostro metodo con quello alle differenze finite sino a 3 cifre significative nelle zone centrali, e uno meno buono (2 cifre) nelle altre zone.

Caso bidimensionale - Risultati analoghi ai precedenti, perchè l'errore % (N eguale sia per l'asse x_1 che per l'asse x_2) nei nostri esperimenti non ha mai superato lo 0,3%. (F_ν sono prodotti di autosoluzione dell'operatore precedente).

Caso tridimensionale: g=2, r=3; D,A,B,C costanti a pezzi, simmetria rispetto ai piani x_1=o, x_2=o. (F_ν prodotti di autosoluzione come prima).

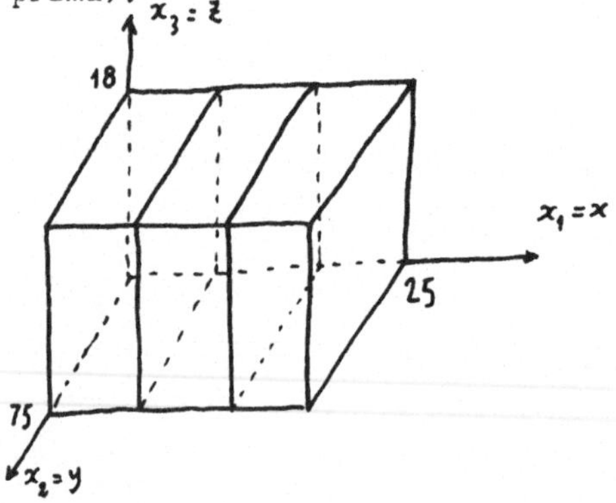

Risultati

a) $N_z = N_y = 1$, $N_x = 3$ e cioè N totale abbastanza piccolo ($N=N_xN_yN_z$): con 6 iterazioni (30" IBM 7090) λ_N approssima λ_{df} entro il 3,3%.

b) aumentando N_y da 3 a 10 l'e% si riduce al 3% (1'47" di macchina).

In generale $\lambda = \lambda_N$ è di tipo monotono (crescente in N), e in 15 iterazioni al più, nei casi considerati, si ha l'autovalore con la approssimazione cercata (1%) mentre è ben noto che con il metodo delle differenze finite il numero delle iterazioni sale, in genere, almeno a circa 50÷60.

u_1, u_2 (sulle rette y=6, Z=8 in figura) sono in buon accordo con i valori ottenuti a differenze finite, tranne nelle interfacce ove lo scarto è ∿ 2,3%.

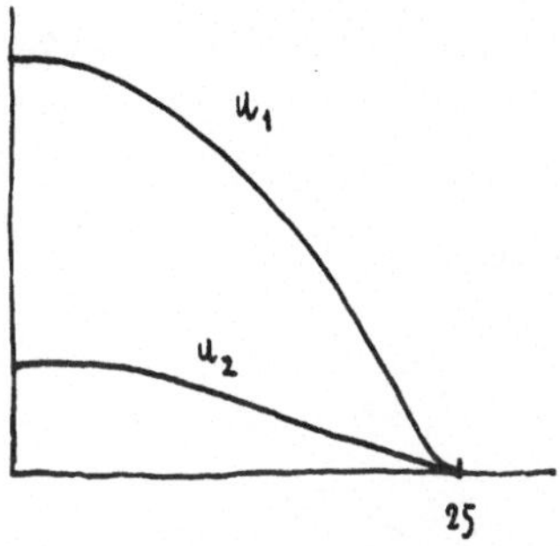

Osservazione - Il metodo iterativo variazionale fornisce l'autovalore massimo in modo assai soddisfacente, sia per precisione che per tempi di calcolatore (rispetto al metodo delle differenze finite), ma invece è inferiore a quest'ultimo metodo per la precisione della tabulazione della soluzione specie per quel che riguarda l'andamento della u_2 che può presentare dei "picchi" nelle zone non centrali (e vicino alle interfacce), picco a volte assai mal descrivibile col metodo variazionale.

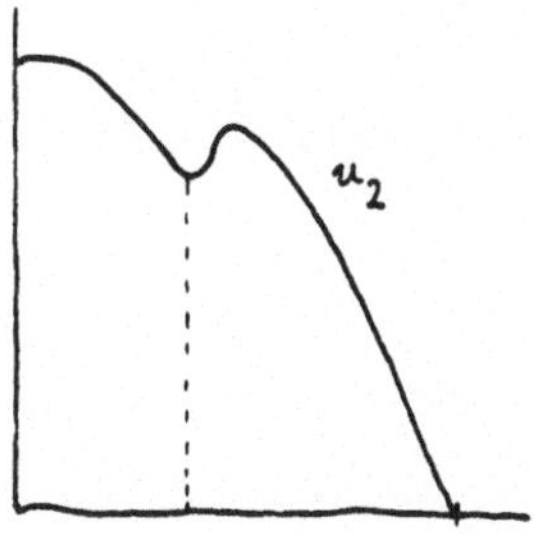

PARTE 2ª - Teoria a multigruppi dipendente dal tempo

§ 1 - E' ben noto che nella teoria della diffusione dei neutroni nell'approssimazione a più gruppi g di velocità, che supporremo due per semplicità, l'evoluzione del tempo dei flussi veloce e lento, rappresentati da u_1, u_2 e della concentrazione dei così detti "neutroni ritardati" rappresentata da C, è retta dal seguente sistema:

$$
(1)\qquad
\begin{aligned}
&\frac{1}{v_1}\frac{\partial u_1}{\partial t} - D_1\,\Delta u_2 + a_{11}\,u_1 - a_{12}\,u_2 - \lambda C = f_1\\
&\frac{1}{v_2}\frac{\partial u_2}{\partial t} - D_2\,\Delta u_2 + a_{22}\,u_2 - a_{21}\,u_1 = f_2\\
&\frac{\partial C}{\partial t} + \lambda C - h\,u_2 = f_3
\end{aligned}
$$

essendo assegnate:

1) i coefficienti (funzioni misurabili e limitate essenzialmente >0) e le funzioni di "sorgente" $f_{1,2,3}$,
2) le condizioni (Dirichlet) per le u_1, u_2 al contorno $\partial\Omega \times]0,T[$, $0<t<T$, $\Omega\subset R_n$,
3) le condizioni di "trasmissione" relative ad u_1, u_2 rispetto una interfaccia γ (una sola per semplicità) che rappresenta la frontiera comune a due subregioni Ω_1,Ω_2

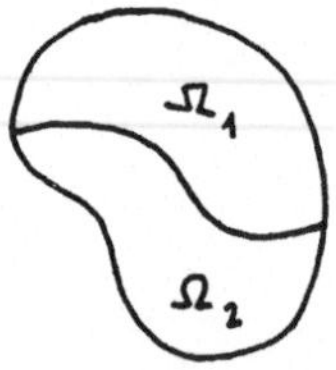

4) la condizione iniziale di Cauchy per tutte le incognite:

$u_{1,2}(x,o) = u^o_{1,2}(x)$

$C(x,o) = C^o(x)$

Il sistema stazionario associato ad (1) è ellittioo secondo

Nirenberg-Douglis, ma non secondo Petrowsky, e pertanto essenzialmente diverso, ad esempio, del sistema ellittico del tipo elasticità. Infatti, l'operatore matriciale associato è del tipo:

$$(2) \qquad A = \begin{vmatrix} -D_1\Delta + a_{11} & -a_{12} & -\lambda \\ -a_{21} & -D_2\Delta + a_{22} & 0 \\ 0 & -h & \lambda \end{vmatrix}$$

da cui si vede subito che sulla diagonale principale l'ordine degli operatori sono rispettivamente 2, 2, 0.
Risulta quindi <u>naturale</u> assegnare le condizioni al contorno (e di trasmissione) solo per u_2, u_2.

Inoltre essendo il sistema a coefficienti discontinui i risultati di Agmon-Douglis-Nirenberg per i sistemi ellittici (che esigono la continuità dei coefficienti) non si applicano. Appare perciò naturale, pensare ad una impostazione di tipo variazionale.
Per questo vedi un lavoro di prossima pubblicazione di Albertoni-Daneri-Geymonat. [5]

Non presenteremo, pertanto, qui risultati teorici circa (1) essendo le ricerche in corso, ma considereremo solo il problema discretizzato che si ottiene da (1) mediante il metodo delle linee (discretizzazione solo sulle variabili spaziali) ottenendo:

$$(2) \qquad V^{-1}\frac{d\phi}{dt} + B\phi = 0 \quad ; \quad \frac{d\phi}{dt} = Q\phi \quad , \quad Q = -VB \quad ,$$

essendo ϕ = vettore di R_{2N+N_1} ove N è il numero dei punti del reticolo spaziale (esclusa la frontiera) e N_1 è il numero dei punti ove $h(x) \neq 0$. Inoltre è:

$$V^{-1} = \begin{pmatrix} V_2^{-1} & 0 & 0 \\ 0 & V_2^{-1} & 0 \\ 0 & 0 & E \end{pmatrix} \qquad \text{ove:} \begin{cases} V_2^{-1} & \text{matrice diagonale NXN} \\ V_2^{-1} & \text{matrice diagonale NXN} \\ E & \text{identità in } R_{N_1} \end{cases}$$

$$B = \begin{pmatrix} L_1+A_{11} & -A_{12} & -\lambda E_2 \\ -A_{21} & L_2+A_{22} & 0 \\ 0 & -H & \lambda E \end{pmatrix}$$

essendo:

L_1	$\rightarrow$	$D_1\Delta$	una matrice pentadiagonale NXN ,
L_2	$\rightarrow$	$D_2\Delta$	una matrice pentadiagonale NXN ,
A_{11}	$\rightarrow$	a_{11}	una matrice diagonale NXN a coefficienti >o ,
A_{12}	$\rightarrow$	a_{12}	una matrice diagonale NXN a coefficienti >o ,
A_{21}	$\rightarrow$	a_{21}	una matrice diagonale NXN a coefficienti >o ,
A_{22}	$\rightarrow$	a_{22}	una matrice diagonale NXN a coefficienti >o ,
H	$\rightarrow$	h	una matrice N_1XN a coefficienti >o ,
E_1	$\rightarrow$	λ	una matrice NXN_1 a coefficienti >o .

Osservazione I - Lo studio del problema discreto (2) è egualmente molto importante in analisi numerica, perchè, a differenza di quanto accade nei casi parabolici, concernenti in generale più o meno la diffusione del calore, la matrice può avere autovalori >0, il che comporta a volte una crescenza molto rapida di ϕ cosa sempre delicata da controllare dal punto di vista numerico.

Inoltre le $v_{1,2}$ (in generale costanti) sono $\stackrel{\sim}{\sim} 10^6$, mentre gli altri coefficienti sono $\stackrel{\sim}{\sim} 1$, e pertanto dal punto di vista numerico si incontrano difficoltà simili a quelle che si hanno nei problemi di "boundary layer" connessi con equazioni differenziali contenenti piccoli parametri nelle derivate più alte.
Usando schemi impliciti (per ragioni di stabilità) si generano da (2) "grossi" sistemi lineari per i quali occorrono metodi iterativi la cui convergenza, che era da indagare, è stata verificata in. [3]

§ 2 - Proprietà del sistema (2).

In [3] sono stati ottenuti i seguenti risultati:

1) Q è irriducibile;

2) Q è essenzialmente >o;

3) Q possiede un autovalore $\omega_0 > -\lambda$ cui corrisponde un autovettore $v>o$, tale che se α_i è un qualsiasi altro autovalore è: $R\ \alpha_i < \omega_0$;

4) dalle 1), 2), 3), seguendo Birhoff-Varga [7], si dimostra $(t \to \infty)$: $\phi(t) = K\, e^{\omega_0 t} v + O\,(e^{\mu t})$ con $R\ \alpha_i < \mu < \omega_0$ (K dipende da $\phi(o)$);

5) le matrici di Jacobi e di Gauss-Seidel associate alla matrice $\alpha I - Q$ sono convergenti per $\alpha > \omega_0$.

§ 3 - Metodi di risoluzione di (2).

a) Metodo Esplicito: $\phi(t) = (I + \Delta t\ Q)\ \phi(t - \Delta t)$. Tale metodo è stato scartato nei nostri casi perchè ha una soglia di stabilità troppo bassa (Δt troppo piccolo).

b) Metodo Implicito: $\phi(t) = (I - \Delta t\ Q)^{-1} \phi(t - \Delta t)$.

Ad ogni passo temperale c'è da risolvere un sistema del tipo:

$$\left(\frac{I}{\Delta t} - Q\right) x = K \quad \text{(noto)}.$$

Se $\Delta t\ \omega_0 < 1$ allora, in base alla proprietà 5 del §2, i metodi di Jacobi e Gauss-Seidel relativi sono convergenti.

Osservazione - Quando si ha un "transiente" molto rapido si è trovato che anche il metodo implicito (e pure quelli di Crank-Nicolson e si Saulyev [4]) risultano molto imprecisi. Pertanto si pone il problema di trovare qualche metodo meno impreciso. La valutazione (4) ha fornito l'idea base per il seguente metodo che chiameremo metodo ω.

Nel metodo ω s'è pensato di esprimere la soluzione nella forma:

$$(3) \qquad \Phi(t) = e^{\omega_o t}\, \psi(t)$$

Allora la (2) si trasforma in:

$$(4) \qquad V^{-1}\frac{d\psi}{dt} + (B + \omega_o V^{-1})\psi = 0 \quad ; \quad \psi(o) = \psi(o)\ .$$

In ogni caso però c'è il problema di determinare ω_o che non è determinabile con procedimento tipo "metodo delle potenze Rayleigh-Kellog" non essendo l'autovalore quello di modulo massimo. Questa questione è abbastanza difficile dal punto di vista numerico, perchè è vero che si può tentare di "translare" lo spettro al fine di condursi ad un problema di autovalore di massimo modulo, ma così facendo, dovendo poi sottrarre il passo di traslazione, se questo è molto grande si può perdere ogni significato.

Allora posto $B=M-N$:

$$M = \begin{pmatrix} L_{11}+A_{11} & 0 & 0 \\ -A_{21} & L_2+A_{22} & 0 \\ 0 & -H & \lambda E \end{pmatrix} ; \quad N = \begin{pmatrix} 0 & A_{12} & \lambda E_1 \\ 0 & 0 & 0 \\ 0 & 0 & 0 \end{pmatrix}$$

ci si riduce, come equazione agli autovalori per ω_o (essendo ω_o autovalore di $-VB$) alla (5): $(M+\omega_o V^{-1})\, x = Nx$.
Introdotto un parametro fittizio μ si dimostra che l'equazione $(M+\omega V^{-1})x = \frac{1}{\mu} Nx$ possiede un autovalore di massimo modulo cui corrisponde un autovettore $>o$ ottenibile col metodo iterativo delle potenze. Quello che si dimostra è che $\mu = \mu(\omega)$ è monotona decrescente, e che $\mu=1$ individua ω_o.

Il sistema (4) è poi risolto con il metodo implicito Crank-Nicolson ed i relativi metodi iterativi risultano convergenti. [3]

§ 4 - Metodo dei passi frazionari

Recentemente, [5] , abbiamo pure sperimentato il metodo di Marchuck ed altri per la (1) (g=1) assumendo C=o e come decomposizione dell'operatore A una del tipo (vedi § 5) :

$$
\begin{aligned}
A &= A_1 + A_2 \\
A_1 &= \frac{\partial}{\partial x}\left(D \frac{\partial}{\partial x}\right) + \tfrac{1}{2} b \qquad (7)\\
A_2 &= \frac{\partial}{\partial y}\left(D \frac{\partial}{\partial y}\right) + \tfrac{1}{2} b
\end{aligned}
$$

Lo schema alternato per il passaggio da $t_n \to t_{n+1},\ t \in [t_n, t_{n+1}[$ è il seguente:

$$
\begin{aligned}
&\frac{1}{v}\frac{d\,u^{n+\frac{1}{2}}}{dt} + A_1\, u^{n+\frac{1}{2}} = 0 \\
&u^{n+\frac{1}{2}}(t_n) = u(t_n) \\
&\frac{1}{v}\frac{d\,u^{n+1}}{dt} + A_2\, u^{n+1} = 0 \qquad (8)\\
&u^{n+1}(t_n) = u^{n+\frac{1}{2}}(t_{n+1})
\end{aligned}
$$

fornente $u(t_{n+1}) = u^{n+1}(t_{n+1})$. La discretizzazione della (8) ci da poi sistemi del tipo:

$$
\begin{aligned}
&V^{-1}\frac{du}{dt} = \alpha_1 u \ ; \qquad V^{-1} \ \text{matrice diagonale;} \\
&V^{-1}\frac{du}{dt} = \alpha_2 u \ ; \qquad \alpha_1, \alpha_2 \ \text{matrici tridiagonali} \qquad (9)
\end{aligned}
$$

Osservazione - Un primo vantaggio del metodo è che abbiamo ora a che fare con matrici tridiagonali (invece di pentadiagonali) invertibili anche con metodi diretti.

§ 5 - Risultati numerici

Si è sviluppato un caso "test" per il quale si conosce la soluzione esatta e si sono fatti confronti con i metodi a), b), c)(§ 3).

Problema "test": $g = 1$ (e senza equazione in C).

$$\frac{1}{v}\frac{\partial u}{\partial t} = \mathrm{div}(d \; \mathrm{grad}\; u) + bu \equiv Au$$

$u|_{\Gamma} = 0$ e condizioni di trasmissione su γ_{12}

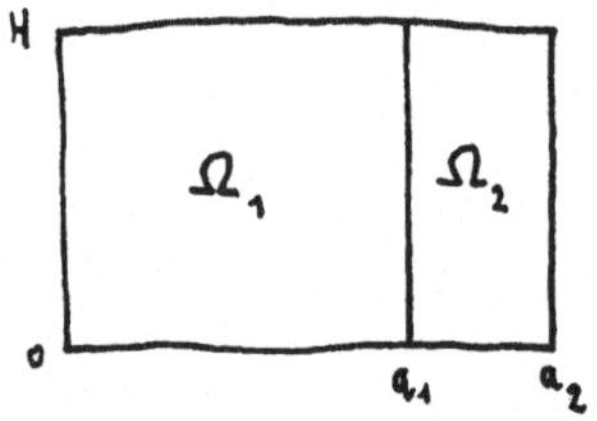

essendo

$$v = \begin{cases} v_1 \; ; \; x \in \Omega_1 \\ v_2 \; ; \; x \in \Omega_2 \end{cases} \quad ; \quad d = \begin{cases} d_1 \; ; \; x \in \Omega_1 \\ d \; ; \; x \in \Omega_2 \end{cases} \quad ; \quad b = \begin{cases} b_1 \; ; \; x \in \Omega_1 \\ b_2 \; ; \; x \in \Omega_2 \end{cases}$$

Se

$$\frac{d_1}{d_2} = \frac{v_1}{v_2} \; ; \quad b_2 = \{b_1 v_1 + (\tfrac{\pi}{H})^2 (d_2 v_2 - d_1 v_2)\}/v_2 \quad ; \; a = a_1 + (a_2 - a_1)\, \frac{d_1}{d_2}$$

allora:

$$u_j = \mathrm{sen}\, \frac{\pi}{H}\, y \cdot \begin{cases} \mathrm{sen}\, \frac{\pi}{a}\, jx \quad ; \quad x \in \Omega_1 \\ \mathrm{sen}\, \frac{\pi}{a}\, j(\frac{d_1}{d_2}\, x + (1 - \frac{d_1}{d_2}) a_1) \quad ; \quad x \in \Omega_2 \end{cases}$$

sono delle autosoluzioni di A, e $\omega_j = \{a_1 - d\, ((\frac{\pi}{a} j)^2 + (\frac{\pi}{H})^2)\}\, v_1$ sono i corrispondenti autovalori. Inoltre se $u(x,y,o) = \sum_{1}^{N} j\; C_j\, u_j$; si ha la soluzione esatta:

$$(6) \qquad u(x,y,t) = \sum_{1}^{N} j\; C_j\, e^{\omega_j t}\, u_j$$

L'ω_0 calcolato come indicato alla fine del § 3 risulta $\omega_0 = 63{,}69$, mentre il valore esatto è 63,21. (Le differenze finite sovrastimano l'autovalore e, a nostra conoscenza, ci sono risultati teorici per questa stima solo nel caso di coefficienti continui in $\bar{\Omega}$).

I risultati numerici sono riportati in tabella dove $\bar{u}$ è il valore medio su Ω della soluzione esatta, $\bar{u}_{M\omega}$ è la soluzione approssimata ottenuta con la trasformazione ω ed il metodo di Marchuck, $\bar{u}_M$ l'analoga soluzione senza trasformazione ω, $\bar{u}_{IM\omega}$ quello ottenuto con il metodo implicito che fa seguito alla trasformazione ω, mentre la $\bar{u}_{IM}$ è quella che si riferisce al metodo implicito diretto.

Come si vede in ogni caso la trasformazione ω, comunque sia associata ad altre tecniche, dà i risultati migliori, ed il metodo dei passi frazionari si è rivelato superiore, nei casi fatti, al metodo implicito.

TABELLA ($\Delta t = 10^{-3}$)

t	Soluzione esatta $\bar{u}$	$\bar{u}_{M\omega}$	$\bar{u}_{M}$	$\bar{u}_{IM\omega}$	$\bar{u}_{IM}$
.025	1.976	2.000	2.039	2.024	2.132
.030	2.708	2.749	2.815	2.780	2.960
.035	3.714	3.779	3.886	3.821	4.111
.040	5.094	5.200	5.366	5.252	5.711
.045	6.986	7.153	7.411	7.220	7.934
.050	9.583	9.838	10.23	9.926	11.02
.055	13.14	13.53	14.14	13.65	15.31
.060	18.03	18.62	19.52	18.76	21.28
.065	24.73	25.61	26.96	25.80	29.57
.070	33.92	35.23	37.24	35.47	41.08
.075	46.54	48.47	51.44	48.77	57.09
.080	63.83	66.69	71.05	67.06	79.32
.085	87.56	91.77	98.14	92.20	110.2
.090	120.1	126.68	135.22	126.8	153.1
.095	164.7	173.68	187.22	174.3	212.9
.100	226.0	238.89	258.59	239.7	295.6

Bibliografia

(1) S.Albertoni: "Metodi variazionali per certi sistemi di equazioni a derivate parziali" - Ist. Lombardo Scienze e Lettere Vol. 100-1966.

(2) S.Albertoni - M.Lunelli - G.Maggioni: "Metodi iterativi variazionali per problemi ellittici nella teoria dei reattori nucleari" Atti Seminario Mat. e Fis. Univ.Modena Vol.XIV,1965.

(3) A.Daneri - A.Daneri - I.Galligani: "A numerical approach to the time dependent neutron diffusion equations" EUR 3742e - 1968.

(4) I.Galligani: "Numerical solutions of the time dependent diffusion equations using the alternative method of Saul'yev" - Calcolo Vol.2, Suppl. 1, 111 - 1965.

(5) S.Albertoni - A.Daneri - G.Geymonat: "Existence and approximation theory for general differential equations of the multigroup diffusion reactor theory" (in corso di pubblicazione).

(6) I.Marek: "Iterations of linear unbounded operator in non self-adjoint eigenvalores problems and Kellog iteration processes" (Cech.Meth.Journal 12 (1962)).

(7) G.Birkoff - R.S.Varga: "Reactor Criticality and non negative matrices" J.Soc.Indust.Appl.Math. 6, 354-377 (1958).

(8) G.I. Habetler - M.A. Martino:"The multigroup diffusion equations of reactor physics, KAPL, 1886, July 28, 1958

(9) G.Stampacchia: "Su un problema relativo alle equazioni di tipo ellittico del 2° ordine" Ricerche di Mat., Vol. V (1956).

CENTRO INTERNAZIONALE MATEMATICO ESTIVO

(C.I.M.E.)

I. BABUŠKA

"PROBLEMS OF OPTIMIZATION AND NUMERICAL STABILITY IN COMPUTATIONS"

Corso tenuto ad Ispra dal 3-11 Luglio 1967

PROBLEMS OF OPTIMIZATION AND NUMERICAL STABILITY IN COMPUTATIONS [1)]

by

I. Babuška (Praga)

1. Introduction

Computer Science is a new scientific discipline. An important part of this discipline is the numerical mathematics. The "Art of Computation" is becoming science ; new questions and problems become important.

A typical problem is the problem of the creation of numerical methods, the determination of their "worth" and, in general, the choice of the most suitable method for the given purpose.

For example, the program-library in a computing centre contains mostly many algorithms for solving single mathematical problems. Opinions on the expedience of these algorithms are usually quite different and subjective. This statement is still more apparent when a method of applied mathematics is to be appreciated, especially in the field of scientific-technical computations. These scientific - technical computations are that part of the computer science in which I have some experience.

My paper will deal with questions which are more or less associated with this kind of computations.

I think that these computations may be characterized as a mathematical and constructive way of processing (transformation) of the given information to the required one [2)] . I am sure that in scientific-techical computations it is necessary to emphasize the knowledge of information which we may collect and the appreciation of its reliability. Further it is

1) In this paper some results obtained recently in Prague will be given

2) Henrici 23 defines numerical analysis as the theory of constructive methods in mathematical analysis (with emphasis on the word "constructive") .

necessary to formulate clearly the required information on the given problem . The necessity of a mathematical and constructive way of this processing is obvious here .

The "clarity" of the given and required information is an important part for a successful solution of a technical problem . Numerical mathematics are the rudiments of this constructive processing of information.

Numerical method generates (in a constructive manner) a mapping, from the class (space) of the given information to the class of the required one. It is important that this mapping is defined on the entire class of information. This class will be the domain of definition of the given method (mapping) .

Numerical process is an exact constructive law (prescription) of creation of the given mapping.

Computation is a concrete realisation of the numerical process in the given case. We shall talk about exact realisation when we compute without round-off errors and about a realization (or disturbed realisation) in a real computation.

It is obvious that there are many different manners of a constructive creation of one given mapping, i.e. many processes exist which transform the given information to the requested one and solve the same mathematical problem. It is evident that the question of choosing a process is very important.

It is clear that the choice and every optimization must necessarily be relative to the given information. This does not mean, however, that some methods might not be advantageous in a certain generality.

The manner in which we appreciate the method is of great importance. My experience is that, from the practical point of view, it is very important to respect an incredulity of the given information . This incredulity can be of different kinds . Some of them will be shown in the next part of the paper . It is essential that the method -and in general all conclusion - be stable with respect to these incredulities. I think that this stability is one of the most important points when choosing a method in practice.

In the next part I shall point out some aspects of these questions.

2. The problem of quadrature formulas [1)]

In this section I shall show some aspects of ideas, which I mentioned previously, in a simple case of quadrature formulas.

Let our task be to determine numerically

$$J(f) = \frac{1}{2\pi} \int_0^{2\pi} f(x)\, dx \, . \tag{2.1}$$

We shall suppose that we know the following about the integrated function $f(x)$:

1. The function $f(x)$ is a continuous periodic function with the period 2π .

2. We can evaluate only the function $f(x)$ (i.e. compute the values of $f(x)$) .

In this case, the simpliest quadrature formula $T_n(f)$ is mostly used in practice, with

$$T_n(f) = \frac{1}{n} \sum_{j=1}^{n} f\left(\frac{2\pi}{n} j\right) . \tag{2.2}$$

This formula is the well known trapezoid formula.

I will now analyse the question, if there are any reasons for selecting

---

1) In this part we are not dealing with the problems of the round-off errors.

the trapezoid formula ; we may ask e.g. why the Simpson-formula isn't better than the formula previously mentioned. Some arguments for choosing the trapezoid formula (in this case of integration of a periodic function) are included in some papers, e.g. Milne (25), Davis (18) and others.

The error bounds for the trapezoid formula are studied in many papers. See (4), (5), (21), (24), and others. We will now analyse the problem of the choice of the quadrature formula according to the information we mentioned previously. In our considerations we shall confine the class of possible formulae to the linear one.

The choice of the quadrature formula means, in our case, to determine of the sequence of linear functionals I_n in the form

$$(2.3)\qquad J_n(f) = \sum_{k=1}^{n} a_{k,n}\, f(x_{k,n}), \qquad 0 < x_{k,n} \leqslant 2\pi$$

with the requirement that $J_n(f) \to J(f)$ (weak) for all functions $f(x)$ of the given class of functions.

We shall measure the amount of work in using a formula by the number of evaluations of the integrated function.

Let us now assume that B is a Banach space. Then we can define

$$(2.4)\qquad \omega(n, B) = \inf_{\substack{a_j, y_j \\ j=1,\dots,n}} \ \sup_{\|f\|_B \leqslant 1} \left| \sum_{j=1}^{n} a_j f(y_j) - J(f) \right|$$

and

$$(2.5)\qquad \varrho(n, B) = \inf_{a_j} \ \sup_{\|f\|_B \leqslant 1} \left| \sum_{j=1}^{n} a_j f\left(\frac{2\pi}{n} j\right) - J(f) \right|$$

$\omega(n, B)$ is the minimal possible error under the assumption that we know only that $\|f\|_B \leqslant 1$. $\varrho(n, B)$ has analogous meaning when we confine ourselves to use equidistant points in the quadrature formula.

We shall further introduce

$$(2.6) \qquad \Lambda(n,B) = \sup_{\|f\|_B \leq 1} |T_n(f) - J(f)|.$$

$\Lambda(n,B)$ is evidently the error-bound of the trapezoid formula in the space B. An objective measure of convenience of the given formula is given here by the comparison of $\Lambda(n,B)$ with $\omega(n,B)$, $\varrho(n,B)$ resp. This appreciation is obviously relative to the space B.

The choice of the space B is very problematical in practice. In majority of cases there is a large incredulity as to whetherit is convenient to take the integrated function as an element of a certain space - B. If the conclusion on the suitability of a formula is strongly dependent on the choice of B, then the conclusion is not "stable" and it is not advantegeous to use that formula in practive. Further we shall see that this "unstability" will appear in the case of the optimal formula, i.e. when we use the formula whose error equals $\varrho(n,B)$ or $\omega(n,B)$ then the results will strongly depend on the space B. Conversely, a formula will be advantageous in practice if its error is nearly equal to $\varrho(n,B)$ or $\omega(n,B)$ but more or less independent of the space B.

Later we shall see that only the trapezoid formula which is not an optimal one, has this property. We now introduce a class of Banach spaces of periodic functions.

Definition 2.1 The Hilbert space H (over complex numbers) will be said to be periodic if:

1) Every $f \in H$ is a 2π periodic, continuous function.

2) Let $\|f\|_c$ signify the norm in the space C, then

$$(2.7) \qquad \|f\|_c \leq C(H)\,\|f\|_H.$$

3) If $f \in H$, then $g(x) = f(x+c) \in H$ for every real c and $\|f\|_H = \|g\|_H$.

The space H will be said to be strongly periodic if it is periodic and if :

4) $e^{ikx} \in H$, $k = \dots, -1, 0, 1, \dots$ and $\| e^{ikx} \|_H = \| e^{-ikx} \|_H$.

5) If $|j| \geq |k|$, then $\| e^{ijx} \|_H \geq \| e^{ikx} \|_H$.

6)

$$\| e^{i[n\alpha]x} \|^2_H \sum_{t=0}^{\infty} \| e^{i([n\alpha]+ tn)x} \|_H^{-2} \leq D , \tag{2.8}$$

for $0 \leq \alpha \leq 2$

and D does not depend on n.

At the beginning of this section it was said that f(x) is a periodic function. It is obvious that this information is insufficient. However, I think it is convenient to assume that the function f(x) is an element of a periodic or strongly periodic space H .

It is evident that now too we have a large incredulity as regards the concrete selection of the space H. The importance of this incredulity is well seen in the next theorem and example.

Theorem 2.1 Let H be a strongly periodic space with the norm

$$\| f \|^2 = \int_0^{2\pi} (|f|^2 + A |f'|^2) \, dx , \quad A > 0 . \tag{2.9}$$

Then the error-bound of the formula

$$R_n^{(A)}(f) = C(n, H) \, T_n(f) \tag{2.10}$$

where

$$C^{-1}(n, H) = 1 + \frac{2}{n^2} \sum_{t=1}^{\infty} \frac{1}{(tA)^2 + \frac{1}{n^2}}$$

is equal to $\varrho(n, H)$.

The theorem 2.1 affirms that the formula (2.10) is an optimal one if we are using the equidistant net. Now we shall introduce the following example :

Example 2.1 Let $f(x) = e^{\alpha \sin x}$, $= 3, 10$.

Then $J(f) = \frac{1}{2\pi} \int_0^{2\pi} f(x)\, dx = 4,88079258586502408\ldots$

resp . 2815,71662846625447.

In Tab. 2.1 we show the result obtained by the trapezoid formula $R_n^{(A)}$ for $A = 1$. From this table we see that an optimal formula used in an inconvenient space may give bad results. We see that the conclusion of the convenience of the optimal formula is very "unstable" with respect to the choice of H) . From this table we also see that the trapezoid formula (C = 1) gives very good results ; however, the following theorem is true :

Theorem 2.2 For every periodic space H

(2.11) $$\Lambda(n,H) > \varrho(n,H) .$$

This theorem shows that the trapezoid formula cannot be optimal in a periodic space. Nevertheless this formula is very advantageous in practice. The explanation of this fact can be seen in the following statement:

Theorem 2.3 . Let H be a periodic space. Then

(2.12) $$\lim_{n \to \infty} \frac{\Lambda(n,H)}{\varrho(n,H)} = 1 .$$

No other formula has the property that the left-hand side of (2.12) is bounded for all periodic spaces (except for a finite number of indices of n)

This theorem shows that the efficiency of the trapezoid formula is

Number of points	$T_n(f)$, $f = e^{\alpha \sin x}$		$R_n^{(1)}(f)$, $f = e^{\alpha \sin x}$	
n	$\alpha = 3$	$\alpha = 10$	$\alpha = 3$	$\alpha = 10$
8	4,88241999058958100	3047,90959481962441	4,64604604 ...	2900,35030...
16	4,88079258593666173	2815,77672896656761	4,81902223 ...	2780,14081...
24	4,88079258586502408	2815,71662897903758	4,85310536 ...	2799,74394...
Exact value	4,88079258586502408	2815,71662846625447		

Table 2.1

roughly the same as that of an optimal formula. This statement is now "stable" with respect to the choice of H. The asymptotic optimality (in the sense of (2.12)) of the trapezoid formula is valid only when the equidistant net is used, i.e. when we compare $\Lambda(n, H)$ with $\rho(n, H)$.

The situation becomes more complicated when we compare $\Lambda(n, H)$ with $\omega(n, H)$. What happens will be seen from the following theorems:

Theorem 2.4 For every sequence $\xi_1, \xi_2, \ldots, \xi_i, > 0$ there exists a periodic space H such that

$$\text{(2.13)} \qquad \limsup_{n \to \infty} \frac{\Lambda(n, H)}{\omega(n, H)\, \xi_n} = \infty$$

Theorem 2.5. Let H be a strongly periodic space. Then we have

$$\text{(2.14)} \qquad \limsup_{n \to \infty} \frac{\Lambda(n, H)}{\omega(n, H)\sqrt{n}} < \infty$$

Theorems 2.4 and 2.5 show that it is reasonable to demand the universal efficiency relative to the set of strongly periodic spaces. The necessity to confine ourselves to some kind of incredulity is natural. This is a general statement when dealing with every kind of incredulity. I think, however, that the previously mentioned theorems show well the role of the trapezoid and optimal formulas and the role of the incredulity.

We have been dealing with the analysis of the convenience of the trapezoid formula $T_n(f)$. Now let us mention the error of $T_n(f)$. Practically all the known error estimations are based on the choice of the space H (or on the more general space B).
The choice may be carried out a priori i.e. before the computation or a posteriori i.e. the choice is made with respect to the results obtained during the computation.

The error estimate is then

(2.15) $$\varepsilon_n(f, H) = \Lambda(n, H) \| f \|_H .$$

The norm $\| f \|_H$ has to be estimated (a priori or a posteriori).

There are many papers dealing with the estimation of $\Lambda(n, H)$ for different spaces, e.g. (5), (2), (24), (6), and others. Many results have been gathered in special books. See e.g. (26), (34) and others. With a suitable choice of H (resp B) we may obtain the estimates in C_n, the "derivative-free" estimates (22); for the estimation of $\omega(n, H)$ and $\varrho(n, H)$ see e.g. (5), (8), (21), (36). Moore's important and principal results (30), (31), (32) may also be understood as an a posteriori choice of B and an a posteriori estimation of the norm $\| f \|_B$. This a posteriori choice is made here by the computer. I want to emphasize that a choice of an a priori given class of possible spaces is given here. The problem of the optimal choice of a space in connection with the error estimation will now be discussed.

Let f(x) be a 2π-periodic continuous function and let $\not\subset \neq$ $\neq \mathcal{H}(f) = E(H; H \text{ periodic}, f \in H)$.

Then the following theorem is valid:

Theorem 2.6. Let $f(x) = \sum_{k=-\infty}^{\infty} a_k e^{ikx}$.

Then

(2.16) $$\varepsilon_n(f) = \inf_{H \in \mathcal{H}(f)} \Lambda(n, H) \| f \|_H = \sum_{\substack{t=-\infty \\ t \neq 0}}^{\infty} |a_{tn}|$$

There is often a simple way of obtaining this "best" estimation: We introduce

<u>Definition 2.2.</u> The function $g(x) = \sum_{k=-\infty}^{\infty} b_k e^{ikx}$ will be called an overfunction to the function $f(x) = \sum_{k=-\infty}^{\infty} a_k e^{ikx}$ if $b_k \geq |a_k|$, $k = \dots -1, 0, 1 \dots$

This will be denoted by $g \circledcirc f$.

The overfunction can often be simply constructed. E.g. $e^{\sin x}\sin^2 x \circledless \frac{1}{2} e^{\cos x}(1+\cos 2x)$. With this knowledge of the overfunction it is possible to get an estimation of $\varepsilon_n(f)$ in (2.16).

Theorem 2.7 Let $g \circledgtr f$ and g''' be a continuous function. Then

$$\varepsilon_n(f) \leqslant \eta_n(f) = \frac{1}{n^2} T_n(g'') \tag{2.17}$$

The efficiency of this estimation will be shown in the following example:

Example 2.2. Let $f(x) = e^{\alpha \sin x}, \alpha > 0$. Then $f(x) \circledless e^{\alpha \cos x}$. In table 2.2 we introduce the value $\eta_n(f)$ and $\sigma_n = |J(f) - T_n(f)|$ for $\alpha = 10$, $\alpha = 50$. From table 2.2 a good accordance between the estimation and the real error may be seen. This error estimation is closely related to the ideas of Dahlquist (19), (20).

We dealt with the analysis of the computation of (2.1). Similar ideas can be used for the computation of the Fourier coefficients

$$J_p(f) = \frac{1}{2\pi} \int_0^{2\pi} f(x)\, e^{ipx}\, dx. \tag{2.18}$$

We obtain a principal new problem when we want to compute simultaneously the values J_{p_j} $j = 1, \dots k$. Obviously the simplest way is to compute these values independently. There is a question if it is possible to gain something when we make the computations simultaneously.

I will show it in the simplest case. Let us assume that we will compute both values J_o and J_1 simultaneously. Put

$$\Omega^{(n)}_{o,1}(H) = \inf_{\substack{a_j, j=1,..n \\ g(p), p=0,1}} \max_{p=0,1} \sup_{\|f\|_H \leq 1} \left| \sum_{j=1}^{n} a_j\, g(p)\, f\left(\frac{2\pi}{n} j\right) - J_o(f) \right| \tag{2.19}$$

Number of points n	$f = e^{10 \sin x}$		
	$T_n(f)$	$J_n(f) - T_n(f)$	$\eta_n(f)$
8	3047, 909594819624415	232, 192966353369944	232, 373271956578 7845
16	2815, 776728966567611	0, 0601005003131402	0, 0601005003142606
24	2815, 716628979037584	0, 0000005127831140	0, 0000005127831167
32	2815, 716628466254842	0, 0000000000003720	0, 0000000000003743
	$f = e^{50 \sin x}$		
8	0, 6480887567505754520 +21	0, 3548333783656418203 +21	0, 5063081633888883995 +21
16	0, 3384555456320188415 +21	0, 0452001672473682524 +21	0, 0452789214681998668 +21
24	0, 2951999264551136014 +21	0, 0019445480701799697 +21	0, 0019445486761790910 +21
32	0, 2932816292532110631 +21	0, 0000262508682774314 +21	0, 0000262508682775479 +21
40	0, 2932554985285131181 +21	0, 0000001201435794864 +21	0, 0000001201435794864 +21
48	0, 2932554985285131181 +21	0, 0000000002019997072 +21	0, 0000000002019997072 +21

Table 2.2

$\Omega^{(n)}_{o,1}$ is apparently the minimal possible error in a simultaneous computation.

We shall analyse what can be gained by this kind of computation.

Let

$$(2.20) \qquad \Pi^{(n)}_{o,1}(H) = \max\left(\Lambda(n,H), \| J_1 \|_H\right)$$

This is the error if we compute J_o with the trapezoid formula and if we put $J_1(f) = 0$.

The following theorem may be proved :

<u>Theorem 2.8</u>. Let H be strongly periodic. Then

$$\lim_{n \to \infty} \sup \frac{\Pi^{(n)}_{o,1}}{\Omega^{(n)}_{o,1}(H)} \leq \sqrt{2}\ .$$

The theorem shows that we can gain practically nothing while performing a simultaneous computation. Theorem 2.8 is a special case of theorems which have been proved by P. Přikryl (33).

We analysed the case if only the function values were used in computing. All I said can be done if we use also the values of k derivatives. Here we shall assume besides (2.7) the the following :

$$(2.7) \qquad \| f^{(s)} \|_c \leq C_s(H) \| f \|_H \, , \quad s = 0, \ldots, k\ .$$

and

$$(2.8') \qquad \| e^{i[n\alpha]x} \|^2_H \sum_{t=o}^{\infty} \| e^{i([n\alpha]+tn)x} \|^{-2}_H (t+\alpha)^{2k} \leq D$$

$$0 \leq \alpha \leq 2k+2$$

and D does not depend on n.

In this case the space H will be said k-periodic or k-strongly periodic.

Analogously to (2.5) we now have

$$(2.22)\quad \rho_k(n,B) = \inf_{\substack{a_j^{(s)} \\ j=1,\dots,n \\ s=0,\dots,k}} \sup_{\|f\|_B \leq 1} \Big| \sum_{s=0}^{k} \sum_{j=1}^{n} a_j^{(s)} f^{(s)}\Big(\frac{2\pi}{n} j\Big) - J(f) \Big|$$

Now I shall mention a special result of K. Segeth (see (35)) who studied this field of problems. One of the problems here is roughly speaking, the following :

Is it better to use more values of a functions in the quadrature or is it better to compute and use the values of the derivatives?

An answer to this is given by the comparison between ρ_o and ρ_k . It can be shown that for 2-periodic spaces

$$(2.23)\quad \rho_2(n,H) = \inf_{\substack{a_j^{(s)} \\ j=1,\dots n \\ s=0,2}} \sup_{\|f\|_H \leq 1} \Big| \sum_{s=0}^{2} \sum_{j=1}^{n} a_j^{(s)} f^{(s)}\Big(\frac{2\pi}{n} j\Big) - J(f) \Big| \quad s \neq 1$$

Let us assume that the amount of work needed for the evaluation of f(x) is equal to 1 and that for the derivative is α . Then the whole work with the use of n points will be $n(1+\alpha)$. This value will be the measure of the "work" when using the given formula with n points .

$$(2.24)\quad S(\alpha, H) = \limsup_{n\to\infty} \frac{\rho_2(n,H)}{\rho([n(1+\alpha)], H)}$$

gives now the required answer (relatively to the space H) .

Thus, for example, the following theorem is true :

Theorem 2.9 .Let H be a 2-strongly periodic space. Let $\alpha \geq 1$. Let $\|e^{inx}\|_H^2 = g(n^2)$ where g is an entire function. Then $S(\alpha, H) > 1$. If g is not a polynomial then $S(\alpha, H) = \infty$ for $\alpha > 1$ and $S(1, H) = 3$.

Theorem 2.9 shows more or less that if the amount of work needed for the evaluation of derivatives is not less than that needed for the evaluation of the function, it is not advantegeous to use the formula with derivatives.

Previously in this section we dealt with the trapezoid formula T_n. An analogous role is played here by the formula

$$T_n^{(2)}(f) = \frac{1}{n}\sum_{k=1}^{n} f\left(\frac{2\pi}{n}k\right) + \frac{1}{n^3}\sum_{k=1}^{n} f''\left(\frac{2\pi}{n}k\right) \tag{2.25}$$

There is also a theorem analogous to Theorem 2.9 for the use of (2.25), given more exactly and in detail more in (25). As an illustration I shall give the following example:

Example 2.3. Compute also $J(f) = \frac{1}{2\pi}\int_0^{2\pi} f(x)\,dx$ for $f(x) = e^{\beta \sin x}$, $\beta = 10, 50$. Let us assume $\alpha = 1$. In table 2.3 we see the error when using the formulas T_n and $T_n^{(2)}$ in dependence on the amount of work (i.e. on n resp. $n(1+\alpha)$). We see that the computation without the sue of derivatives is more advantageous. This agrees fully with the theoretic investigations. In accordance with the theorem the error of the formula with derivatives is nearly three times larger than that of the formula without derivatives. All we said was connected with the computation of (2.1), and (2.18) respectively.

Now I shall briefly speak about the computation of

$$J_g(f) = \frac{1}{2\pi}\int_0^{2\pi} f(x)\,g(x)\,dx, \quad g \in \dot{L}_2 \tag{2.26}$$

We shall not analyse all the problems associated with this computation. All can be done analogously. The formula which plays the same role here as the trapezoid formula is the following (see(6)) :

$$T_n^{(g)}(f) = \frac{1}{n}\sum_{j=1}^{n} S_n\left(\frac{2\pi}{n}j\right) f\left(\frac{2\pi}{n}j\right) \tag{2.27}$$

where

$$S_n(x) = \sum_{k=-\left[\frac{n}{2}\right]+1}^{\left[\frac{n}{2}\right]} b_k^{(n)} e^{ikx}$$

$$b_k^{(n)} = b_k \quad \text{for} \quad k < \left[\frac{n}{2}\right]$$

$$b_{\frac{n}{2}}^{(n)} = \frac{1}{2}\left(b_{\frac{n}{2}} + b_{-\frac{n}{2}}\right), \quad \frac{n}{2} = \left[\frac{n}{2}\right]$$

Amount of work n	$f(x) = e^{10 \sin x}$ Error of the formula without derivatives (T_n)	$f(x) = e^{10 \sin x}$ Error of the formula with derivatives ($T_n^{(2)}$)	$f(x) = e^{50 \sin x}$ Error of the formula without derivatives (T_n)	$f(x) = e^{50 \sin x}$ Error of the formula with dirivatives ($T_n^{(2)}$)
16	0,60100 -1 0, 37 -12	0, 18030 0 0, 11200 -11	0, 45200+20 0, 26250+17 0, 20199+12	0, 15147+21 0, 78754+17 0, 60599+12

Table 2.3

$$g(x) = \sum_{k - \infty}^{\infty} b_k \; e^{ikx}$$

The error estimation by an overfunction can be made. As an illustrastion I shall show

Example 2.4. Compute

$$(2.28) \qquad L = \int_{-\frac{\pi}{2}}^{+\frac{\pi}{2}} e^{\alpha \sin x} \cos x \, dx, \quad \alpha = 1,5$$

Apparently this integral may be written like this :

$$(2.29) \qquad L = \int_{-1}^{+1} e^{\alpha x} \, dx.$$

In the table 2.4 we have shown the errors of (2.27) in comparison with the Rombergs integration (see (16) (17)). In the table there is also shown the error obtained by the use of the overfunction $e^{\cos x} \cos x$. The computation was made with computer ICT 1905 in double precision.

In a simple case of quadrature we have shown some aspects of incredulity with respect to the given information and the meaning of "stability" of a conclusion. It is possible to generalize these ideas in different ways . A possibility of a generalization can be seen in (28) . I shall, however, not deal with it here.

3. Boundary-value Problems for Ordinary Differential Equations

In section 2 we showed one kind of incredulity as regards the given information and how to deal with it. I shall now mention some other aspects of incredulity. A simple problem will again be analised. Let us solve the following boundary-value problems

$$(3.1) \qquad (p(x)\, y')' - q(x) y = f(x)$$

Number of evaluations of function n	$f = e^{\sin x}$ Error of (2.27) for (2.28)	$f = e^{\sin x}$ Estimation based on overfunction	$f = e^{x}$ Error of Romberg formula of (2.29)
9	0,634 -8	0,171 -5	-0,421 -10
17	0,271 -19	0,245 -16	-0,416 -14
25	0	0,437-18	0,407 -19
33	0		
	$f = e^{5 \sin x}$	$f = e^{5 \sin x}$	$f = e^{5x}$
9	0,510 -2	0,381 0	-0,844 -3
17	0,254 - 8	0,126 -5	-0,208 -5
25	0,492-16	0,801-13	-
33	-0,217-18	0,268-16	0,128 -8

Table 2.4

with the boundary conditions

(3.2) $y(0) = a \qquad y(L) = b.$

We assume that p(x), g(x) and f(x) are sufficiently smooth and $p(x) > \alpha > 0$, $q(x) \geq 0$.

The functions p, q, f have a physical meaning. Nevertheless, we know them only approximately in practice.

Let the possible disturbances (incredulities) of p, q, f be σ, ϱ, φ respectively. From the physical point of view these perturbances are small in a certain sense (norm). They may also have further properties. Such perturbances will be called admissible disturbancies. We shall assume that small admissible disturbances result in a small change in the solution.

It is well known that a numerical process cannot be realized with an absolute exactness. Every realization of a process by computation is disturbed (by round-off errors). We can mostly imagine, however, this disturbed realization as an exact one (without disturbance) but with the disturbed given information. We shall speak about replaced disturbances (of information) in this case [1]. It is reasonable to speak about a suitable numerical process if the replaced disturbances

a) are admissible

b) the order of disturbances is the same as the order of error in the individual operations.

[1] The method of replaced disturbances (backword-method) was used with large success by Wilkinson. See (42), (43).

Bauer (13), (14), (15) used a similar approach in his investigations of numerical processes in algebraic problems.

There are suitable and non suitable processes. I shall show them by the process of solving (3.1) and (3.2).

Example 3.1. The method of combination of solutions leads to a non suitable process. This method, as known, consists in solving two initial-value problems for the initial conditions $y(0) = 0$, $y(0) = \gamma_j$, $j = 1, 2$ and the required solution (3.1) (3.2) is determined by a suitable combination. Let $p(x) = (1 + x)$, $q(x) = 500$, $f(x) = \pi \cos \pi x - (500 + \pi^2 (1 + x)) \sin \pi x$, $L = 1$, $a = b = 0$. We solve the initial problem by the Runge-Kutta-Gill method of the 4th order for step $h = 0,025$ (computer LGP 30) See (12). The results obtained are given in Table 3.1.

Example 3.2. The factorization methods leads to a suitable process. By this method (see e.g. (12)) we solve the following system

$$\psi' + q\psi^2 = \frac{1}{p}, \quad \psi(0) = 0,$$

$$u' + q\psi_u = f, \quad u(0) = -a,$$

$$\psi y' - \frac{y}{p} = \frac{u}{p}, \quad y(L) = b.$$

Let us solve the same problem as in Example 3.1 by this method. The initial problems are also solved with Runge-Kutta-Gill method with $h = 0,025$. We obtain the results mentioned in Table 3.1.

I have said that we can mostly consider the disturbed realization of a process as an exact realization with the disturbed input (i.e. given) information. In this case in the method of factorization the replaced disturbances are small in the following norms: σ, ρ in C norm and φ in the norm $\|\varphi\| = \left\| \int_0^x \varphi \, dx \right\|_C$. It may be seen that

x	y (x) by method of combination	Exact solution	y (x) by method of factorization
0, 100	0, 3090103	0, 3090170	0, 3090018
0, 400	0, 9510075	0, 9510565	0, 9510461
0, 500	1, 005031	1, 000000	0, 9999897
0, 700	0, 8577343	0, 809017	0, 8090081
0, 750	1, 374171	0, 7071068	0, 7070985
0, 800	0, 0000000	0, 5877852	0, 5877778
0, 900	9, 700032	0, 3090170	0, 3090119

Table 3. 1

these disturbances are admissible.

It is obvious that the questions of existence of a suitable numerical process for the solution of the given problem is very important. The method of factorization may be generalized to a general boundary (or multipoint) problem for the system

$$x'(s) - A(s)\, x\,(x) = f(s). \tag{3.4}$$

J. Taufer, see (38), (39), has investigated in detail the replaced disturbances for a concrete kind of factorization and has shown that his factorization method is suitable in the previously mentioned sense, Another kind of factorization method, sometimes called method of the transfer of boundary conditions, was investigated in recent years, for example, by Abramov (1), (2) who also briefly mentioned the possibility of showing the suitability of this process for the general case (3.4). See (3).

In (7) and (12) the stability of the differential equations of the factorization method in special cases has been studied.

Example 3.3

As an example I shall show the computation of a continuous beam of 20 fields built in at the end and constantly loaded. In practice, the method of transfer of matrices which is very similar to the method of combination of solutions is very often used. See e.g. (45).

In the following table 3.2 there are shown the moments at some supports computed by the usual method as well as by Taufer s factorization method. The previously mentioned factorization method can also be used in solving the eigenvalue problem. See (40).

4. Stability of numerical processes.

In the previous sections we dealt with some aspects of incredulity as to the choice of a numerical process.

Number of support	Exact moment at the support	Computed moment at the support by mentioned method	Computed moment at the support by factorization method
4	5,000000 -3	4,9999999 -3	4,999999999 -3
12	5,000000 -3	4,9492238 -3	5,000000006 -3
19	5,000000 -3	1,618765 -1	5,000000004 -3
20	5,000000 -3	7,790814 -1	4,999999999 -3

Table 3.2

In this section we shall deal with a quantitative characterization of the numerical stability of a given numerical process. See (9), (10), (11), (12) . In computations of problems of mathematical analysis, the existence of a subscript n (e.g. number steps) is typical so that we obtain the required result only for $n \to \infty$. In section I we introduced a numerical process. Here we shall define it more exactly .

Definition 4.1 . Let there be given a sequence of normed vector spaces

$$X^{(n)}_{-p_n}, \; X^{(n)}_{p_n+1}, \; \dots , \; X^{(n)}_{o}, \; X^{(n)}_{1}, \dots \; X^{(n)}_{N_n}, \qquad n = 1, 2, \dots$$

and a sequence of continuous operators :

$$A^{(n)}_{i}, \quad i = 0, 1, \dots , \; N_n - 1, \qquad n = 1, 2, \dots$$

mapping the Cartesian product

$$X^{(n)}_{-p_n} \times X_{-p_n+1} \times \dots \times X^{(n)}_{i} \text{ into } X^{(n)}_{i+1} .$$

Further let the sets

$$M^{(n)}_{k} \subset X^{(n)}_{k} \quad \text{for} \quad k = -p_n, -p_n + 1, \dots, 0$$

be given. Then the sequence of equations

$$x^{(n)}_{i+1} = A^{(n)}_{i} (x^{(n)}_{-p_n}, x^{(n)}_{-p_n+1}, \dots, x^{(n)}_{i})$$

$$i = 0, 1, \dots, N_n - 1 , \; x^{(n)}_{k} \in X^{(n)}_{k}, \; k > 0$$

$$x^{(n)}_{k} \in M^{(n)}_{k} , \; k \leqslant 0 , \; n = 1, 2, \dots$$

will be called a numerical process. The set $M_k^{(n)}$ will be called the set of input data and the elements $x_k^{(n)}$ $k = 1, 2 \ldots N_n$ will be called the solution corresponding to input elements $x_k^{(n)}, k=-p_n, \ldots, 0$.

In practice, the numerical processes as by Definition 4.1 cannot be solved exactly by the computer (round-off errors). Hence we introduce the following definition:

Definition 4.2.

Let there be given a numerical process in the sense of Definition 4.1. Let there be given the input elements $x_k^{(n)}$, $k = -p_n, \ldots, 0$ and a sequence of numbers $\left\{a_j^{(n)}\right\}^k = \xi^{(n)}$ $j = -p_n, \ldots, N_n$; $n = 1, 2, \ldots$ and denote $\tilde{x}_i^{(n)} \in X_i^{(n)} i=-p_n, \ldots, 0, 1, \ldots$ $\ldots, N_n$ the elements satisfying the equations:

$$(4.2) \quad \tilde{x}_{i+1}^{(n)} = A_i^{(n)} (\tilde{x}_{-p_n}^{(n)}, \tilde{x}_{-p_{n+1}}^{(n)}, \ldots \tilde{x}_i^{(n)}) + \vartheta_{i+1}^{(n)}, \quad i = 1, \ldots N_n - 1 .$$

$$(4.3) \quad \tilde{x}_i^{(n)} = x_i^{(n)} + \vartheta_i^{(n)}, \quad \tilde{x}_i^{(n)} \in M_i^{(n)}, \quad i = -p_n, \ldots, 0.$$

The solution of the given numerical process corresponding to input elements $x_k^{(n)}$,

$k = -p_n, \ldots, 0$ and to the sequence $\xi^{(n)}$ will be called

β_s - solution if

$$(4.4) \quad \lim_{\Delta \to 0} \sup \frac{1}{\Delta} \sup_{|\vartheta_i^{(n)}| \leq a_i^{(n)} \Delta, \; i=-p_n, \ldots, N_n} \sup |\tilde{x}_i^{(n)} - x_i^{(n)}| \leq C n^s$$

and C does not depend on n.

We will speak about B_s -solution if $S_0 = \inf s$.

The investigations of concrete given processes have been done in the previously mentioned way in many cases. See e.g. (12), (27), (32), (41), (44), and others.

I shall now give some examples explaining the meaning of the previous definitions. Let us solve the initial problem for an ordinary differential equation

$$y' = f(x,y) \;, \quad y(a) = y \;. \tag{4.5}$$

The Runge-Kutta method can be written as follows

$$y_{i+1}^{(n)} = y_i^{(n)} + h^{(n)} \Phi_f(x_i^{(n)}, y_i^{(n)}, h^{(n)}.)$$

A slight change will be made to simplify the notation in Definitions 4.1 and 4.2. We shall investigate two processes

I. $y_{i+1}^{(n)} = y_i^{(n)} + h^{(n)} \Phi_f(x_i^{(n)}, y_i^{(n)}, h^{(n)}) \;, \quad i = 0, 1, 2, \ldots,$

$$y_o^{(n)} = y \;, \quad x_{i+1}^{(n)} = x_i^{(n)} + h^{(n)} \quad i = 0, 1, 2, \ldots, \; x_o^{(n)} = a \;. \tag{4.6}$$

$$h^{(n)} = \frac{C}{n}$$

II. $y_{i+1}^{(n)} = y_i^{(n)} + h^{(n)} \Phi_f(x_i^{(n)}, y_i^{(n)}, h^{(n)}) \;, \quad i = 0, 1, 2, \ldots$

$$y_o^{(n)} = y \;, \quad x_{i+1}^{(n)} = a + ih^{(n)} \quad i = 0, 1, 2, \ldots, \; x_o^{(n)} = a, \tag{4.7}$$

$$h^{(n)} = \frac{C}{n} \,.$$

Let further

$$y_i^{(n)} \in Y_i^{(n)} \,, \quad h^{(n)} \Phi_f \in Z_i^{(n)} \,, \quad x_i^{(n)} \in X_i^{(n)} \,, \quad h^{(n)} \in H^{(n)}. \tag{4.8}$$

The spaces $X_i^{(n)}$, $Y_i^{(n)}$, $Z_i^{(n)}$, $H^{(n)}$ are spaces of real numbers with the norm $|x|$. The meaning of the previously mentioned mapping is evident in this ease. The numerical process is also clear. The disturbed process is as follows

$$\text{(4.9)} \qquad \tilde{y}_{i+1}^{(n)} = \tilde{y}_i^{(n)} + \tilde{z}_i^{(n)} + {}^{I}\vartheta_{i+1}(Y, n),$$

$$\text{(4.10)} \qquad \tilde{x}_{i+1}^{(n)} = \tilde{x}_i^{(n)} + \tilde{h}^{(n)} + {}^{I}\vartheta_{i+1}(X, n),$$

$$\text{(4.11)} \qquad \tilde{z}_i^{(n)} = \tilde{h}^{(n)} \Phi(\tilde{x}_i^{(n)}, \tilde{y}_i^{(n)}, h^{(n)}) + {}^{I}\vartheta_i(Z, n),$$

$$\text{(4.12)} \qquad \hat{h}^{(n)} = \frac{C}{n} + {}^{I}\vartheta(H, n).$$

II. The equations (4.9) (4.11) (4.12) remain unchanged. (4.10) now has the following form

$$\text{(4.10')} \qquad \tilde{x}_{i+1}^{(n)} = a + i\, h^{(n)} + {}^{II}\vartheta_{i+1}(X, n)$$

We can compute these processes in a different manner. These computations differ in disturbances. The following mathematical models can be assumed

a) Fixed point computation

$$|{}^{j}\vartheta_i(Y, n)| \leq \Delta, \quad |{}^{j}\vartheta_i(X, n)| \leq \Delta,$$

$$|{}^{j}\vartheta_i(Z, n)| \leq \Delta, \quad |{}^{j}\vartheta(H, n)| \leq \Delta, \quad j = I, II.$$

b) Scaled point (floating point) computation

$$|{}^{j}\vartheta_i(Y, n)| \leq \Delta, \quad |{}^{j}\vartheta_i(X, n)| \leq \Delta,$$

$$|{}^{j}\vartheta_i(Z, n)| \leq h^{(n)}\Delta, \quad {}^{j}\vartheta(H, n) \leq h^{(n)}\Delta, \quad j = I, II.$$

b') Normalized floazing point computation

$$|{}^{j}\vartheta_i^{(Y,n)}| \leqslant |\tilde{y}_i^{(n)}|\Delta, \quad |{}^{j}\vartheta_i^{(X,n)}| \leqslant |\tilde{x}_i^{(n)}|\Delta,$$

$$|{}^{j}\vartheta_i^{(Z,n)}| \leqslant |\tilde{z}_i^{(n)}|\Delta, \quad |{}^{j}\vartheta^{(H,n)}| \leqslant \tilde{h}^{(n)}\Delta,$$

$$j = I, II.$$

c) Normalized floating point computation with computation (4.9) in the process II in double precision

$$|{}^{II}\vartheta_i^{(Y,n)}| = 0, \quad |{}^{II}\vartheta_i^{(X,n)}| \leqslant |\tilde{x}_i^{(n)}|\Delta, \tag{4.13}$$
$$|{}^{II}\vartheta_i^{(Z,n)}| \leqslant |\tilde{z}_i^{(n)}|\Delta, \quad |{}^{II}\vartheta_i^{(H,n)}| \leqslant h^{(n)}\Delta.$$

The sequence $\xi^{(n)}$ is obvious and I shall not describe it. The following theorem may be proved.

Theorem 4.1. The previously mentioned processes are

a) B_1 solution,

b), b') B_1 solution,

c) B_0 solution.

I shall now show the meaning of this theorem by means of the following example :

Example 4.1. We shall solve the initial problem for the equation

$$y' = x(x+2)y^3 + (x+3)y^2 \tag{4.14}$$

$$y(a) = -\frac{2}{a(a+2)}, \tag{4.15}$$

with the standard Runge-Kutta method of the 4th degree.

$y(a+\frac{1}{2})$ is to be solved. Here we obviously have $C = \frac{1}{2}$. Our task is to estimate $|\tilde{y}_n^{(n)} - y_n^{(n)}|$ in dependence on n. Since the solution of (4.14) and (4.15) is $y(x) = -\frac{2}{x(x+2)}$ and we do not know $y_i^{(n)}$

we shall use $y_n^{(n)} \rightarrow y(a+\frac{1}{2})$ and put $\varepsilon_n^{(n)} = |\tilde{y}_n^{(n)} - y(a+\frac{1}{2})|$

In the following figures there are the outcomes of computations. In fig. 4.1 a are the results for the process I.b, a = 0,5, obtained with MINSK 22.[1] The parameter n has been selected as a decadid value.

It is interesting to ask what happens if we investigate $\eta_n^{(n)} = |\tilde{y}_n^{(n)} - y(\tilde{x}_n^{(n)})|$. It may be shown that this is also a B_1 solution. In fig. 4.1 b we see the results.

A further interesting question is what happens if we use n diadic. In fig. 4.2 we also see the results $\varepsilon_n^{(n)}$ for n diadic. We see that this computation has a different character. This is more or less an accident.

In computation II.b there is no difference between diadic and dedacid n as mentioned above.

For a diadic n exactly the same results for computation I.b and II.b . are obtained. In fig. 4.3 and 4.4 we have the results for I.b n dedadic and II.b .

In fig. 4.5 we have the results for II.c and a = 300 .

From the mentioned example we can clearly see that in computations there are different kinds of importance with respect to the stability. E.g. we have seen that the floating point makes the round-off smaller but the results remain unchanged. Such a kind of considerations may be very valuable in practice, yet we cannot deal with it here.

It is obvious that the knowledge of stability, especially the B_s stability is an important factor in a suitable choice of method.

---

[1] This computer is a diadic one.

There is a question if it exist a B_o solution for computation in simple precision normalized fleating point. The answer is positive. For the sake of simplicity, I will show one of them on the example of the quadrature formula T_n. Let $n=2^k$. Let the process of the computation of T_n be the following

$$P_s = 2^{-s} \sum_{j=1}^{2^{s-1}} \left(2 f\left(\frac{2\pi}{2^s}(2j-1)\right) - f\left(\frac{2\pi}{2^s}(2j-2)\right) - f\left(\frac{2\pi}{2^s} 2j\right)\right)$$

$$P_o = \frac{1}{2} (f(0) + f(2\pi))$$

$$k_{n+1} = k_n + P_{k-n} , \quad k_o = 0 .$$

$$n = 0, 1, \dots, k .$$

$$k_{k+1} = T_n .$$

Then the computation is a B_o solution provided that we compute with normalized floating point and simple precision.

We shall show an other interesting example. Let us solve the initial problem for the differential equation

$$y'' = f(x.y) \tag{4.8}$$

The usual difference method leads to the following formula

$$y_{n+3} - y_{n+2} + y_{n+1} = \frac{1}{12} h^2 (13 f_{n+2} - 2 f_{n+1} + f_n) \tag{4.9}$$

This formula can be written in the following form

$$z_{n+1} - z_n = h f_n \tag{4.10}$$

$$y_{n+2} - y_{n+1} = \frac{1}{12} (13 z_{n+2} - 2 z_{n+1} + z_n)$$

and the following theorem is true :

Theorem 4.2. The numerical process based on (4.9) resp. (4.10) is a B_2, B_1 process respectively for $\xi^{(n)} = \{1, 1, \dots\}$

This example (see (12) (41)) shows the possibilities of getting a better stability through simple changes in the method. The question when it is possible to write a formula in a forma having a botter stability is solved in (41) .

As a further example I shall show the numerical stability of the numerical process of overrelaxation for usual finite-differential equations (see (32)) . We put $N_n = \infty$ in the definition 4.1 . Let us measure the error of the result in the norm $\eta^{(n)} = \frac{1}{m} \sum_{j=1}^{m} |\varepsilon_j^{(n)}|$

where $\varepsilon_j^{(n)}$ is the error in one point of the net; m is the numer of the net-points and $h = \frac{6}{n}$ is the step. Let the matrix A of finite-difference equations have the form

$$A = \begin{pmatrix} I, B \\ \\ B^T, I \end{pmatrix}$$

where I is the unit matrix. Then the following theorem holds.

<u>Theorem 4.3.</u> Let the previous assumptions hold. Then the numerical process is a B_2 process if $0 < \quad \leq \omega \leq 2 - C . h, \ C > 0$.

Evidently a special case of theorem 4.3 is when ω is independent of h or ω is the optimal overrelaxation parameter.

We shall introduce an example.

<u>Example 4.2.</u> Let us solve the one dimensional proble $y'' = 1, y(0) = y(1) = 0$ with the finite-difference method and overrelaxation. Because of the round-off error the iterations do not, in general converge to the required solution. They will "quasi-converge" in a more or less well known sense. In Fig. 4.6 we see $\eta^{(n)}$ $n = \frac{1}{h}$ in dependence on h.

We see a good agreement with theorem 4.3. It is possible to formulate the theorem 4.3 for a $0 < \omega \leq 1$ in a more general form. See (12) .

Further processes have also been investigated. I shall mention here the stability of the Kellog process for the determination of eigenvalue (see (27)) and a numerical process for solving a problem of the theory of reactors (44) and the process for computation of conform mapping . (see also (12)) .

I have shown a few different aspects of incredulity with regard to the given information which appear in computations. I think that this kind of investigations is very important when choosing an algorithm in general.

Fig. 4. 1. a.

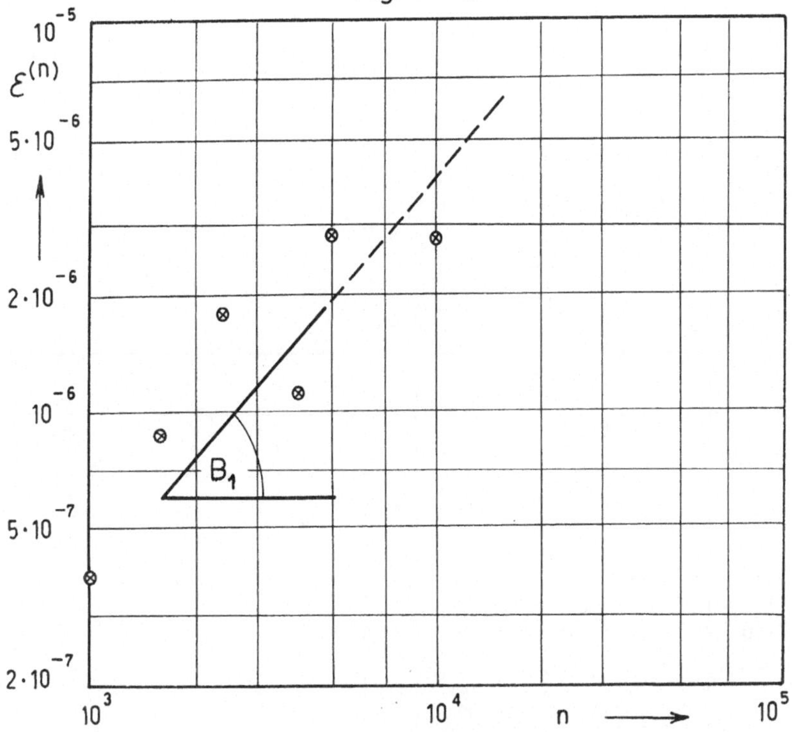

⟨ 0.5 , 1 ⟩

I. b decadic

Fig. 4. 1. b

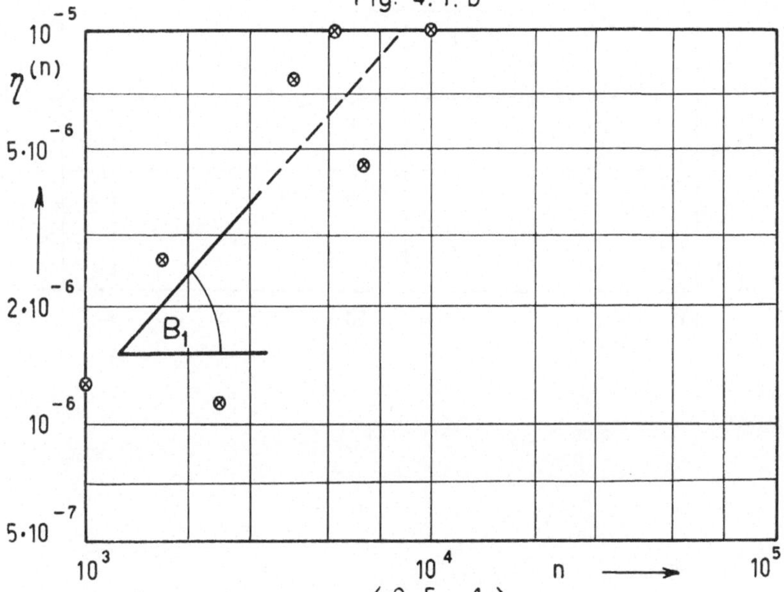

⟨ 0.5 , 1 ⟩

I. h decadic

Fig. 4.2.

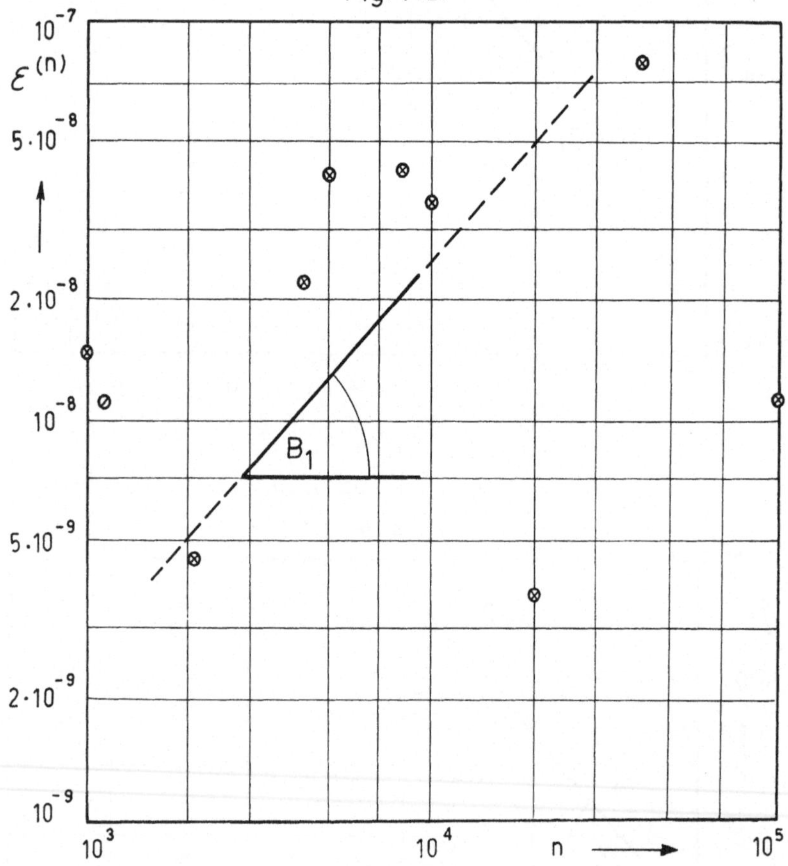

⟨ 0.5 , 1 ⟩

II. b

Fig. 4.3.

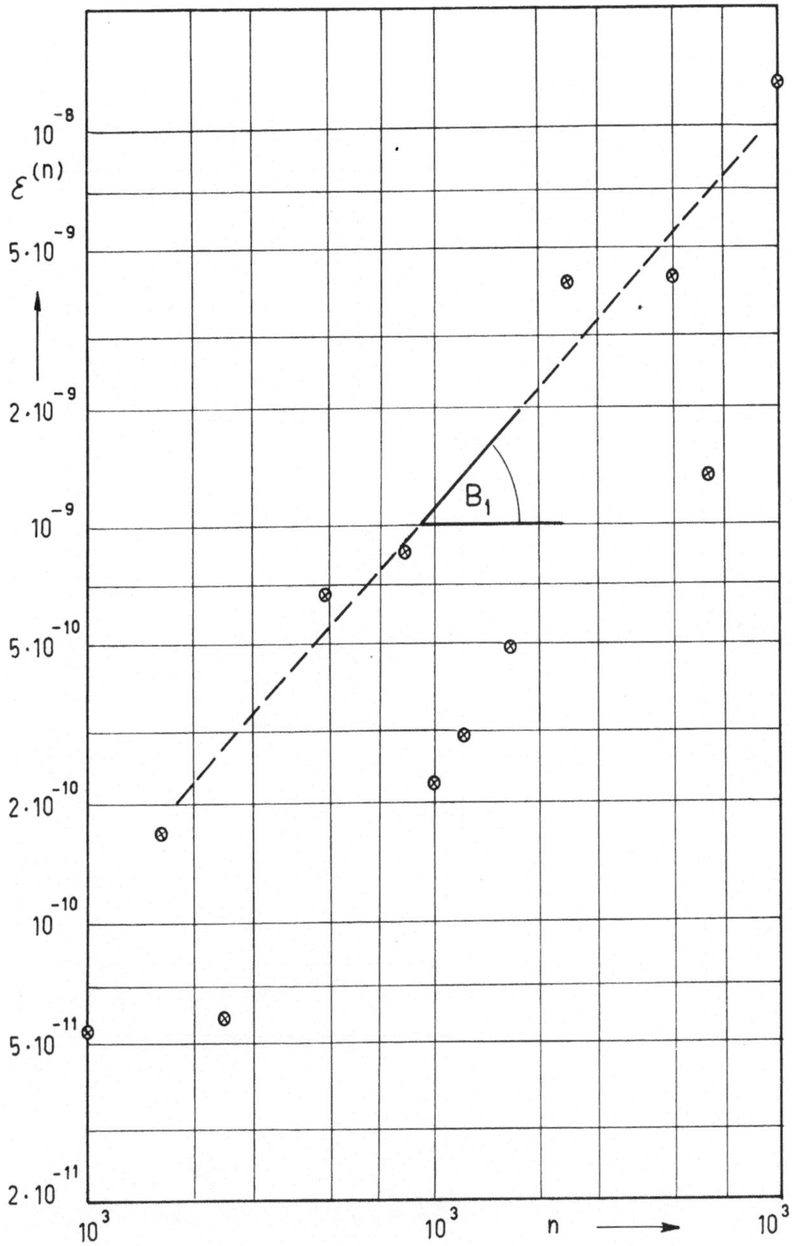

⟨300, 300.5⟩

I, b

Fig. 4.4.

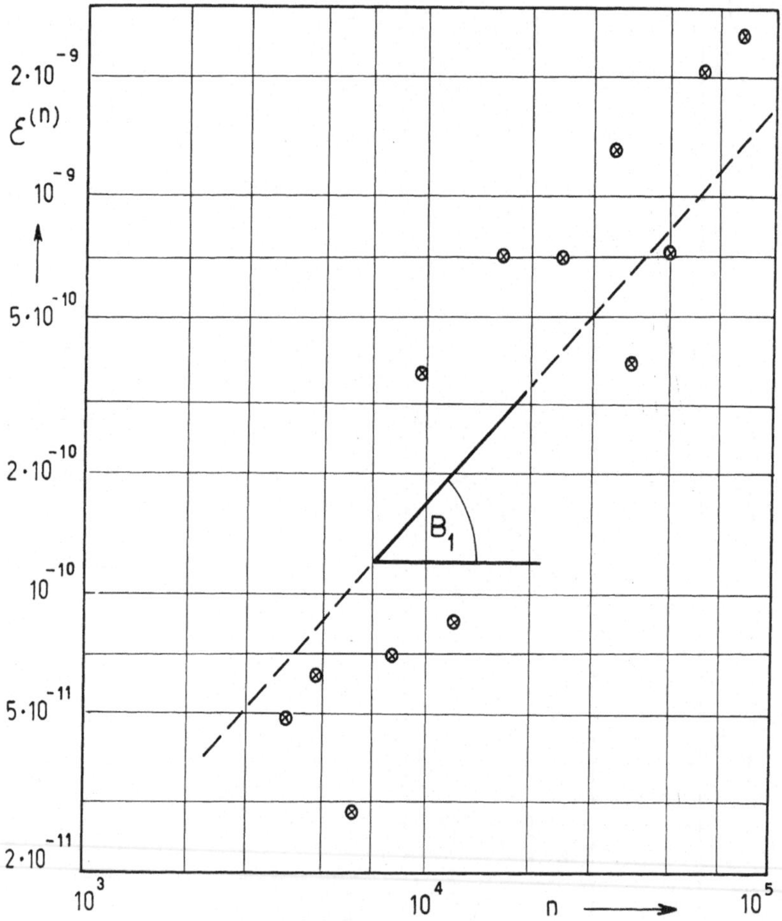

⟨ 300, 300.5 ⟩

II. b

Fig. 4.5.

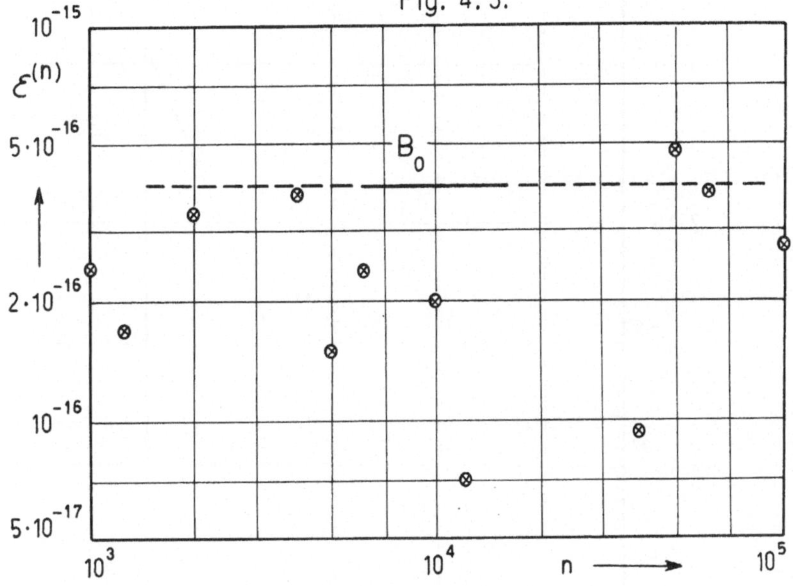

⟨ 300 , 300.5 ⟩

II. c

Fig. 4.6.

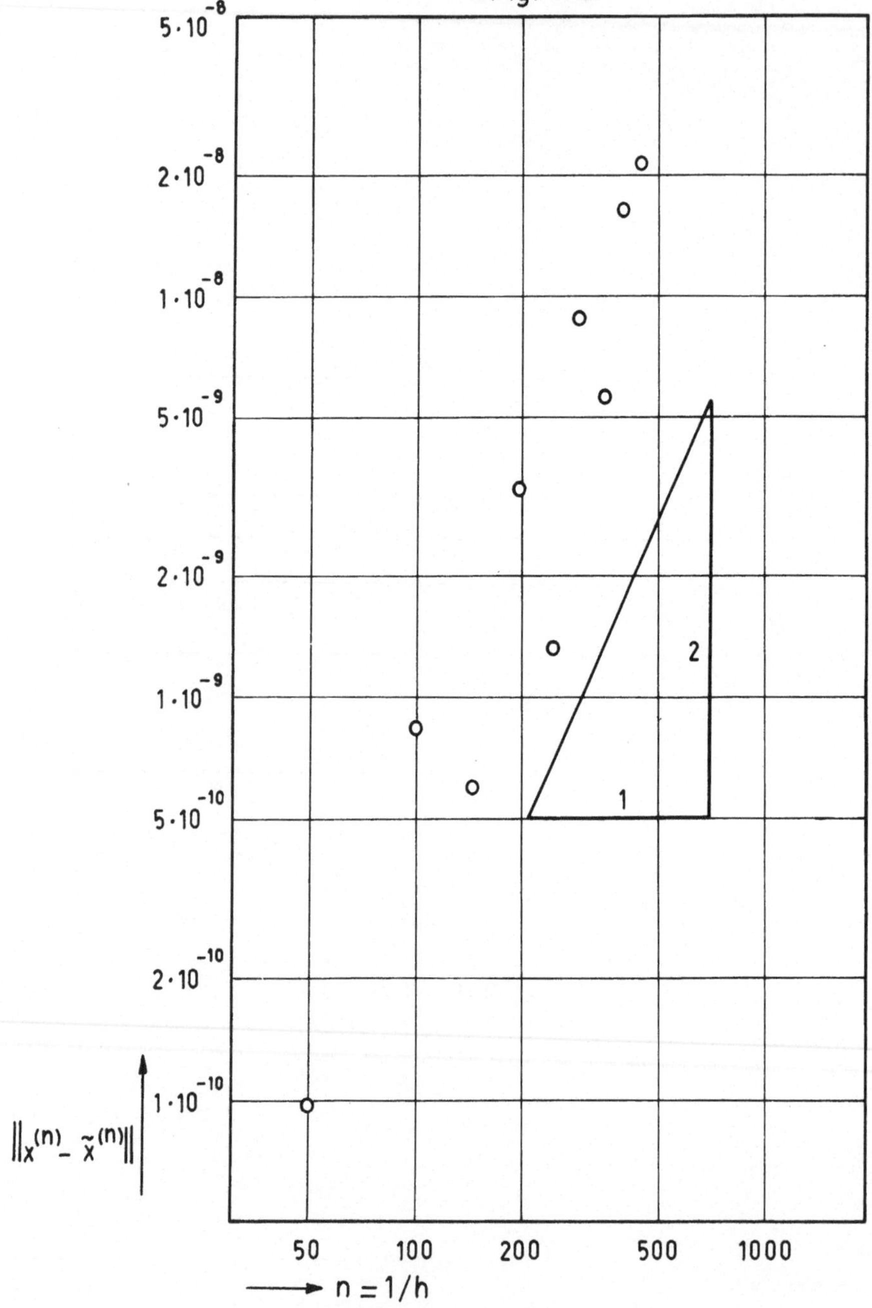
5·10^-8
2·10^-8
1·10^-8
5·10^-9
2·10^-9
1·10^-9
5·10^-10
2·10^-10
1·10^-10
$\|x^{(n)} - \tilde{x}^{(n)}\|$
2
1
50
100
200
500
1000
n = 1/h

I. Babuška

References

[1]   А.А. Абрамов: О переносе граничных условий для систем линейных
      обыкновенных дифференциальных уравнений.Ж.выч.мат. и мат.
      физ. 1961, 1, 542-545.

[2]   А.А. Абрамов: Вариант метода прогонки. Ж.выч.мат. и мат. физ.
      1961, 1, 349-351

[3]   A.A. Abramov : Transfer of boundary conditions for system of
      ordinary linear differential equations. Proc. of IFIP Congress 65,
      p.420

[4]   С.А. Агаханов: О точности некоторых квадратурных и кубатурных
      формул. Сиб.мат.журнал Х1, 1965, 1-15

[5]   I. Babuska : Über die optimale Berechnung der Fourierischen Koeffi-
      zienten. Apl. Mat. 11, 1966, 113-122.

[6]   I. Babuska : Über universelloptimale Quadraturformeln. Apl. Mat. 1968

[7]   I. Babuska , M. Prager : Numerisch stabile Methoden zur Lösung
      von Randwertaufgaben. ZAMM 1961 , H. 4-6

[8]   И.Бабушка, С.Л.Соболев: Оптимизация численных методов.
      Apl.mat. 10, 1965, 96-129

[9]   I. Babuska, M. Práger, E. Vitásek; Numerische Stabilität von
      Rechenprozessen. Wiss. Z. Techn. Hochsch. Dresden 1963, 12,
      101-110.

[10] I. Babuska, M. Práger, E. Vitásek; Numerické řešení differenciál-
      nich rovnic 1964 , SNTL

[11] I. Babuska, M. Práger, E. Vitásek: Stability of Numerical Processes,
      Proc. of IFIP 65 , 602-603 .

[12] I. Babuška, M. Práger, E. Vitásek: Numerical Process in Differen-
      tial Equations. Interscience Publishers 1966

[13] F.L. Bauer: Numerische Abschätzung und Berechnung   von Eigen-
      werten nichtsymmetrischen Matrizen. Apl. Mat. 10, 1965,
      178-189.

[14] F.L. Bauer et al. Moderne Rechenanlagen , Stuttgart 1965, p. 64.

[15] F.L. Bauer: Genauigkeitsfragen bei der Losung linearer Gleichungssy-
      steme. ZAMM 46, 1966, 409-421.

[16] F.L. Bauer, H. Rutishauser, E. Stiefel : New Aspect in Numerical
      Quadrature. Proc. of Symp. in Appl. Mat. 1963 , XV, 199-218.

[17] R. Bauman : Algol Manual der Alcor-Gruppe, Sonderdruck aus Elektronischen Rechenanlagen H 5/6 (1961) H 2 (1962) R. Oldenburg, Munchen.

[18] P. J. Davis: On the Numerical Integration of Periodic Analytic Functions. Proceedings of Symposium Madison 1959.

[19] G. G. Dahlquist : On Rigorous Error Bounds in the Numerical Solution of Ordinary Differential Equations. Numerical Solution of Nonlinear Differential Equations. Wiley 1966, 89-96 .

[20] G. G. Dahlquist: Private communication.

[21] H. Ehlich : Untersuchungen zur numerischen Fourier analyse. Math. Zeitschr. 91 (1966) , 380-420 .

[22] G. Hammerlin: Uber ableitungsfreie Schranken für Quadraturfehler. Numerische Mathematik 5, 1963 , 226-233; 7, 1965 , 232-237.

[23] P. Henrici: Elements of numerical Analysis. J. Wiley Sons, Inc New York-London-Sydney, 1964.

[24] D. Jagerman : Investigation of Modified Mid-Point Quadrature Formula, Math. of Comp. 20 1966 , 78-89.

[25] G. Kowallewski : Interpolation und genährte Quadratur. Leipzig 1930 , p. 130

[26] В.И.Крылов: Приближенное вычисление интегралов. Москва 1959.

[27] I. Marek: Numerische Stabilität der Prozesse vom Keloggschen Typus. Liblice 1967 Apl. Mat. 13, 1968.

[28] J. Milota ; Universal Almost Optimal Formulae Solutions of Boundary - Value Problems for Ordinary Differential Equations. Liblice 1967 , Apl. Mat. 13, 1968.

[29] R. E. Moore: The automatic Analysis and Control of Error in Digital Computation. Vol. 1, 61-130 . Proceedings of a seminar University of Wisconsin, Madison Octobre 5-7, 1964.

[30] R. E. Moore; Interval Analysis. Prentice Hall 1966.

[31] R. E. Moore: Practical Aspect of Interval Computation. Liblice 1967. Apl. Mat. 13, 1968 .

[32] M. Prager : Numerical Stability of the Method of Overrelaxation. Liblice 1967. Apl. Mat. 13, 1968.

[33] P. Prikryl: On Computation of Fourier Coefficients in Strongly Periodic Spaces. Liblice 1967, Apl. Mat. 13, 1968.

[34] A. Sard : Linear Approximation. Providence 1963.

[35] K. Segeth : On Universally Optimal Quadrature Formulae Involving Values of Derivatives of Integrand. Liblice 1967 , Apl. Mat. 13, 1968.

[36] С.Л.Соболев: Лекции по теории кубатурных формул. Новосибирск 1965.

[37] H. J. Stetter : Numerical Approximation of Fourier-Transform. Num. Math. 8, 1966 , 235 - 249.

[38] J. Taufer: On Factorization Method. Apl. Mat. 11, 1966, 427-452 .

[39] J. Taufer : Faktorisierungsmethode für ein Randwertproblem eines linearen Systems von Differentialgleichungen, Liblice 1967 . Apl. Mat. 13, 1968.

[40] J. Taufer: Faktorisierungsmethode für ein Eigenwertproblem eines linearen Systems von Differentialgleichungen , Liblice 1967

[41] E. Vitásek; Numerical Stability in Solution of ordinary Differential Equations of Higher Order, Liblice 1967, Apl. Mat. 13, 1968.

[42] J. H. Wilkinson : Rounding errors in algebraic processes. London H. M. S. O. 1963

[43] J. H. Wilkinson : A. Survey of Errors Analysis of Matrix Algorithms. Liblice 1967, Apl. Mat. 13, 1968.

[44] R. Zezula : Numerische Stabilität eines Algorithmus zur Berechnung des Eigenparameters eines Matrizenoperator mit Hilfe der Reduktionsmethode und der Banachschen Iterationene. Liblice 1967, Apl. Mat. 13, 1968.

[45] R. Zurmuhl : Matrizen und ihre technische Anwendungen. Berlin 1964.

CENTRO INTERNAZIONALE MATEMATICO ESTIVO

(C.I.M.E.)

J.H. BRAMBLE

ERROR ESTIMATES IN ELLIPTIC BOUNDARY VALUE PROBLEMS

Corso tenuto ad Ispra dal 3-11 Luglio 1967

INTRODUCTION

Error Estimates in Elliptic Boundary Value Problems

J.H. Bramble (University of Maryland)

In these lectures, I will discuss some methods of obtaining error estimates for finite difference approximations to solutions of elliptic differential boundary value problems. Because of the limited time, I shall restrict my attention to the Dirichlet problem, although some of the methods are easily carried over to other boundary conditions. The first part will be devoted to second order problems. In fact, in order to illustrate the methods, I will restrict my attention to the Dirichlet problem for Poisson's equation, with "zero boundary values" and the classical Dirichlet problem for Laplace's equation. The last part will be devoted to some results on higher order elliptic equations. In most cases, I will not give more than a sketch of the proof, indicating the details. Instead of choosing to discuss a general class of operators and a corresponding general class of difference schemes, I shall consider a specific operator and certain specific difference schemes so as not to obscure the essential points of the method of analysis.

I will not, during these talks, make extensive references to related work but shall include in the bibliography a number of closely related papers. I shall restrict the references to the specific results under discussion.

I choose to formulate the first problem in a weak form. Let R be a bounded open set in E_n with boundary ∂R and let Δ be the Laplace operator

$$\Delta = \sum_{i=1}^{N} \left(\frac{\partial^2}{\partial x^2} \right) .$$

We need the following class of functions V defined on R :

$$V = \{ \phi \mid \phi \in H_o^1 \; ; \; \Delta \phi \in C_o^\infty \} ,$$

where C_o^∞ is the class of infinitely differentiable functions with support contained in R and H_o^1 is the Hilbert space obtained by completing C_o^∞ with respect to the norm

$$\|v\|_1^2 = \int_R u^2 \, dx + \sum_{i=1}^{N} \int_R \left(\frac{\partial v}{\partial x_i} \right)^2 dx .$$

All functions for the present will be assumed to be real valued. We now state problem I.

<u>Problem I:</u> Given $F \in (L_\infty)'$, find $u \in Lp$, $1 \leq p < \frac{N}{N-2}$ such that

$$\int_R u \Delta \phi = \langle \phi , F \rangle \; , \forall \phi \in V .$$

Here $(L_\infty)'$ is the space of continuous linear functionals defined on L_∞
The quantity $\langle \phi , F \rangle$ is the value of the functional F at the point ϕ .

For example, if "$F \in L_1$" then $\langle \phi , F \rangle = \int \phi F$. The space L_p is the completion of the C^∞ functions in R with respect to the norm

$$\|u\|_{L_p} = \left(\int_R |u|^p \, dx \right)^{1/p} .$$

We will discuss this problem later but let me first remark that when F and ∂R are sufficiently smooth Problem I is just the classical problem

$$\Delta u = F \quad \text{in} \quad R$$
$$u = 0 \quad \text{on} \quad \partial R \quad .$$

The second problem is the classical Dirichlet problem

<u>Problem II:</u> Given $f \in C^o(\partial R)$, find $u \in C^o(\overline{R})$ such that $\Delta u = 0$ in R .

We will refer to well known results (eg. on regularity) for this problem when needed. We can immediately state the following theorems.

<u>Theorem 1:</u> There exists a unique solution to problem I.

<u>Theorem 2</u>: If ∂R is such that at each point of ∂R there exists a barrier then problem II has a unique solution.

Theorem 2 is classical. As we will see Theorem 1 (existence part) can be proved by means of a difference method.

Let us now formulate corresponding difference problems and investigate their properties.

Let E_{Nh} be the set of mesh points in E_N , i.e., points of the form $(i_1 h, \cdots, i_N h)$, for $h > 0$ and $i_1, \cdots, i_N$ integers.

We make some definitions.

a) $\bar{R}_h = R \cap E_{Nh}$

b) $S_\rho(x) = \{ y \mid |x-y| \leq \rho \}$

c) $R_h = \bar{R}_h \cap \{ x \mid S_{\sqrt{N}h}(x) \subset R \}$

d) $\partial R_h = \bar{R}_h - R_h$.

e) $\partial_i V(x) = \dfrac{V(x+he_i) - V(x)}{h}$; $\bar{\partial}_i V(x) = \dfrac{V(x) - V(x-he_i)}{h}$

(e_i is the vector with 1 in the <u>ith</u> position and 0 in the others)

f) $\Delta_h V(x) = \sum_{i=1}^{N} \partial_i \bar{\partial}_i V(x)$

We can immeidately state the following.

<u>Lemma 1</u>: Suppose $-\Delta_h V \geq 0$ in R_h , $V \geq 0$ in $E_{Nh} - R_h$. Then $V \geq 0$ in E_{Nh} . From this it immediately follows that the discrete problem

$$\Delta_h V(x) = F(x) , x \in R_h$$

(1)

$$V(x) = g(x) , x \in \partial R_h$$

has always one and only one solution for any given F and g . Thus we can introduce the discrete Green's function $G_h(x,y)$ defined as

$$\Delta_{h,x}\, G_h(x,y) = -\, h^{-N}\, \delta(x,y) \ , \ x \in R_h \ , \ y \in \overline{R}_h$$

(2)

$$G_h(x,y) = \delta(x,y) \ , \ x \in \partial R_h \ , \ y \in \overline{R}_h \quad .$$

We can now make some statements about G .

Lemma 2: $G_h(x,y) \geq 0$.

Lemma 3: F or any V defined on $\overline{R}_h$

(3) $$V(x) = -\, h^N \sum_{y \in R_h} G_h(x,y)\, \Delta_h\, V(y) + \sum_{y \in \partial R_h} G_h(x,y)\, V(y) \quad .$$

This follows from uniqueness in the discrete problem.

We next introduce the following function. Let

(4) $$V(x) = \begin{cases} 1/\gamma_2 \ \ln\left[\dfrac{d_o^2 + \alpha_h^2}{|x|^2 + \alpha h^2}\right] , & N = 2 \ , \\[2ex] \dfrac{1}{(N-2)\,\gamma_N} \left[\, |x|^2 + \alpha h^2 \,\right]^{\frac{2-N}{2}} , & N \geq 3 \ . \end{cases}$$

Then we can prove

Lemma 4: For suitably chosen γ_N , α, and d_o (independent of h)

$G_h(x,y) \leq V(x-y)$, $y \in R_h$.

The proof of lemma 4 consists in showing that it is possible to choose γ_N , α and d_o in such a way that for $y \in R_h$

$$
\begin{aligned}
& - \Delta_{h,x} V(x-y) = h^{-N} \quad , \quad x = y \\
(5) \qquad & - \Delta_{n,x} V(x-y) \geq 0 \quad , \quad x \neq y \\
& V(x-y) \geq 0 \quad \text{all } x \quad .
\end{aligned}
$$

Then it follows from lemma 1 applied to $V(x-y) - G_h(x,y)$ that lemma 4 is true. From lemma 4 we can prove

<u>Lemma 5:</u> Let $1 \leq p < \frac{N}{N-2}$. Then there is a constant C_p independent of h and x such that

$$\left(h^N \sum_{y \in R_h} G_h^p(x,y) \right)^{1/p} \leq C_p \quad .$$

After using lemma 4 the sum is estimated by comparing the sum with corresponding analogous integrals using the fact that $|x-y_o|^{-\beta}$ is subharmonic, $\beta \geq N-2$, $x \neq y_o$. For the case $N \geq 3$, for example, we can obtain the estimate

$$(6) \qquad h^N \sum_{y \in R_h} G_h^p(x,y) \leq C \int_S \frac{dx}{|x-y_o|^{p\left(\frac{N-2}{2}\right)}} \quad ,$$

where S is a sufficiently large sphere containing R and having center at an arbitrary point $y_o \in R$. The integral is convergent if $p < \frac{N}{N-2}$.

By taking $V \equiv 1$ in lemma 3 we obtain

Lemma 6: $$\sum_{y \in \partial R_h} G_h(x,y) = 1 \quad .$$

Now it is a simple matter to give a convergence theorem. First we define the discrete problem.

Problem II_1:

$$\Delta_h \, u_h(x) = 0 \qquad x \in R_h$$

$$u_h(x) = f(\overline{x}) \;, \; x \in \partial R_h \;, \quad \overline{x} \in \partial R$$

$$\overline{x} \in S_{\sqrt{N}h}(x)$$

Of course u_h exists and is unique. Now we have

Theorem 3: Let u be the solution of problem II and suppose that $u \in C^2(\overline{R})$. Then $u_h \to u$ uniformly as $h \to 0$. We apply lemma 3 to $u - u_h$.

$$u(x) - u_h(x) = -h^N \sum_{y \in R_h} G_h(x,y) \, \Delta_h \, u(y) + \sum_{y \in \partial R_h} G(x,y) \, (\, u(y) - f(\overline{y}) \,) \quad .$$

Hence by lemmas 5 and 6,

$$(7) \qquad \left| u(x) - u_h(x) \right| \leqslant C \sup \left| \Delta_h u \right| + \sup_{\substack{y, \overline{y} \in \overline{R} \\ y \in S_{\sqrt{N}\,h}(\overline{y})}} \left| u(y) - u(y) \right| \quad .$$

Clearly the right hand side tends to zero as $h \to 0$.

Immediately from (7) we can deduce

Theorem 4: If $u \in C^{2,\alpha}(\bar{R})$ $0 < \alpha \leq 1$, then

$$\sup_{\bar{R}_h} \left| u - u_h \right| \leq Ch^{\alpha} .$$

The proof is obvious. Since u is harmonic, $|\Delta_u u| \leq Ch^{\alpha}$ in $\bar{R}$ and $|u(y) - u(\bar{y})| \leq Ch$.

Now no matter how smooth u is in the closure of R the best possible result is with $\alpha = 1$. The last term on the right of (7) prevents us from obtaining a higher order estimate. As we will see, however, theorems 3 and 4 can both be improved in the sense that if we place some restrictions on ∂R then we can obtain similar theorems with u less regular up to the boundary.

Theorem 5: Suppose ∂R is such that u can be approximated uniformly by a sequence of functions each of which is harmonic in $\bar{R}$. Then $\tilde{u}_h \to u$ uniformly as $h \to 0$. ($\tilde{u}_h$ is the extension of u_h as a constant on each cube $C_h(x) = \{y \mid x_i - h/2 < y_i \leq x_i + h/2 , i = 1 , \cdots , N\}$)

Remark. Conditions on ∂R have been given by several authors,(c.f. Brelot [8] and Walsh [11] such that every function continuous on ∂R can be approximated uniformly by functions harmonic in $\bar{R}$. This is not necessary for our theorem as we can see by the following example. Let R be star-shaped with respect to some point which, without loss, we call the origin. Then if u exists, it can be uniformly approximated by a sequence each member of which is harmonic in $\bar{R}$. For, define

$$u_n(x) = u(p_n x)$$

where $p_n \uparrow 1$ as $n \to 0$. Now $\Delta u(p_n x) = 0$ in $\overline{R}$. Clearly since $u \in C^0(\overline{R})$, $\tilde{u}_n \to u$ uniformly in $\overline{R}$. Thus starshapedness is a sufficient condition for the convergence of u_h to u whenever u exists.

The proof of theorem 5 is immediate. We have only to estimate $u_h - u_n$ for h small and n large. But from lemma 3

$$|u_h - u_n| \leq (\sup_{R_h} |\Delta_h u_n| + \sup_{\substack{\overline{x},\, x \in \overline{R} \\ x \in S_{\sqrt{N}h}(x)}} |u(x) - u_n(\overline{x})|$$

which we can clearly make as small as we wish by first taking n large and then h small. We also can prove the following.

<u>Theorem 6:</u> Let $\partial R \in C^2$ (piecewise, with no reentrant cusps if $N = 2$). Suppose that u is Hölder continuous with exponent $0 < \lambda \leq 1$. Then, for every $\varepsilon > 0$

$$\sup_R |u - \tilde{u}_h| \leq K(\varepsilon)^{\lambda - \varepsilon}$$

where $K(\varepsilon)$ is a constant which depends on ε, R and u but not on h .

The proof of this is based on the following two lemmas.

<u>Lemma 7:</u> Let $\partial R \in C^2$ and u be harmonic in R and λ - Hölder continuous in $\overline{R}$. Let $d(x)$ be the distance from the point $x \in R$ to the boundary (which is well defined in a strip S_δ of fixed width δ) . Then for every $\varepsilon > 0$ there is a $K(\varepsilon)$ such that

$$|\Delta_h u(x)| \leq K(\varepsilon)\, h^{\lambda - \varepsilon}\, d(x)^{\varepsilon - 2}$$

for $x \in R_h \cap S_\delta$.

This lemma follows from the mean value theorem for harmonic functions. One obtains an estimate for the Hölder continuity of the second derivatives which depends on the distance to the boundary.

The next lemma is crucial.

Lemma 8: For every $\varepsilon > 0$ there exists $K(\varepsilon)$ such that, if $\partial R \in C^2$

$$h^N \sum_{y \in S_\delta \cap R_h} G_h(x,y)\, d(y)^{\varepsilon-2} \leq K(\varepsilon) \quad .$$

The proof of this lemma is tedeous, long and involved, but the motivation is the following. We consider formally

$$\psi(x) = \int_R G(x,y)\, d(y)^{\varepsilon-2}\, dy$$

where G is the continuous Green's function and we have assumed that d has been suitably extended to R. Then, formally,

$$- \Delta\psi = d^{\varepsilon-2} \quad \text{in} \quad R$$

$$\psi = 0 \quad \text{on} \quad \partial R \quad .$$

It is not difficult to see using the maximum principle that for $\partial R \in C^2$

$$\psi(x) \leq C\, d^{\varepsilon}(x)$$

Hence the procedure for proving lemma 8 is based on constructing a suitable comparison function and using the maximum principle, lemma 1.

We shall now turn our attention, for a time to problem I. First I will briefly sketch a proof of Theorem 1. For uniqueness it suffices to show that if $\int v\Delta\phi = 0$, $\forall\phi\varepsilon V$ then $v = 0$. But one can show that if $\psi \varepsilon C_a^\infty$ then there exists a $\phi \varepsilon V$ such that $\Delta\phi = \psi$. Uniqueness follows. The existence follows once the inequality

(8) $$\|\phi\|_{L_\infty} \leq C \|\Delta\phi\|_{L_q} , q > \frac{N}{2}$$

is established. (One used the Hahn-Banach theorem.)

Since we shall be interested in the convergence properties of related difference schemes, we shall sketch the proof of (8) by the difference method.

For any integrable function f we define

$$(f)_h(x) = \frac{1}{h^N} \int_{C_h(x)} f\, dy$$

where $C_h(x)$ is the cube with center x , side h and sides parallel to axes, defined previously.

We let ϕ_h be the solution of

$$\Delta_h \phi_h(x) = (\Delta\phi)_h(x) , x \varepsilon R_h$$

$$\phi_h(x) = 0 , x \varepsilon E_{Nh} - R_h$$

Then it follows from lemmas 3 and 5 and Hölder's inequality that

(9) $$\|\bar{\phi}_h\|_{L_\infty} \leq C\ \|\Delta\phi\|_{L_q} \quad , \ q > N/2$$

where $\bar{\phi}_h$ is the extension of ϕ_h as a constant in each cube C_h .

What we would like to show is

<u>Lemma 9</u>: Let $\phi \in V$. Then $\bar{\phi}_h \to \phi$ as $h \to 0$, weakly in L_p , $1 < p < \infty$.

This can be done by using (9) and the fact that Lp , $1 < p < \infty$ is reflexive and hence bounded sets are weakly compact. We then choose a sequence $\{h_n\}$, $h_n \to 0$ as $n \to \infty$ such that $\bar{\phi}_{h_n} \to \bar{\phi} \in L_p$, weakly as $n \to \infty$. We then show that $\bar{\phi} = \phi$ and that every sequence converges to ϕ .

Clearly (8) follows from lemma 9 and Theorem 1 is then proved.

We next want to pose a discrete problem analogous to problem I .

<u>Problem I_1 :</u>

$$\Delta_h\, u_h(x) = \langle M_h(x) \, , F \rangle \quad , \quad x \in R_h$$

$$u_h(x) = 0 \quad , \quad x \in E_{Nh} - R_h \quad .$$

where $M_h(x)$ is the function of y for fixed x defined by

$$M_h(x) = \begin{cases} h^{-N} & , y \in C_h(x) \\ 0 & , y \notin C_h(x) \quad . \end{cases}$$

Note that if $F \in L_1$ then

$$<M_h(x) , F> = (F)_h(x) .$$

It is now rather easy to prove

Theorem 7: Let $F \in L_1$ and u and u_h be the solutions to problems I and I_1, respectively. Then $\tilde{u}_h \to u$ weakly in L_p, $1 \leq p < \frac{N}{N-2}$ as $h \to 0$.

We consider the following identity for arbitrary $\phi \in V$:

$$(10) \qquad \int \tilde{u}_h \Delta\phi = h^N \sum u_h (\Delta\phi)_h = h^N \sum \phi_h F_h = \int \tilde{\phi}_h F .$$

Again, using lemma 5 we can show that

$$\|\tilde{u}_h\|_{L_p} \leq C \|F\|_{L_1} , 1 \leq p < \frac{N}{N-2} ,$$

so that again we can take a sequence $\tilde{u}_{h_n} \to \tilde{u}$ (some element of L_p) weakly in L_p, $1 < p < \frac{N}{N-2}$. Clearly the left hand side of (10) converges to $\int \tilde{u}\Delta\phi$ as $n \to \infty$. Now $\tilde{\phi}_{h_n} \to \phi$ weakly in L_p for $1 < p < \infty$ and by (8) and (9) $\tilde{\phi}_{h_n}$ and ϕ are bounded. Thus it is a simple matter to conclude that

$$\int \tilde{\phi}_{h_n} F \to \int \phi F \quad \text{as} \quad n \to \infty .$$

Hence $\tilde{u}$ is the solution of problem I or $\tilde{u} = u$. But every sub sequence converges to u so that $\tilde{u}_h \to u$.

In order to show the strong convergence we introduce a sequence $F_n \in C_o^\infty(R)$ for each u and such that $\lim_{n\to\infty} \int_R |F-F_n| = 0$. We also introduce the corresponding functions u_n and u_{nh}. Now we have

$$\|\tilde{u}_h - u\|_{L_p} \leq \|\tilde{u}_h - \tilde{u}_{nh}\|_{L_p} + \|\tilde{u}_{nh} - u_n\|_{L_p} + \|u_n - u\|_{L_p}$$

Now the first and last terms are bounded by $C\,\|F-F_n\|_{L_1}$ where C does not depend on h. The middle term can be treated rather easily directly but we can observe that since $F_n \in C_o^\infty(R) \subset L_2$ for each n as a special case of the work of Céa [9] we conclude that $\|\tilde{u}_{nh} - u_n\|_{L_2} \to 0$ as $h \to 0$ for fixed n. But since it is easy to see that $u_n \in V$ and hence is bounded so that in fact $\|\tilde{u}_{nh} - u_n\|_{L_p} \to 0$ as $h \to 0$ for all $p < \infty$. Thus taking n large and then h small we have proved.

<u>Theorem 8</u>: Let $F \in L_1$ and u_h and u be the solutions to problem I and I_1 respectively. Then $\tilde{u}_h \to u$ strongly in L_p, $1 \leq p < \frac{N}{N-2}$ as $h \to 0$.

The next two theorems give information when F is not necessarily in L_1.

<u>Theorem 9</u>: Let $F \in (L_\infty)'$ and let C_p be the class of piecewise continuous functions in R. Suppose that F, when considered as a functional on $C_p \cap L_\infty$ has support Ω contained in R (i.e. Ω is closed and if $\phi \in C_p \cap L_\infty$ and $\phi = 0$ on Ω then $\langle \phi, F \rangle = 0$). Then $\tilde{u}_h \to u$ weakly in L_p, $1 \leq p < \frac{N}{N-2}$ as $h \to 0$.

We easily obtain $\int \tilde{u}_h \Delta\phi = \langle \bar{\phi}_h, F \rangle$ and $\|\tilde{u}_h\|_{L_p} \leq C\,\|F\|_{(L_\infty)'}$.

It then must be shown that $\langle \bar{\phi}_h, F \rangle \to \langle \phi, F \rangle$, as $h \to 0$.

The additional hypothesis is imposed now since F can not necessarily be approximated (strongly) by nice functions. However by using the results of Bramble and Hubbard [5] we can show uniform convergence of $\tilde{\phi}_h$ to ϕ on compact subsets of R.

Instead of puting a condition on F we could impose one on ∂R.

<u>Theorem 10</u>: Let $F \varepsilon (L_\infty)'$ and $\partial R \varepsilon C^2$. Then $\tilde{u}_h \to u$ weakly in L_p, $1 \leqslant p < \frac{N}{N-2}$ as $h \to 0$.

If $\partial R \varepsilon C^2$ then in fact each $\phi \varepsilon V$ will belong to $C^1(\bar{R})$ so that theorem 6 with $\lambda = 1$ can be applied to show the uniform convergence of $\tilde{\phi}_h$ to ϕ in $\bar{R}$. In fact using the estimate of theorem 6 (slightly modified) we obtain a kind of estimate for the rate of weak convergence.

<u>Theorem 11</u>: Let $F \varepsilon (L_\infty)'$ and $\partial R \varepsilon C^2$. Then for any $\psi \varepsilon C_o^\infty$

$$\left| \int_R (\tilde{u}_h - u)\, \psi \, dx \right| \leqslant C\, h\, \|F\|_{(L_\infty)'}$$

where C depends on ψ but not on h.

We next observe that if F is the "delta function" when restricted to V for an arbitrary point x then the solution u of problem I will be the Green's function with singular point x. That is we have the existence and uniqueness of $G(x,y)$ such that

$$(11) \qquad \phi(x) = - \int_R G(x,y)\, \Delta\phi(y)\, dy \quad , \ \forall\, \phi \,\varepsilon\, V \ .$$

Now it is possible to prove the following.

<u>Lemma 10:</u> Let $F \in L_1$. Then

$$u(x) = - \int_R G(x,y) F(y) dy \tag{12}$$

is the solution to problem I.

This lemma is proved by first showing that $G(x,y)$ is symmetric and then applying the basic relation (11) enjoyed by G . The only difficulty arrises in showing that the theorem of Fubini-Tonelli on interchange of integration can be applied. This can be done with the aid of the difference approximations.

Having the representation of lemma 10 we can show

<u>Theorem 12</u>: Let $F \in L_q$, for some $q > N/2$. Then $\tilde{u}_h \to u$ pointwise in R as $h \to 0$.

Clearly from lemma 3

$$u_h(x) = - h^N \sum_{y \in R_h} G_h(x,y) \ (F)_h(y)$$

and hence

$$\tilde{u}_h(x) = - \int_R \tilde{G}_h(x,y) F(y) dy \quad .$$

Combining this with (12) we have

$$\tilde{u}_h(x) - u(x) = \int_R \left(G(x,y) - \tilde{G}_h(x,y)\right) F(y) dy \quad .$$

But from theorem 9 and the continuity of ϕ in R it follows that for each fixed $x \in R$ $\tilde{G}_h(x,\cdot) \to G(x,\cdot)$ weakly in L_p, $1 \leq p < \frac{N}{N-2}$. Since $F \in L_q$, for some $q > N/2$ the theorem follows.

In this same spirit we can prove

Theorem 13: Let $F \in L_q$ for some $q > N/2$ and $\partial R \in C^2$. Then $\tilde{u}_h \to u$ uniformly as $h \to 0$.

To show this we simply approximate F strongly in L_q by a sequence $\{F_n\}$ with $F_n \in C_o^\infty(R)$ for each n. Then

$$\tilde{u}_n(x)-u(x) = \int_R \left(G(x,y)-\tilde{G}_h(x,y)\right)F_n(y)dy + \int_R \left(G(x,y)-\tilde{G}_h(x,y)\right)\left(F(y)-F_n(y)\right)dy$$

so that

$$|\tilde{u}_h(x)-u(x)| \leq |\phi_h(x)-\tilde{\phi}_{nh}(x)| + C \, \|F-F_n\|_{L_q}$$

where C does not depend on x and $\phi_n \in V$. For large n the second term is small and by the previous remarks $\phi_{n_h} \to \phi_n$ uniformly as $h \to 0$.

Finally, if F is smooth and ∂R is smooth we have the analog of theorem 6.

Theorem 14: Suppose $F \in C^{o,\lambda}(\overline{R})$ and $\partial R \in C^2$. Then for problem I we have for every $\varepsilon > 0$

$$\sup_R \left|\tilde{u}_h - u\right| \leq K(\varepsilon)\, h^{\lambda-\varepsilon}$$

where $K(\varepsilon)$ is a constant independent of h.

This is similar to theorem 6 for II. Take u_1 such that $\Delta u_1 = F$, $u_1 \varepsilon C^{2,\lambda}$. Then set $u_2 = u - u_1$. Theorem 14 then follows from theorems 4 and 6.

In order to obtain rate of convergence estimates which are of higher order it is clearly necessary to modify the difference scheme near the boundary. We will for the time being still be considering problem I but with various assumptions on ∂R and F.

We shall now redefine R_h and ∂R_h.

a) $N_h(x)$ is the set of "neighbors" of x with respect to Δ_h.

b) $R_h = E_{Nh} \bigcap R$

c) ∂R_h is the set of points on ∂R which lie on "mesh lines".

d) $R'_h = R_h \bigcap \{ x | N_h(x) \quad \bar{R} \}$

e) $R^*_h = R_h - R'_h$

f) $R_h = R_h \bigcup \partial R_h$.

For V defined on ∂R_h we define

g) $\partial_i V(x) = \dfrac{V(x+\alpha_i e_i h) - V(X)}{\alpha_i h}$; $\bar{\partial}_i V(x) = \dfrac{V(x)-V(x-\alpha_i e_i h)}{\alpha_i h}$

where $\alpha_i h$ is the distance between adjacent points of $\overline{R}_h$ in the x_i direction $0 < \alpha_i \leq 1$.

h) $\Delta_h V(x) = \sum_{i=1}^{N} \partial_i \overline{\partial}_i V(x)$.

Now everything said so far remains true with this definition , with the minor redefinition of $(F)_h$ and $M_h(x)$ regarding problem I_1. Essentially, if one defines F as "zero" outside R then everything is the same. However we can give now an additional estimate for G_h which allows us to obtain second order convergence for smooth solutions.

Lemma 11:

$$h^{N-2} \cdot \sum_{y \in R_h^*} G_h(x,y) \leq 1 \ .$$

This is trivially obtained by taking for V in lemma 3, V = 0 on ∂R_h and V = 1 in R_h . Then

$$\Delta_h V(x) = 0 \quad , \quad x \in R_h'$$

$$-\Delta_h V(x) \geq h^{-2}, \quad x \in R_h^* \ .$$

Let us call the analog of problem I_1, problem I_2. Then we get immediately

Theorem 15: Let u and u_h be the solutions of I and I_2 respectively and $u \in C^4(\overline{R})$. Then

$$\sup_{x \in R_h} \left| u_h(x) - u(x) \right| \leq C h^2 \ .$$

We simply note that

$$|\Delta_h u - \Delta u| \leq \begin{cases} C\,h & \text{on } R_h^* \\ C\,h^2 & \text{on } R_h' \end{cases}$$

so that puting $V = u_h - u$ in lemma 3 we have the result. In fact we note that even though the approximation locally near the boundary is of order h the contribution to the error is a term of order h^3 . This, we will see, will lead to higher order approximations.

Before mentioning other approximations we shall see what can be said if certain more specific information is known about F . Thus we shall consider briefly the case in which F is very smooth except at one point and ∂R is also smooth. Again in this case the Green's function technique is fruitful.

The following theorem gives information about the error in this case.

<u>Theorem 16</u>: Let ∂R be smooth and $u \in C^4(\overline{R}-0)$ where the origin 0 is an arbitrary point of $\overline{R}$ (for convenience always chosen to be the center of a mesh cube.) Suppose that $m + \lambda > 2 - N$ where m is an integer and $0 < \lambda \leq 1$ and

$$(13) \qquad |D^k u(x)| \leq K \begin{cases} 1 & , \ |k| \leq m \\ |x|^{m+\lambda-|k|} & , \ m+1 \leq |k| \leq 4 , \end{cases}$$

where k is a multi-index $k_1 = (k_1 \cdots , k_j)$, $k_1,\dots,k_j$ non-negative integers, $|k| = \sum k_i$ and

$$D^k u = \frac{\partial^{|k|} u}{\partial x_1^{k_1} \cdots \partial x_j^{k_j}} .$$

The for u and u_h the solutions of I and I_2 respectively

$$|u(x)-u_h(x)| \leq K \begin{cases} h^{m+\lambda+N-2-\varepsilon} |x|^{\varepsilon+2-N}, & 2-N<m+\lambda\leq 4-N \\ h^2\left(|x|^{n+\lambda-2} + 1\right), & 4-N<m+\lambda \end{cases} .$$

In this case since we have quite specific knowledge of the behavior of the solution at the origin we obtain an estimate for the error which is point dependent. Thus even if the solution u has the form

$$u(x) = |x|^{2-N+\delta} + \text{regular function}, \quad \delta > 0 ,$$

we get from the theorem that $u_h \to u$ uniformly on every compact subset not containing the origin.

Also note that under the assumption (13) if $u \in C^{2+\delta}$,$\delta > 0$ $(m=2)$ then the convergence is second order. This shows clearly that the usual sufficient condition that $u \in C^4(\bar{R})$ is far from necessary for $O(h^2)$ convergence.

Note also that when $u \in C^\lambda (m=0)$ we obtain a uniform rate of h^λ or $h^{\lambda-\varepsilon}$.

As might be imagined the theorem is proved by using the representation lemma 3 and estimating the resulting expressions. The details are long and technical and are found in [7] but I want to point out the crucial points. First of all the essential ingredient is the majorant (lemma 4) for G_h . This tells us that, as might be expected G_h behaves quite like the fundamental solution for Laplace's equation in the neighborhood of the singularity. Having this we are led to proving a sum relation analogous to a well known integral expression

Lemma 12: If $-N < p , q < 0$ and $x , z \in R_h$ are such that $|x-y| \geq ah , |z-y| \geq ah , a > 0 , \forall y \in R_h$ then

$$h^N \sum_{y \in R_h} |x-y|^p |z-y|^q \leq C \{ |x-z|^{N+p+q} + 1 \} .$$

The proof is done by showing that the sum is majorized by $C \int_R |x-y|^p |z-y|^q dy$.

For this formulation, however, we can get a sharper theorem than either theorem 6 or 15.

Theorem 17: Let $\partial R \in C^2$ (piecewise, with no reentrant cusps if $N = 2$). Suppose that u is the solution of II and u_h is the solution of II_2 (the analog of II_1 for the reformulation). Let $u \in C^{p,\lambda}(\bar{R})$. Then for $\varepsilon > 0$

$$\sup_{x \in R_h} \left| u_h(x) - u(x) \right| \leq K \left[h^{p+\lambda-\varepsilon} + h^2 \right] , \ p = 0,1,2 \ldots .$$

The proof is similar to that of theorem 6. Note that with $u \in C^{2,\lambda}(\overline{R})$ the theorem 4 would show only a rate of h^{λ} and for $C^{1,\lambda}$, $\lambda < 1$ we would conclude nothing. However, we get second order convergence when the second derivatives are Hölder continuous in $\overline{R}$.

Clearly these methods are not restricted to these particular difference formulations for the Dirichlet problem for Poisson's equation. One can treat

a) More general operators (second order).

b) Various boundary conditions.

c) Eigenvalue problems.

d) Various difference approximations.

I shall discuss an example of the last extension, since it brings out the fact that in the transition from the interior to a curved boundary one can (in the Dirichlet problem) take approximations which are of the order of accuracy (locally) worse by a factor of h^2 and still obtain as a global error that of the interior.

For this example I choose $N = 2$ and consider the nine point approximation

$$(14)\quad \Delta_h V(x) = \frac{1}{6h^2}\{4V(x_1+h,x_2) + 4V(x_1-h,x_2) + 4V(x_1,x_2+h)$$

$$+ 4V(x_1,x_2-h) + V(x_1+h,x_2+h) + V(x_1-h,x_2+h) + V(x_1+h,x_2-h)$$

$$+ V(x_1-h,x_2-h) - 20\,V(x_1,x_2)\}$$

If $u \in C^6$

$$|\Delta_h u - [\Delta u + \frac{h^2}{12} \Delta^2 u]| \leq K h^4 ,$$

so that Δ_h is now locally 4th order. Now we take Δ_h to be defined by (14) in R_h at points of R_h (say R_h') where only R_h points are involved in (14). At $R_h^* = R_h - R_h'$ we take a second order approximation to Δ . ∂R_h we take as in the second formulation. One can then show by appropriate modifications of the previous Green's function method that if u_h is the solution of our new problem II_3 (for $\Delta u = 0$) then we have

Theorem 18: Let u be the solution of II and $u \in C^6(\overline{R})$. Then if u_h is the solution of II_3

$$\sup_{x\in R_h} \left| u_h(x) - u(x) \right| \leq C h^4 .$$

But in fact we can lessen considerably the requirement that $u \in C^6(\overline{R})$ and obtain by the methods of theorem 17

Theorem 19: Let $\partial R \in C^2$ and u and u_h the solution of II and II_3 respectively. Then if $u \in C^{p,\lambda}(\overline{R})$

$$\sup_{x\in R_h} \left| u_h(x) - u(x) \right| \leq C \left\{ h^{p+\lambda-\varepsilon} + h^4 \right\} .$$

So far the approximations mentioned have all possessed a common property, i.e., that of being "of positive type." This means that if A^{ij} is the matrix of coefficients of the linear system then

$$A^{ii} > 0$$

(15)

$$A^{ij} \leq 0 \quad i \neq j$$

the second condition possibly failing near the boundary. In fact, it is this condition, together with

(16)
$$-\sum_j A^{ij} \leq A^{ii} \quad , \quad i = 1, \ldots, k \quad .$$

that makes lemma 1 trivial.

To show that this is just a convenience I wish to give another example. Suppose instead of (15) we used the 9-point $0(h^4)$ approximation

$$\Delta_h V(x) = h^{-2}\Big\{-\frac{1}{12}\Big[V(x_1+2h,x_2)+V(x_1-2h,x_2)+V(x_1,x_2+2h)+V(x_1,x_2-2h\Big] +$$
$$+\frac{4}{3}\Big[V(x_1+h,x_2)+V(x_1-h,x_2)+V(x_1,x_2+h)+V(x_1,x_2-h)\Big] - 5V(x_1,x_2)\Big\}.$$

It is possible to show that the error is of the order h^4 when $u \in C^6(\bar{R})$ and probably a theorem like 19 is also true. Since the properties (15) and (16) are not possessed by the resulting system the analysis is much more difficult. It is interesting to note, however, that a corresponding discrete Green's function will still be positive, a fact which is no longer completely trivial.

As is evident the preceding discussion is in many respects special for second order equations, since much use was made of the maximum principle or, what is the same, the positivity of the Green's function. Thus it appears that, in attempting to treat higher order equations, we should work more with norms other than the maximum norm.

I would like first to sketch some results of Thomée[10] on higher order equations and difference approximations.

Consider the differential operator

$$L\,u = L(D)\,u = \sum_{|\beta|=|\gamma|=m} a_{\beta\gamma}\, D^{\beta+\gamma}\, u\ ,\quad a_{\beta\gamma} = a_{\gamma\beta}$$

where β and γ are multi-indices, i.e. $\beta=(\beta_1,\ldots,\beta_N)$, where the β_j's are non-negative integers, $|\beta| = \sum_{j=1}^{n} \beta_j$ and similarly for γ . The $a_{\beta\gamma}$ are real constants and

$$D = \left(i^{-1}\frac{\partial}{\partial x_1}\right)^{\beta_1} \cdots \left(i^{-1}\frac{\partial}{\partial x_N}\right)^{\beta_N} .$$

We assume that L is elliptic, i.e. for real $\xi = (\xi_1,\cdots,\xi_N)$

$$L(\xi) = \sum_{|\beta|=|\gamma|=m} a_{\beta\gamma}\, \xi^{\beta+\gamma} \geq C|\xi|^{2m}, \xi^{\beta} = \xi_1^{\beta_1}\cdots\, \xi_N^{\beta_N}\ ,\ |\xi|^2 = \sum \xi_j^2\ .$$

The Dirichlet problem (III)

$$L\,u = F \quad \text{in} \quad R$$

$$D^{\beta}u = 0\ ,\quad |\beta| \leq m-1 \quad \text{on} \quad \partial R$$

has a smooth solution provided F and ∂R are sufficiently smooth.

Consider approximations of the form

$$L_h u_\xi = h^{-2m} \sum_\alpha C_\alpha u_{\xi+\alpha}$$

where $u_\xi = u(\xi h)$ and the C_α's are complex numbers defined for all α but zero except for a finite number of α's . A point $(\xi+\alpha)h$ will be called a neighbor of ξh if $C_\alpha \neq 0$.

This time R_h will be defined as those points of $R \cap E_{Nh}$ whose neighbors also lie in R . Define $R \cap E_{Nh} - R_h = \partial R_h$. The characteristic polynomial of L_h is defined as the trigonometric polynomial

$$p(\theta) = \sum_\alpha C_\alpha e^{i(\alpha,\theta)}$$

where $\theta = (\theta_1, \cdots, \theta_N)$, $(\alpha,\theta) = \sum \alpha_j \theta_j$. Because of periodicity θ can be taken in the set

$$S = \{\theta \mid |\theta_j| \leq \pi \ , \ j = 1, \cdots, N\}$$

We say that L_h is consistent with L if at an arbitrary point (which we take as the origin)

$$L_h u_o = L\, u(0) + o(1) \quad \text{when} \quad h \to 0 \ .$$

$$(+ 0(h^k) - \text{ consistant of order } k)$$

Now it can be shown

Lemma 13: L_h is consistent with L if and only if

$$p(\theta) = L(\theta) + o\left(|\theta|^{2m}\right) \quad \text{when} \quad \theta \to 0 \ .$$

Now we shall denote the set of complex valued mesh functions on R_h by D_h and

$$(u,v)_h = h^N \sum_{\xi} u_\xi \overline{v}_\xi \ ,$$

$$\|u\|_h^2 = (u,u)_h \ .$$

The sum will always be finite since all functions considered will vanish outside some bounded set.

Define

$$\partial_j u_\xi = (ih)^{-1}\left(u_{\xi+e_j} - u_\xi\right)$$

$$\partial^\beta = \partial_1^{\beta_1} \cdots \partial_N^{\beta_N}$$

and

$$\|u\|_{h,m} = \left(\sum_{|\beta| \leq m} \|\partial^\beta u\|_h^2\right)^{1/2} \ .$$

Now we call the <u>difference</u> operator L_h elliptic if $p(\theta)$ satisfies

$$p(\theta) > 0 \quad \text{for} \quad 0 \neq \theta \varepsilon S \ .$$

In particular if L_h is elliptic $p(\theta)$ is real so that $C_{-\alpha} = \overline{C}_\alpha$.

With this definition, Thomée then gives two a priori inequalities which we state as the next two theorems.

<u>Theorem 20</u>: Let L_h be consistent with L . Then L_h is elliptic if and only if there is a constant C independent of u and h such that

$$\|u\|^2_{h,m} \leqslant C(L_h u,u)_h \ , \ u \ \varepsilon \ D_h \ .$$

The main tool in the proof is the "Fourier transform." For this reason only constant coefficients are treated.

If we define R'_h that part of R_h whose neighbors are in R_h and $R^*_h = R_h - R'_h$ and also

$$L_{h,m} u_\xi = \begin{cases} L_h u_\xi \ , & \xi \ \varepsilon \ R'_h \\ h^m L_h n \ , & \xi \ \varepsilon \ R^*_h \\ 0 & \xi \notin R_h \end{cases}$$

Then it can be shown

<u>Theorem 21</u>: Let ∂R be sufficiently smooth and L_h consistent with L and elliptic. Then

$$||u||_{h,m} \leq C\,||L_{h,m}u||_h \ , \quad u \in D_h \ .$$

This theorem will show that the difference approximation can be cruder near the boundary.

Let $u_{h\xi}$ be the solution of problem III.

$$L_h\, u_{h\xi} = F_\xi \ , \quad u_\xi \in D_h \ .$$

From theorem 20, III_1 has one and only one solution. We have the following convergence estimate of Thomée.

<u>Theorem 22</u>: Let u and u_h be the solutions of III and III_1, respectively. Suppose L_h consistent with L and elliptic and $u \in C^{2m+1}(\bar{R})$. Then if $e_h = u - u_h$ in R_h and 0 outside R_h we have

$$\|e_h\|_{h,m} = 0\left(h^{1/2}\right) \quad \text{as} \quad h \to 0 \ . \tag{17}$$

As interesting examples Thomée gives a number of applications of his theorem to special cases. I want to discuss two of them since in these two one can obtain an additional inequality which together with theorem 22 shows that

$$\|e_h\|_{h,m-1} = 0(h) \quad \text{as} \quad h \to 0 \ .$$

(A) Let

$$L\,u = \sum_{j,k} a^{jk} D^j D^k u \ .$$

and take,

$$L_h u_h = \sum a^{jj} \partial_j \bar{\partial}_j u_h + \frac{1}{4} \sum_{j\neq k} a^{jk} (\partial_j + \bar{\partial}_j)(\partial_k + \bar{\partial}_k)\, u_h \ .$$

(We remark that in this case we can also take u^{ij} to be variable and treat the self adjoint operator

$$L\,u = \sum_{i,j} \frac{\partial}{\partial x_j} \left(a^{ij} \frac{\partial u}{\partial x_i}\right).\ )$$

The matrix a^{ij} is assumed symmetric and to satisfy

$$\sum_{j,k} a^{jk} \xi_j \xi_k \geq a_o \sum |\xi_j|^2$$

for real ξ .

In obtaining (17) the reason for the low power of h is that near the boundary the approximation gave rise to a lower order error term. Thus we want to try to estimate the (m-1) L_2 norm in such a way that the approximations near the boundary are not so important. For motivation we consider a very simple way of obtaining an L_2 inequality in the continuous problem. We suppose that ∂R is smooth and let ψ be the smooth function satisfying

$$-L\,\phi \geq 1 \quad \text{in} \quad R$$

$$\phi = 0 \quad \text{on} \quad \partial R \ .$$

Then by the maximum principle $\phi \leq 0$ in R. Clearly since u and ϕ are zero on ∂R

$$\int u^2\, dx \leq \int u^2(-L\phi)dx = \int \phi(-Lu^2)dx \quad .$$

But

$$- L\, u^2 = 2\, u(-Lu) - 2 \sum_{j,k} a^{jk}\, D^j u\, D^k u$$

$$= 2\, u(-Lu) + 2 \sum_{j,k} a^{jk}(iD^j u)\,(iD^k u)$$

$$\geq 2\, u\,(-Lu)$$

Thus since $\phi \leq 0$ in R

(18) $$\phi(-Lu^2) \leq 2\phi u(-Lu)$$

and hence

(19) $$\int_R u^2\, dx \leq 2 \int_R \phi\, u(-Lu)\, dx \quad .$$

Now the importance of this estimate is that $\phi = 0$ on ∂R and hence Lu is not so influential near the boundary. It is just this type of estimate that would give us something for the difference problem. Unfortunately (18) does not hold in the case of the present difference approximation. However what can be shown in the following. For any V such that $V = 0$ in $E_N - R_h$

(20) $$||V||^2 \leq C\, h^N \sum \phi\, V(-L_h V) + C\, h\, ||V||^2_{h,1} \quad ,$$

where C does not depend on h . The fact to note is that although (18) holds pointwise it was only used in the mean. Now one expects, because of consistency, that an expression for the difference operator, analogous to (18) will hold to within higher order terms. It turns out that these terms can be estimated in the means, hence giving us (20). Thus we conclude

Theorem 23: In the case of example (A) we have

$$\|e_h\|_{h,o} = O(h) \quad \text{as} \quad h \to 0 \ .$$

The one thing that must be used here is an inequality given by Thomée. That is that

$$h^{N-2} \sum_{R_h^*} V^2 \leq C \ \|V\|^2_{h,1} \ .$$

(B) For the second example if we take (N=2)

$$L\, u = \Delta^2\, u$$

and

$$L_h\, u_h = \Delta_h^2\, u_h$$

then we have the first boundary value problem for the biharmonic equation Thomée's result is the first error estimate in this problem for a general domain. He obtains according to theorem 22 an estimate of the order of $h^{1/2}$ for the $m = 2$ norm. We want to look at the $m - 1 = 1$ norm and obtain an order h estimate for the error in this norm.

Again we are motivated by a pointwise differential inequality. Miranda, in obtaining a maximum principle, made use of the fact that

$$\Delta\left[u_x^2 + u_y^2 - u\Delta u\right] \geq u \quad \Delta^2 u$$

(Unfortunately this is special for $N = 2$).

Again we can obtain the analogus expression for the difference operator, which is

$$\Delta_h \left[\frac{1}{2}(V_x^2 + V_{\bar{x}}^2 + V_y^2 + V_{\bar{y}}^2) - V \, \Delta_h \, V \right] \geq$$

$$- V \, \Delta_h^2 \, V + \frac{h^2}{2} \left[(V_{x\bar{x}}^2)_{x\bar{x}} + (V_{y\bar{y}}^2)_{y\bar{y}} \right]$$

where the subscript x and $\bar{x}$ denote the usual forward and backward divided differences. Once again we can introduce a function ϕ such that

$$- \Delta\phi \geq 1 \quad \text{in} \quad R$$

$$\phi = 0 \quad \text{on} \quad \partial R$$

$$\phi \geq 0 \quad \text{in} \quad R \; .$$

Then it is (almost) clear that the last term

$$\frac{h^2}{2} \phi \left[(V_{x\bar{x}})^2_{x\bar{x}} + (V_{y\bar{y}})^2_{y\bar{y}} \right]$$

can be estimated in the mean by

$$C \, h \, \|V\|^2_{h,2} \quad .$$

Thus we are led to the discrete a priori inequality

$$\|V\|^2_{h,1} \leq C\, h^2 \sum \phi\, V\, \Delta_h^2 V + C\, h\, \|V\|^2_{h,2} \quad .$$

This together with theorem 22 leads to the estimate $\|e_h\|_{h,1} \leq C\, h$.

Although it is not true that the maximum norm can be estimated by the Dirichlet integral in the continuous case, we can obtain a meaningful estimate in the discrete case, $N = 2$.

<u>Lemma 14:</u> Let, $V = 0$ in $E_N - R_h$. Then

$$\max_{R_h} |V| \leq C|\ln h|^{1/2} \|V\|_{h,1} \quad .$$

This can be obtained by using the discrete Green's function G for Δ_h . From the representation lemma 3 and partial summation we have

$$V(x) = h^2 \sum_y V_{y_1} G_{y_1}(x,y) + V_{y_2} G_{y_2}(x,y)$$

and by Schwarz's inequality

$$|V(x)| \leq \left(h^2 \sum_y \left[G^2_{y_1}(x,y) + G^2_{y_2}(x,y)\right]\right)^{1/2} \|V\|_{h,1} \quad .$$

But it is easy to see that

$$h^2 \sum_y \left(G^2_{y_1}(x,y) + G^2_{y_2}(x,y)\right) = G(x,x) \leq V(0) \leq C|\ln h| \quad .$$

Hence we obtain the estimate $\max_{R_h} |e_h| \leq C|\ln h|^{1/2}\, h$.

Concerning second order approximations in this problem (B) I have given one in Bramble [4] and Zlámal, in [13] has given one. To my knowledge no second order approximation has been proved in general for (A).

As regards the second order approximation given by Zlámal for problem (B) I wish to mention that it is simplier than the one given by me and also that the technique used by Zlámal holds for more general 4th order equations but for $N = 2$. He shows, essentially, that a certain second order interpolation near the boundary is sufficient to increase the rate of convergence from $h^{1/2}$ for the $m = 2$ - norm to $h^{3/2}$ and in the case of example (B) from h to h^2 for the $m - 1 = 1$ - norm.

BIBLIOGRAPHY

1. Babuška, I., Práger, M., and Vitásek, E. Numerical Processes in Differential Equations. Interscience publishers, New York (1966).

2. Bramble, J. H. "On the convergence of difference schemes for classical and weak solutions of the Dirichlet problem." To appear in the proceedings on Differential Equations and Their Applications II, Bratislava, Czechoslovakia (1966).

3. Bramble, J. H. (editor) Numerical Solution of Partial Differential Equations. Academic Press, New York (1966).

4. Bramble, J. H. "A second order finite difference analog of the first biharmonic boundary value problem" Numerische Mathematik 9, 236-249 (1966).

5. Bramble, J. H. and Hubbard, B. E. "Approximation of derivatives by finite difference methods in elliptic boundary value problems." Contributions to Differential Equations, Vol. III, No. 4 (1964).

6. Bramble, J. H., Hubbard, B. E. "Discretization error in the classical Dirichlet problem for Laplace's equation by finite difference methods." Univ. of Md. Tech. Note BN-484 (1967) (to appear, SIAM Series B).

7. Bramble, J. H., Hubbard, B. E., and Zlámal, M. "Discrete analogs of the Dirichlet problem with isolated singularities." Univ. of Md. Tech. Note BN-475 (1966) (in print).

8. Brélot, M. "Sur l'approximation et la convergence dans la theorie des fonctions harmoniques ou holomorphes." Bull. Soc. Math. France 73, 55-70 (1945).

9. Céa, J. "Sur l'approximation des problemes aux limites elliptiques." Compte rendus 254, 1729-1731 (1962).

10. Thomée, V. "Elliptic difference operators and Dirichlet's problem." Contributions to Differential Equations, Vol. III, No. 3 (1964).

11. Walsh, J. L. "The approximation of harmonic functions by harmonic polynomials and by harmonic rational functions."

12. Zlámal, M. "Asymptotic error estimates in solving elliptic equations of the fourth order by the method of finite differences." SIAM Series B2, 337-344 (1965).

13. Zlámal, M. "Discretization and error estimates for elliptic boundary value problems of the fourth order." (in print).

CENTRO INTERNAZIONALE MATEMATICO ESTIVO

(C. I. M. E.)

G. CAPRIZ

THE NUMERICAL APPROACH TO HYDRODYNAMIC PROBLEMS

Corso tenuto ad Ispra dal 3-11 Luglio 1967

THE NUMERICAL APPROACH TO HYDRODYNAMIC PROBLEMS

by

G. Capriz

(Centro Studi Calcolatrici Elettroniche
del CNR presso l'Università di Pisa,
Pisa, Italy)

1. Introduction

Interest in the numerical solution of hydrodynamic problems has been alive for a long time; the book of ref. 1, for instance, bears the date 1922. The reasons of the interest are obvious: so few explicit solutions of the equations of hydrodynamics are known and great gaps still exist in the knowledge on the qualitative behaviour of general solutions [2, 3] . As ref. 1 already shows, attempts at numerical integration were made before the age of computers: some references to this earlier work can be found in the textbooks of Allen [4] and Thom-Apelt [5] , among others.

Von Neumann called attention repeatedly to this field of research [e.g. 6, p. 236] , suggesting that computers would be the right tool for inquiry. Attention was devoted at first to studies of compressible flows [6 A; 7, vol. 4] ; sometimes through the integration of reduced equations of the boundary layer type [7, vol. 3; 8, and the papers quoted there] .

The heuristic interest of the calculations was soon pointed out and among the first problems tackled were those for which the curiosity of the experimenter had not yet been satisfied by the results of the theoretician [9, 10] .

Incompressible flows are studied now with great zest; there is interest in such flows for analysis of motions with a free surface [11, 12, 13], and of motions of natural convection [14, 15], for wheather analysis and prediction [16]perhaps using a "shallow fluid" approximation [17, 18] , for forecasts of flood waves in rivers [18] and many other questions.

Perhaps the most ambitious goal is pursued by those research workers who try to decide, by a thorough numerical study, as to what extent the Navier-Stokes equations (in a finite difference form) are able to describe phenomena of fluid flow instability and even of transition to turbulence. Interesting results have been already obtained in the description of the formation of Karman vortices behind an obstacle [19, 20] , of the spike and bubble in the Rayleigh--Taylor form of instability for superposed fluids [13, 21] of the Taylor vortices at high Reynolds number in the Couette flow [22, 23]. The calculations are so precise that they can be used to deduce values of the functionals of flow (such as heat transfer coefficients and viscous drag coefficients) much more satisfactorily than by other approximate means. The wealth of results obtained by Harlow and collaborators at the Los Alamos Laboratory are so spectacular that they have found space in Scientific American [24] , Science [25] , Datamation [26] , Sciences [27] .

Attempts have been made to follow, in a fluid flow, the production of small eddies from larger ones in three dimensions but the work was hampered by the occurrence of numerical instability [28, 29] . Similar and other difficulties have limited the range of results obtained in studies on the development of perturbations in laminar plane Poiseuille flow [30, 31, 32] . More detailed are the conclusions of another analysis of transition from laminar to turbulent flow (for a flow over a flat plate) [33] ; proposals have also been made for a direct numerical study of turbulent flows.

2. Finite difference approximations for the Navier-Stokes equations

In almost all the researches quoted in Section 1 the authors have resorted to finite difference analogues of the Navier-Stokes equations. So we introduce now those equations restricting our consideration to the incompressible case

$$\frac{\partial \underline{u}}{\partial t} = - (\text{grad}\ \underline{u}) . \underline{u} - \text{grad}\ \varphi + \nu \Delta \underline{u} + \underline{g}\ , \tag{1}$$

$$D = \text{div}\ \underline{u} = 0\ ; \tag{2}$$

here $\underline{u}$ is speed, φ is the ratio of pressure over (constant) density, ν kinematic viscosity and g applied force per unit mass.

To eqns (1), (2) the appropriate boundary conditions must be added, perhaps on unknown boundaries (flows with a free surface).

For the purposes of a numerical study, discrete equivalents to (1), (2) and the boundary conditions can be used, which are based on a net of points where the relevant quantities must be determined.

The discrete equivalent must have a form which suggests feasible numerical algorithms; they must be sufficiently accurate without leading to cumbersome computations and not be subject to numerical instability. Although the requirements are numerous and stringent there is a variety of procedures that meets them; the choice depends on a not well defined criterion of economy.

To obtain convenient numerical algorithms the differential eqns (1), (2) are not the best starting points, for a number of reasons; first of all one must try to separate the unknowns $\underline{u}$ and φ. If the boundary conditions do not involve φ, this unknown can be eliminated altogether from (1) using (2). In fact, this equation states that $\underline{u}$ is solenoidal; hence it can be expressed as the curl of a vector potential ψ which is itself solenoidal; at the same time φ can be eliminated from (1) by taking the curl of both sides :

$$\begin{aligned} \underline{\chi} &= \text{curl } \underline{\phi} \ , \\ \underline{\phi} &= \text{curl } \underline{u} \ , \\ \underline{u} &= \text{curl } \underline{\Psi} \ , \end{aligned} \tag{3}$$

$$\frac{\partial \underline{\phi}}{\partial t} + \text{curl } (\text{grad } \underline{u} \cdot \underline{u}) = - \nu \text{ curl } \underline{\chi} + \text{curl } \underline{g} \ ; \tag{4}$$

here account was taken of the identity: curl curl $\underline{v}$ = grad div $\underline{v} - \Delta \underline{v}$. Thus, using eqns (3) (4) the conservation of mass is exactly verified. If one wants to work in terms of the variables $\underline{u}$ and φ directly, one can substitute (2) with a consequence of (1), in whose derivation (2) plays a rôle. Here difficulties are met because the very important eqn (2) would thus intervene only indirectly; in practice one finds that great care must be taken in the computation if the approximate values of $\underline{u}$ have to correspond to values of D which are sufficiently small to be accepted. In theory one could rely on the following consequence of (1)

$$\Delta \varphi = \text{div } \underline{g} - \text{div } (\text{grad } \underline{u} \cdot \underline{u}) \ . \tag{5}$$

In practice one finds that the use of discrete equivalents of (1), (5) leads to rapid accumulation of errors and to large values of D, at least where the discretization is based on a relatively coarse net. It is more convenient to substitute (1) with the equation

$$\frac{\partial \underline{u}}{\partial t} + \text{div } (\underline{u} \otimes \underline{u}) = - \text{grad } \varphi - \nu \text{ curl curl } \underline{u} + \underline{g} \tag{6}$$

and (5) by this consequence of (6) :

$$\int_t^{t+\Delta t} \left\{ \Delta \varphi + \text{div} \left[\text{div } (\underline{u} \otimes \underline{u}) - \underline{g} \right] \right\} dt + D(t) = 0. \tag{7}$$

This equation implies that D (t + Δt) vanishes though the "starting value" D (t) may be different from zero; by such a device errors introduced at one stage tend to be reduced in the next (*).

The form given to the non-linear term in (6) is more convenient than the form of the corresponding term in (1) for our purposes. In fact, one aims at trasforming the differential equation into a difference equation (spacewise) through the following steps:

i) integrate over a mesh-element V, transforming all volume integrals containing space derivatives into surface integrals over the boundary S of V.

ii) approximate surface integrals using only the values of the functions at the meshpoints.

For step (1) eqn (6) is directly fit [so, of course, are eqns (3), (4)] ; precisely we get ($\underline{n}$, unit vector of the exterior normal)

$$\int_V \frac{\partial \underline{u}}{\partial t} dV = - \int_S \underline{u} (\underline{u} \cdot \underline{n}) dS - \int_S \varphi \, \underline{n} \, dS + \\ - \nu \int_S (\underline{n} \times \mathrm{curl}\, \underline{u}) dS + \int_V \underline{g} \, dV . \tag{8}$$

Similarly from (7) it follows

$$\int_t^{t+\Delta t} dt \int_S \left\{ (\mathrm{grad}\, \varphi) \cdot \underline{n} + \left[\mathrm{div}\, (\underline{u} \otimes \underline{u}) - \underline{g} \right] \cdot \underline{n} \right\} dS \\ + \int_S \underline{u}(t) . \underline{n} \, dS = 0. \tag{9}$$

(*) The "penalty method" or the "method of artificial derivatives" described by Professor Lions could also have been used.

We quote here also the integrated versions of eqns (3), (4)

$$\begin{aligned} \int_V \underline{\chi}\, dV &= \int_S \underline{\phi} \times \underline{n}\, dS , \\ \int_V \underline{\phi}\, dV &= \int_S \underline{u} \times \underline{n}\, dS , \\ \int_V \underline{u}\, dV &= \int_S \underline{\Psi} \times \underline{n}\, dS , \end{aligned} \tag{10}$$

$$\begin{aligned} \frac{d}{dt} \int_V \underline{\phi}\, dV + \int_S (\underline{u} \cdot \text{grad}\, \underline{u}) \times \underline{n}\, dS &= \\ = - \nu \int_S \underline{\chi} \times \underline{n}\, dS + \int_S \underline{g} \times \underline{n}\, dS . \end{aligned} \tag{11}$$

Numerical quadratures must now be introduced to approximate the integrals in (8), (9) or (10), (11). For the sake of simplicity we consider here only the case of a regular cubic mesh. It is easy to realize then, (although we not enter here into details) that, for the simplest and relatively most precise approximation of eqns (8),... (11), one must introduce a cubic lattice with the following condition: If $\underline{\phi}$ for instance is supposed to be known on one lattice point P_o, then $\underline{u}$ must be known on the six nearest points $P_1 \ldots P_6$ and conversely.

As a very simple example, consider the second eqn (10): we have (h, mesh-size)

$$h\, \underline{\phi}\, (P_o) \cong \sum_1^6{}_i \quad \underline{u}\, (P_i) \times \underline{n}\, (P_i).$$

Then, the structure of the system (8), (9) and (10), (11) is such that, to achieve best approximations, it is convenient to take V successively as coincident with different but overlapping cubic cells. For instance, with reference to eqns (10), (11) notice that $\underline{\psi}$ must be known in the centres of the faces of the first cell V_1 whereas $\underline{u}$ must be known at the centre of the cell itself. The cells of type V_2 must be such that $\underline{u}$ is known at the centres of the faces, whereas $\underline{\phi}$ is known at the centre of the cell itself, and so on.

The procedure can thus be organized so that no interpolation is required except for the approximate expression of the non-linear terms. We must introduce now a discretization in the time variable. Leaving without a superscript the values at the end t_k of the k-th time step and using the superscript k+1 for values at the end t_{k+1} of the (k+1)-th step, the simplest finite difference approximations to (8), (9) are

$$\left[\int_V \underline{u}\, dV\right]^{k+1} = \int_V \underline{u}\, dV - \delta\left\{\int_S \underline{u}\,(\underline{u}\cdot\underline{n})\, dS + \int_S \varphi\,\underline{n}\, dS + \right. \tag{12}$$

$$\left. + \nu\int_S \underline{n} \times \operatorname{curl} \underline{u}\, dS - \int_V \underline{g}\, dV\right\},$$

$$\delta\int_S\left\{\underline{n}\cdot \operatorname{grad} \varphi + \left[\operatorname{div}(\underline{u}\otimes\underline{u}) - \underline{g}\right]\cdot n\right\} dS + \tag{13}$$

$$+ \int_S \underline{u}\cdot\underline{n}\, dS = 0\,, \qquad \delta = t_{k+1} - t_k\,.$$

These approximations are very rough, but have the great advantage of leading to explicit formulae. The care taken in writing the condition which implies conservation of mass and also the special form given to the term measuring the diffusion of momentum is justified now: it allows the acceptance of the rough formulae above.

Leaving aside for the moment the question of the boundary conditions, the process to follow is this.

Assume that the initial values of $\underline{u}$ be given at t=0. The corresponding distribution of φ is determined through eqn (13). This step can be accomplished through one of the many methods available for the integration of Laplace equation, for instance through an iterative overrelaxation procedure [34] . Successively, the right-hand side of eqn (12) is computed and new values of $\underline{u}$ are determined. The process is then repeated.

Attention was confined so far to time dependent flows. There is interest of course also in the study of steady flows; for such a study some of the remarks still apply. The difficulties in respect to diffusion of dilatation do not occur; we find instead problems of convergence in the schemes of successive approximation that must be introduced to deal with the non-linear terms.

3. Boundary conditions

Boundary conditions for the approximate analysis of our problems must not be lightly stated. For instance: is the usual condition of no slip at a wall always justified?

Only a reference to physical circumstances allows one to give a satisfactory answer to this question. Whether or not slippage is to be allowed depends upon the thickness of the boundary layer that one would expect to develop in the true fluid. If this is much less than the dimensions of a lattice cell and one is not interested in the detail of the boundary flow then a free slip condition is appropriate; if the boundary layer is much larger than one cell, then a no-slip condition is required. For intermediate cases, the proper condition to use depends upon the exact circumstances, and in some cases it is appropriate to try both ways and compare the results.

Another point one must emphasize: sometimes it is convenient for computational purposes to introduce fictitious mesh-points outside the boundary. If such a device is used, one must be sure that the finite difference approximation to D vanishes also at the exterior fictitious cells so that no diffusion of D inside the boundary occurs.

All these warnings are of course of an experimental character and are connected not to any inadequacy in principle of the finite difference approximations, but rather to the need to operate with a relatively small number of cells.

Even more delicate is the question of writing adequate approximations to conditions at a free surface Σ.

Over Σ conditions on stress components must be stated; for instance if the externally applied stress is a pressure $\wp_a$, we should use, in a system of cartesian coordinates, the conditions

$$\begin{aligned} \wp &= \wp_a + \nu N_i N_j (u_{i,j} + u_{j,i}) , \\ 0 &= T_i^{(h)} N_j (u_{i,j} + u_{j,i}) , \; h = 1,2 ; \end{aligned} \tag{14}$$

where $\underline{N}$ is the exterior normal to Σ and $\underline{T}^{(1)}$, $\underline{T}^{(2)}$ are two orthogonal tangential vectors. These conditions are very difficult to set up satisfactorily on a computer and workers in the field have resorted to conditions such as D=0, $\wp = \wp_a$ to balance equations and unknowns. The first choice is justified on the grounds that the gravest source of errors is diffusion of dilatation D through the boundaries. The second choice is motivated by the remark that often viscous effects are small when compared with a directly imposed stress. On the other hand the local orientation of the surface can be usually determined only very roughly, so that a more precise use of eqns (14) is not justified.

It remains to follow the changes of the free surface with time. This is accomplished by introducing marker particles on the free surface (actually in the marker-and-cell method the marker particles are distributed throughout the fluid, though, for analytical

purposes, they are essential only at the boundary).
The speed of the particles is determined by interpolation or extrapolation from nearest mesh points; finally their movement is followed step by step.

A practical procedure is as follows.

One builds up in the computer a picture of the fluid set in a wider field of cells where the free surface can impinge. There are markers to show which cells are occupied (at least in part) and which are free. Pressure and velocity fields are determined over all the occupied cells, boundary conditions intervening in the boundary cells. To avoid ambiguities (i.e. a wrong labeling of internal cells as empty) at least four marker particles for cell are distributed at time t=0 in all occupied cells with further provisions for exceptional cases.

4. Numerical instability; accuracy

Phenomena of numerical instability have been mentioned already; it is well known that their onset depends critically on meshsize, and time-step size. It is also common experience that explicit algorithms such as that embodied in formulae (12), (13) are usually more prone to the disease than more complex implicit algorithms.

If we reduce eqns (12), (13) to a non-dimensional form by introducing a typical velocity $\mathcal{U}$ a typical body force per unit mass G, the time step δ and the meshsize h (assuming for simplicity that the lattice is cubic though phenomena of instability may be quenched sometimes by introducing meshes with appropriate side-ratio [31]) we see that the solution of the finite difference equations depends locally on the following parameters

$$\alpha_1 = \frac{\delta\mathcal{U}}{h}, \quad \alpha_2 = \frac{G\delta}{\mathcal{U}}, \quad \alpha_3 = \frac{\nu\delta}{h^2}$$

For those who are physically inclined we remark that α_1 can be construed as a Strouhal number of the flow based on the numerical time step and meshsize. Similarly α_2 and α_3 can be combined with α_1 to express numerical Froude and Reynolds numbers

$$F_N = \frac{\alpha_1}{\alpha_2} = \frac{\mathcal{U}^2}{Gh}, \quad R_N = \frac{\alpha_1}{\alpha_2} = \frac{\mathcal{U}h}{\nu}.$$

Conditions of numerical stability can then be expressed through limitations on α_1, F_N, R_N. The choice of the relevant values of $\mathcal{U}$ and G will depend on the problem in hand, of course. In the study of flows with a free surface $\mathcal{U}$ can be taken as the speed of surface waves: using shallow fluid theory

$$\mathcal{u} = \left(\frac{g}{k} \tanh k H \right)^{\frac{1}{2}} ,$$

k , wave number; H, depth of fluid.

In the experiments of Harlow and collaborators typical stability conditions were found to be, experimentally, [11, p.28]

$$\alpha_1 < 1 \qquad , \qquad \alpha_3 < 1/4 . \tag{15}$$

In other cases the local velocity intervenes [5, p.137] in studies on the behaviour of a perturbation in a steady flow the excess speed due to the perturbation seems to have relevance. In all cases it was found by experiment or was suggested by heuristic arguments that R_N must be of order of unity if instability has to be avoided. Although the value of $\mathcal{u}$ that must be used is not known exactly in advance, rough evaluations are usually possible. Then the condition just mentioned can give an idea of the size of the problem in hand from a computational point of view.

If conditions such as (15) are satisfied the results of a computation are likely to look reasonable, i.e. not wildly wrong, but they may still be far from accurate. It would be nice to have some tests for accuracy. A check on the value of D must always be kept with automatic stop when D reaches an unacceptable level. If the condition of incompressibility is satisfactorily approximated the measure of the domain occupied by the fluid (as shown by marker particles) must be constant. In the marker-and-cell method a check can therefore be made by comparing the number of cells $\bar{N}$ containing at least one particle with the number of boundary cells (since these cells are constantly controlled in a program, the check is simple).

The experimental value of $\overline{N}$ can be compared with theoretical estimates

One such estimate [12] for plane problems is

$$\overline{N} = \frac{A}{\delta^2} + \frac{2}{\pi\delta}(1 - \lambda)P$$

where A is the constant area of a cross-section of the region occupied by the fluid, P is the length of the boundary of the cross-section and λ the ratio of particle spacing to cell size.

Further checks are sometimes made on the basis of evaluations of total kinetic energy.

3. <u>Numerical analysis of hydrodynamic stability of steady flows.</u>

I mentioned already that a good deal of research effort is applied to the numerical study of stability of certain classical flows: the Poiseuille flow, the Couette flow, the flow over a flat plate, etc.

In these cases the boundaries are fixed and one can make use conveniently of eqn. (3) , (4) ; the time-independent functions describing a fundamental flow are supposed to be known:

$$\underline{\chi}_0 \quad , \quad \underline{\phi}_0 \quad , \quad \underline{u}_0$$

and details on the behaviour of perturbations $\underline{\tilde{\chi}} = \underline{\chi} - \underline{\chi}_0$, $\underline{\tilde{\phi}} = \underline{\phi} - \underline{\phi}_0$, $\underline{\tilde{u}} = \underline{u} - \underline{u}_0$ are required.

The equations are :

$$\underline{\tilde{\chi}} = \text{curl}\ \underline{\tilde{\phi}} \quad , \quad \underline{\tilde{\phi}} = \text{curl}\ \underline{\tilde{u}} \quad , \quad \underline{\tilde{u}} = \text{curl}\ \underline{\tilde{\psi}} \ ,$$

$$\frac{\partial \underline{\tilde{\phi}}}{\partial t} + \text{curl}\left[\text{grad}\ \underline{\tilde{u}} \cdot (\underline{u}_0 + \underline{\tilde{u}}) + \text{grad}\ \underline{u}_0 \cdot \underline{\tilde{u}}\right] = - \nu \ \text{curl}\ \underline{\tilde{\chi}}$$

Because the field of $\tilde{\underline{\Psi}}$ can be taken to be solenoidal, this equation can be written in term of $\tilde{\underline{\psi}}$ only

$$\frac{\partial \Delta \tilde{\underline{\psi}}}{\partial t} - \text{curl}\left[\text{grad curl}\, \tilde{\underline{\psi}} \cdot (\underline{u}_0 + \text{curl}\, \tilde{\underline{\psi}}) + \text{grad}\, \underline{u}_0 \cdot \text{curl}\, \tilde{\underline{\psi}} \right] =$$

$$= \nu \Delta \Delta \tilde{\underline{\psi}} \, .$$

It is convenient to write immediately this equation also in a non-dimensional form using a typical velocity $\mathcal{U}$ and a typical dimension of the domain L , introducing the notation

$$\underline{\psi}^* = \tilde{\underline{\psi}} / \mathcal{U}\, L \; , \quad \underline{s} = \underline{u}_0 / \mathcal{U} \; , \quad \tau = \nu \; t / \, L^2$$

and a physical Reynolds number

$$R = \frac{\mathcal{U}\, L}{\nu}$$

and presuming now that the operators Δ and curl act over non-dimensional space variables

$$\frac{\partial}{\partial \tau} \Delta \underline{\psi}^* - R\, \text{curl} \left[\text{grad curl}\, \underline{\psi}^* \cdot (\underline{s} + \text{curl}\, \underline{\psi}^*) + \text{grad}\, \underline{s} \cdot \text{curl}\, \underline{\psi}^* \right] =$$

$$= \Delta \Delta \underline{\psi}^* \, . \qquad (16)$$

Usually one wants to know the solution of eqn (16) for a sufficiently ample interval of time and over a domain for the space variables which is not bounded, though sometimes the expected phenomenon is periodic in one or more space variables and a reduction to a bounded domain ensues.

On the part of the boundary that represents walls (fixed or in steady motion) $\tilde{\underline{u}} = 0$; often one can conclude from this that all components of $\underline{\psi}$ and their normal derivatives vanish.

There may be conditions at infinity and, on other parts of

the boundary, periodicity conditions may apply; besides the initial conditions must be known.

When one is interested in the decay of an istantaneous disturbance or in the speeding of self amplified perturbation, these are the only conditions that apply. In other cases perturbations may be continuously fed from outside; then $\underline{\Psi}$ and derivatives are assigned on portions of the boundary as known functions of time.

Because the choice $\underline{\Psi} = 0$ corresponds to the fundamental solutions of eqn (16) the interest centres at first on the small perturbations. Although a precise statement can be made only in one special case, it is generally presumed that the behaviour of a perturbation of small amplitude can be qualitatively decided on the basis of the linearized equation

$$\frac{\partial(\Delta \underline{\Psi}^*)}{\partial \tau} - R \operatorname{curl}\left[\operatorname{grad}\operatorname{curl}\underline{\Psi}^* \cdot \underline{s} + \operatorname{grad}\underline{s} \cdot \operatorname{curl}\underline{\Psi}^*\right] = \Delta\Delta\underline{\Psi}^* \quad . \qquad (17)$$

We come thus to a rather complex linear diffusion problem; in the plane case, where $\underline{\Psi}^*$ has only one non-vanishing component eqn (3) has been the object of many classical studies, for instance those related to the stability of Poiseuille flow, or the flow on a flat plate.

Because the coefficients of eqn (17) are independent of time, the solutions can be written as linear combinations of functions of the type

$$\underline{\Psi}^* = e^{k\tau}\,\underline{A} \;,$$

where $\underline{A}$ is a function of the space variables only and k is a complex constant. The equation that follows for $\underline{A}$, from (17),

$$k\Delta \underline{A} - R \operatorname{curl} \left[\operatorname{grad} \operatorname{curl} \underline{A} \cdot \underline{s} + \operatorname{grad} \underline{s} \cdot \operatorname{curl} \underline{A} \right] = $$
$$= \Delta\Delta \underline{A} \qquad (18)$$

and the associated boundary conditions add to an eigenvalue problem depending on the positive parameter R_c . It is essential to decide which is the infimum R_c of the set of values of R for which one eigenvalue K(R) at least has a positive real part. In this field the early work of Thomas must be quoted [35] .

It happens sometimes that the value of k corresponding to R_c vanishes; this analytical fact is related to the physical existence of non-trivial steady flows. In such cases the eigenvalue problem (18) is further simplified. Besides, a search for non-trivial solutions of the non linear problem

$$\Delta\Delta \underline{\psi}^* = - R \operatorname{curl} \left[\operatorname{grad} \operatorname{curl} \underline{\psi}^* (\underline{s} + \operatorname{curl} \psi^*) + \right.$$
$$\left. + \operatorname{grad} \underline{s} \cdot \operatorname{curl} \psi^* \right] \qquad (19)$$

with the associated boundary conditions, can be attempted. For a special case of this problem we have definite results due to Velte, Kirchgässner and others research workers at Freiburg [36, 37, 38] . The special case is examined in some detail later. Mention must be briefly made here of the numerical techniques used to tuckle eqns (16), (17), (18), (19) with the associated boundary conditions.

A trivial explicit method can be used in connection with eqn (16), (17); but more often, to lessen phenomena of numerical instability, it is more convenient to evaluate the term under the biharmonic operator as the average of the values at times τ and $\tau + \Delta\tau$, mantaining for the other terms the evalutation at time τ [31, 33] .

If such technique is adopted a matrix representing the

discrete equivalent of a linear combination of the operators $\Delta\Delta$ and Δ must be inverted. Even when use is made of an explicit method a matrix inversion (although simpler) is required. Techniques of direct inversion or iterative methods must be called for.

Direct inversion though cumbersone may be attractive because it is needed only once for all time steps. The economy of the procedure is much enhanced in cases where the solution is periodic in one or more space variables because the matrices involved are then circulant in submatrices which may even be circulant in their turn. Formulae for the inversion of circulant or block circulant matrices are quoted in the next section [39, 40].

For the solution of the problem (19) with the associated boundary conditions an iterative procedure is always called for, to deal with the non-linear terms. Starting with a reasonable guess, one can make use of the discrete equivalent of the iteration

$$\Delta\Delta\,\underline{\Psi}^{*(k)} = -\,R\,\mathrm{curl}\Big[\mathrm{grad\ curl}\,\underline{\Psi}^{*(k-1)}\cdot(\underline{s} + \mathrm{curl}\,\underline{\Psi}^{*(k-1)}) + \mathrm{grad}\,\underline{s}\cdot\mathrm{curl}\,\underline{\Psi}^{*(k-1)}\Big]\,. \qquad (20)$$

Here again, if the boundary conditions express periodicity at least in one variable, techniques of inversion of circulant matrices may be of use. Both in the analysis of the time-dependent case and during the iteration (2o) phenomena of numerical instability may occur. A word of warning is necessary here; a mild form of numerical instability in diffusion problems may be wrongly taken sometimes as indicative of hydrodynamic instability. The study of the same problem with two different meshsizes (one rectangular and one square for instance, in the plane case) is recommended. "Numerical" eddies change then wavelength so as to cover the same number of cells (the typical wavelength of "numerical" eddies is ten cells).

6. Block-circulant matrices

We have mentioned that in the numerical solution of our problems under periodicity conditions block-circulant matrices appear. To show that simple devices can save at times a lot of work, the property of these matrices is recalled here, that allows an easier inversion.

Let

$$A = \left\{ A_0 , A_1 , \ldots A_{m-1} \right\} = \begin{vmatrix} A_0 & A_1 & \ldots & A_{m-1} \\ A_{m-1} & A_0 & \ldots & A_{m-2} \\ \cdot & \cdot & \cdots & \cdot \\ A_1 & A_2 & \ldots & A_0 \end{vmatrix}$$

be a block-circulant matrix, where the A_i are blocks of order n. Let I be the identity matrix of order n and $\alpha_0 , \alpha_1 , \ldots \alpha_{m-1}$ the m-th roots of unity and put

$$V = \begin{pmatrix} I & \ldots & I \\ \alpha_0 I & \ldots & \alpha_{m-1} I \\ \cdot & \cdots & \cdot \\ \alpha_0^{m-1} I & \ldots & \alpha_{m-1}^{m-1} I \end{pmatrix}.$$

Then

$$V^{-1} = \frac{1}{m} \begin{pmatrix} I & \alpha_0^{m-1} I & \ldots & \alpha_0 I \\ \cdot & \cdot & \cdots & \cdot \\ I & \alpha_{m-1}^{m-1} I & \ldots & \alpha_{m-1} I \end{pmatrix}$$

and

$$V^{-1} A V = \Delta = \begin{pmatrix} \phi(\alpha_0) & 0 & \ldots & 0 \\ 0 & \phi(\alpha_1) & \ldots & 0 \\ \cdot & & \cdots & \cdot \\ 0 & 0 & \ldots & \phi(\alpha_{m-1}) \end{pmatrix}$$

where

$$\phi(x) = A_o + A_1 x + \ldots A_{m-1} x^{m-1}$$

If the matrices $\phi(\alpha_i)$ are not singular also A is non singular and

$$A^{-1} = V \Delta^{-1} V^{-1} .$$

So A^{-1} is also block-circulant; precisely

$$A^{-1} = \frac{1}{m} \left\{ \sum_{o}^{m-1}{}_k \phi^{-1}(\alpha_k), \sum_{o}^{m-1}{}_k \alpha_k^{m-1} \phi^{-1}(\alpha_k), \ldots, \sum_{o}^{m-1}{}_k \alpha_k \phi^{-1}(\alpha_k) \right\} .$$

Therefore the inversion of the matrix A of order $m \cdot n$ is reduced to the inversion of m matrices of order n. If, besides, the matrix A is block-symmetric ($A_1 = A_{m-1}$, $A_2 = A_{n-1}$, ...), then we need invert only $[m/2]$ matrices; the successive algebraic manipulations are also simpler then because the inverse matrix is also block-symmetric.

7. A simple analytical scheme for the study of the stability of Couette flow.

To illustrate with one example the analytical and numerical problems that arise in the study of the stability of a steady flow, we examine now in some detail the behaviour of perturbations introduced in the circumferential flow between two concentric coaxial rotating cylinders (Couette flow). An analysis of this problem has interest for many reasons:

(i) the Couette flow is one of the very few steady flows of a viscous fluid for which one has a precise analytical description.

(ii) the stability of the Couette flow can be studied in the laboratory through relatively simple experiments. There have been precise experimental studies of Taylor, Conelly and more recently of Coles [41, 42, 43] .

(iii) the Couette flow is subject to a form of hydrodynamic instability that lends itself to an analytical treatment, through linearisation of the perturbation equation, with forecasts amply confirmed by experiments [41, 44] .

(iv) the special type of hydrodynamic instability lends itself to a rigorous analytical treatment also through a study of the complete non-linear equations [37] .

The fundamental reason for (ii), (iii), (iv) is the fact that instability is in most cases due to transition to other forms of steady flow (Taylor vortices - for which axial symmetry still holds - or Coles wavy vortices) rather than to transition towards turbulence (as happens instead for Poiseuille flow).

Apart from its analytical-experimental interest, the Taylor vortex flow has importance in practice at least for two reasons: the flow in plain bearings of large rotors (turbines, alternators) is roughly a Couette flow, subject at high speeds to Taylor instability; when Taylor vortices appear the viscous losses in the lubricant become much higher than is forecast on the assumption of Couette flow; hence the interest of a precise understanding of the Taylor instability. There have been also attempts to use the Taylor vortices as seals in bearings. Secondly, boundary layers along concave walls are subject to instability of a similar kind, that can be studied by similar means (Görtler vortices) [45] .

The starting point for a numerical analysis of the (axisymmetric) Taylor vortices are equation (16) and the appropriate boundary conditions. Naturally the peculiarities of the problem allow simplification; $\underline{s}$ is in the circumferential direction, $\underline{\psi}^*$ is assumed not to depend on θ but only on the radial and axial coordinates r and z. We can use the gap $r_2 - r_1$ between cylinders

as typical dimension and put $\xi = (r - r_1) / (r_2 - r_1)$,
$\zeta = z / (r_2 - r_1)$ together
with $\tau = \nu t / (r_2 - r_1)^2$.

Then the radial and axial components of speed are related to the derivatives of the transverse component of $\underline{\psi}^*$ and it would be uneconomical to introduce the other two components of $\underline{\psi}^*$ simply to define the transverse component of speed; the simplest scheme derives from the use of second component ψ of $\underline{\psi}^*$ and the transverse component of speed. Reasons of simplicity connected with special features of the problem (such as the axial symmetry) suggest a slight modification of the usual formulae and the adoption of the following ones, which are self explanatory (Ω_1 , Ω_2 angular speeds of internal and external cylinder respectively , $\lambda = \Omega_2 / \Omega_1$)

$$r v_r = - \frac{r_1 \nu}{r_2 - r_1} \frac{\partial \psi}{\partial \zeta} \qquad r v_z = \frac{r_1 \nu}{r_2 - r_1} \frac{\partial \psi}{\partial \xi}$$

$$v_\theta - v_\theta^{(o)} = \frac{1}{2} (\Omega_1 r_1 + \Omega_2 r_2) \quad (1 - \lambda) v .$$

In writing the two scalar equations a formal complication follows from the use of cylindrical coordinates. To make this exposition as simple as possible reference is made here only to an asymptotic case: that of small clearance $(r_2 - r_1) / r_2 \ll 1$.

The equations valid in that case are

$$\Delta\Delta \psi = \frac{\partial (\Delta \psi)}{\partial \tau} + T \left[1 - \xi + \lambda \xi + \frac{1}{2} (1 - \lambda^2) v\right] \frac{\partial v}{\partial \zeta} + \frac{\partial (\psi, \Delta\psi)}{\partial (\xi, \zeta)} ,$$

$$\Delta v = \frac{\partial v}{\partial \tau} + \frac{2}{1-\lambda} \frac{\partial \psi}{\partial \zeta} + \frac{\partial (\psi, v)}{(\xi, \zeta)}$$

(see 39 for the derivation); here T is a mean Reynolds'number

(Taylors' number)

$$T = \frac{(r_1\Omega_1 + r_2\Omega_2)^2 \, (r_2 - r_1)^2 \, (\Omega_1 - \Omega_2)}{r_1 \nu^2 \, (\Omega_1 + \Omega_2)} \, .$$

The boundary conditions express: vanishing of mean axial flow, periodicity in the axial direction (with arbitrary period 2 q), vanishing of the perturbation on the cylinders

$$v = 0 \, , \; \psi = 0 \, , \; \frac{\partial \psi}{\partial \xi} = 0 \, , \text{ at } \xi = 0, \; \xi = 1 \; ;$$

$$v(\xi, q) = v(\xi, -q) \; ; \; v_\zeta(\xi, q) = v_\zeta(\xi, -q) \; ;$$

$$\psi(\xi, q) = \psi(\xi, -q) \; ; \; \left(\frac{\partial^i \psi}{\partial \xi^i}\right)_{\zeta = q} = \left(\frac{\partial^i \psi}{\partial \zeta^i}\right)_{\zeta = -q} \, , \; i = 1,2,3 \, . \qquad (22)$$

A further simplification can be obtained in (21) by choosing to put $\lambda = 1$; it may seem that the simplification denies physical significance to the resulting problem. In fact it is found that the problem is interesting and certain consequences (such as the critical Taylor number) deduced in the special case can be applied with good approximation also for λ in the closed interval (0,1). We consider then in the following sections this problem: find in the rectangle $(0 \leq \xi \leq 1 \, , -q \leq \zeta \leq q \, , \; \tau \geq 0)$ a vector (ψ, v) satisfying the equations

$$\Delta\Delta\psi = \frac{\partial(\Delta\psi)}{\partial\tau} + T \frac{\partial v}{\partial \zeta} + \frac{\partial(\psi, \Delta\psi)}{\partial(\xi, \zeta)} \, ,$$

$$\Delta v = \frac{\partial v}{\partial \tau} + \frac{\partial \psi}{\partial \zeta} + \frac{\partial(\psi, v)}{\partial(\xi, \eta)} \, , \qquad (23)$$

the boundary conditions (22) and given initial conditions.

8. Some results regarding the differential problem.

We begin the study of our example with an analysis of properties of boundary and initial value problems related to problem (22), (23). To begin with, it is convenient to consider from a partly formal point of view the solution of the linear system

$$\Delta\Delta\psi = T\frac{\partial v}{\partial \zeta}, \qquad \Delta v = \frac{\partial \psi}{\partial \zeta}, \tag{24}$$

with the boundary conditions (22). Obviously if $\psi(\xi,\zeta)$, $v(\xi,\zeta)$ is a solution of this problem so is also $\psi(\xi,\zeta+b)$ $v(\xi,\zeta+b)$ (b, any constant); we require of ψ to be odd and v to be even in ζ. Then we separate variables, looking for solutions ψ_n, v_n of the type

$$\psi_n = A_n(\xi)\sin n\pi(\zeta/q), \qquad v_n = B_n(\xi)\cos n\pi(\zeta/q).$$

The functions A_n, B_n satisfy the system of equations ($a_n = n\pi/q$)

$$\left(\frac{d^2}{d\xi^2} - a_n^2\right)^2 A_n = -Ta_n B_n, \qquad \left(\frac{d^2}{d\xi^2} - a_n^2\right) B_n = a_n A_n \tag{25}$$

and the boundary conditions

$$A_n = A_n' = B_n = 0 \qquad \text{at} \qquad \xi = 0, \ \xi = 1 \tag{26}$$

Supposing that an eigenvalue $\hat{T}_n$ and a corresponding eigenvector $(\hat{A}_n, \hat{B}_n)$ exist for problem (25), (26), then multiplying both sides of the first eqn (25) by the complex conjugate $\hat{A}_n^*$ of $\hat{A}_n$

and integrating over (0,1) one gets

$$I_1 = - \hat{T}_n a_n I_3 \tag{27}$$

where $$I_1 = \int_0^1 \left[|\hat{A}''_n|^2 + 2 a_n^2 |\hat{A}'_n|^2 + a_n^4 |\hat{A}_n|^2 \right] d\xi ,$$

$$I_3 = \int_0^1 \hat{A}_n^* \hat{B}_n \, d\xi .$$

Similarly from the second eqn (25) one gets

$$I_2 = - a_n I_3^* \tag{28}$$

with

$$I_2 = \int_0^1 (|B'_n|^2 + a_n^2 |\hat{B}_n|^2) \, d\xi .$$

Because a_n is a real positive number, such are also I_1, and I_2; these two numbers vanish only on the trivial solution of (25), (26).

Hence $I_3 = I_3^*$ is real negative (from eqn (28)), and the eigenvalue $\hat{T}_n$ is real positive (from eqn (27)).

As a consequence the associated eigenvector can be taken to have real components. From the formulae above it follows that, if $\hat{A}_n$, $\hat{B}_n$ do not vanish,

$$- (\hat{T}_n + a_n^4) I_3 = \frac{1}{a_n} I_1 + a_n^3 I_2 > a_n^4 \int_0^1 (\frac{1}{a_n} \hat{A}_n^2 + a_n \hat{B}_n^2) \, d\xi .$$

On the other hand

$$- I_3 = - \int_0^1 \hat{A}_n \hat{B}_n \, d\xi \leq \frac{1}{2} \int_0^1 (\frac{1}{a_n} \hat{A}_n^2 + a_n \hat{B}_n^2) \, d\xi ;$$

hence

$$\frac{1}{2}\,(\hat{T}_n + a_n^4)\int_0^1 \left(\frac{1}{a_n}\hat{A}_n^2 + a_n \hat{B}_n^2\right) d\xi >$$

$$> a_n^4 \int_0^1 \left(\frac{1}{a_n}\hat{A}_n^2 + a_n \hat{B}_n^2\right) d\xi ,$$

and finally

$$\hat{T}_n > a_n^4 . \qquad (29)$$

The equation, which the eigenvalues satisfy, is easily found. Notice that $\hat{A}_n$ and $\hat{B}_n$ are both solution of the equation in y

$$\left[\left(\frac{d^2}{d\xi^2} - a_n^2\right)^3 + a_n^2 T_n \right] y = 0 .$$

Hence, both $\hat{A}_n$ and $\hat{B}_n$ can be expressed as linear combinations of functions $e^{z\xi}$ with

$$z = \left[a_n^2 + (- a_n^2 T_n)^{1/3} \right]^{1/2} .$$

Let us indicate with z_1, $-z_1$, z_2, $-z_2$, z_3, $-z_3$ the six distinct determinations of z [Notice that inequality (29) excludes the occurrence of multiple roots] .

By imposing the boundary conditions the equation for the eigenvalues can be found. It is expressed by putting equal to zero the determinant Δ of a 6x6 matrix whose first three lines are

$$\begin{vmatrix} 1 & e^{z_1} & 1 & e^{z_2} & 1 & e^{z_3} \\ z_1 & -z_1 e^{z_1} & \alpha z_2 & -z_2 e^{z_2} & \beta z_3 & -\beta z_3 e^{z_3} \\ z_1^2 & z_1^2 e^{z_1} & z_2^2 & z_2^2 e^{z_2} & z_3^2 & z_3^2 e^{z_3} \end{vmatrix} \qquad (30)$$

$$\alpha = e^{2\pi i/3} \qquad\qquad \beta = e^{2\pi i/3}$$

and the other three are formed with the same columns in the order 2, 1, 4, 3, 6, 5.
It can be checked that the determinant Δ is equal to the difference of the squares of two sums S_1 and S_2, where S_1 is the sum of the determinants of the matrices of order 3 obtained by extracting the columns 1, 3, 5; 2, 3, 6; 2, 4, 5; 1, 4, 6 of the matrix (30) and S_2 by a similar sum where the columns 1, 3, 6; 1, 4, 5; 2, 3, 5; 2, 4, 6 are involved. Easy developments lead to the result

$$\sum_{0}^{2} n\, z_n \left(z_{n+1}^2 - z_{n+2}^2\right) e^{\frac{2n\pi}{3} i} \left(1 - e^{z_n}\right) \left(1 + e^{z_{n+1}}\right) \left(1 + e^{z_{n+2}}\right) = 0$$

from which an implicit multivalued function $\hat{T}_n = \hat{T}_n (a_n)$ can be computed. On one branch of this function the relation $\hat{T}_n = a_n^4$ is satisfied (because then the functions $e^{\pm z_i t}$, $i = 1,2,3$ are not independent), but that branch is without interest for computing eigenvalues in view of inequality (29).

A graphical representation of the two branches over which the value of $\hat{T}_n$ corresponding to a given of a_n is the lowest and next lowest are shown in the figure.

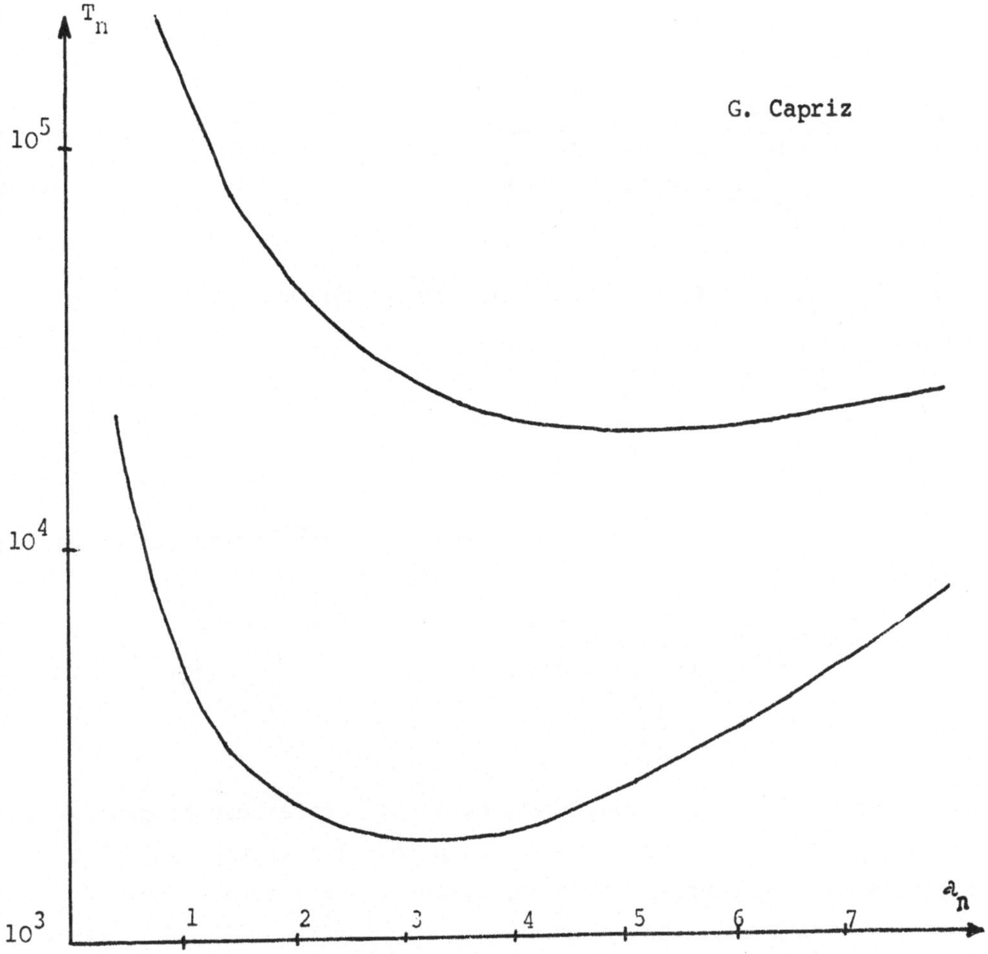

The numerical experiments then show that there are real positive eigenvalues of our problem and that appropriate values of q can be found such that to $a_n = \pi/q$ there corresponds an eigenvalue $\hat{T}_1$ which is not an eigenvalue when a is chosen equal to $j\pi/q$ ($j = 2, 3, \ldots$). To such T there corresponds then only one eigenvector of the type sought in this Section: $\hat{T}_1 \neq \hat{T}_j$ ($j = 2, 3 \ldots$).

These results, based here simply on numerical evidence, can be reached without recourse to experiments [see 36], through reference to properties of solutions of a variational problem based on the equality

$$T = \frac{\int_{-q}^{q} a\zeta \int_0^1 \left\{ |\mathrm{grad}\, \psi_x|^2 + |\mathrm{grad}\, \psi_y|^2 \right\} d\xi}{\int_{-q}^{q} d\xi \int_0^1 |\mathrm{grad}\, v|^2 \, d\xi} \tag{31}$$

which follows from (24) and the boundary conditions (22).

We recall now some results regarding the linear diffusion problem

$$\begin{aligned} \Delta\Delta\psi &= \frac{\partial(\Delta\psi)}{\partial\tau} + T\frac{\partial v}{\partial\zeta}, \\ \Delta v &= \frac{\partial v}{\partial\tau} + \frac{\partial\psi}{\partial\zeta}, \end{aligned} \tag{32}$$

with the boundary conditions (22). We do not quote here properties of special solutions corresponding to particular initial data but rather state the general behaviour of solutions depending on T .

Remark that the general solution of (32), (22) is a linear combination of functions of the type $\psi = e^{\gamma\tau}\psi_1(\xi,\zeta)$, $v = e^{\gamma\tau} v_1(\xi,\zeta)$ where γ is a real number and ψ_1, v_1 satisfy the differential system

$$\Delta\Delta\psi_1 = \gamma\,\Delta\psi_1 + T\frac{\partial v_1}{\partial\zeta},$$

$$\Delta v_1 = \gamma\, v_1 + \frac{\partial\psi_1}{\partial\zeta},$$

with the usual boundary conditions.

By reference to this eigenvalue problem it is possible to prove see[36, pp.109-111] that:

(i) when T is smaller than the smallest eigenvalue $\hat{T}_1$ of the problem of Sect. 8, γ is necessarily negative;

(ii) when T is in an appropriate interval $\hat{T}_1$, $\hat{T}_1 + \delta (\delta > o)$ there are solutions of (22), (32) exponentially increasing with time.

9. The non-linear problem.

We consider here the non-linear problem, whose solution represents the Taylor vortices within our approximation :

$$\Delta\Delta\psi = T\frac{\partial v}{\partial \zeta} + \frac{\partial(\psi, \Delta\psi)}{\partial(\xi,\zeta)} ,$$

$$\Delta v = \frac{\partial \psi}{\partial \zeta} + \frac{\partial(\psi,v)}{\partial(\xi,\eta)} , \qquad (33)$$

with the boundary conditions (22).

We give precise sense to this problem by stating the set where we seek a non-trivial solution: it is a subset $\mathcal{J}$ of a Banach space B of vectors (ψ, v) obtained thus.

Consider the set $\mathcal{Y}_m$ of functions $\varphi\ (\xi,\zeta)$ defined in a strip S_1 larger than the strip S $: 0 \leq \xi \leq 1$, of class C^m in S_1, periodic in ζ with period 2q; introduce in $\mathcal{Y}_m$ the norm

$$\|\varphi\|_n = \|\varphi\|_0 + \sum_{1}^{m}{}_{|i|} \|D^i\varphi\|_0 ,$$

where

$$\|\varphi\|_0 = \left(\int_0^1 d\xi \int_{-q}^{q} \varphi^2 d\zeta\right)^{\frac{1}{2}} ,$$

and let $\bar{\mathcal{Y}}_m$ be the closure of $\mathcal{Y}_m$ with reference to this norm.

Then B is the Banach space of vectors (ψ, v) with ψ in $\bar{\mathcal{Y}}_3$, v in $\bar{\mathcal{Y}}_1$ and the norm

$$\|\psi, v\| = \|\psi\|_3 + \|v\|_1 .$$

We consider also the set of all functions $\varphi(\xi,\zeta)$, C^∞ in S , periodic in ζ with period 2q, which vanish in a strip along the boundary of G. The space obtained by closure of the set with reference to the norm $\|\varphi\|_m$ will be indicated with $\overset{\circ}{H}_m$.

Then $\mathcal{Y}$ is the set of vectors (ψ, ϑ) of B with $\psi \in \overset{\circ}{H}_2$ and $\vartheta \in \overset{\circ}{H}_1$; in fact vectors such that $\psi \in \overset{\circ}{H}_2$, $\vartheta \in \overset{\circ}{H}_1$ satisfy, in a generalized sense, the boundary conditions at $\xi = 0$, $\xi = 1$.

We will look then for solutions of our problem (33) in $\mathcal{Y}$. It is possible to show first of all that there are no solutions except the trivial one, for $T < \hat{T}_1$ (for a proof, see 39, pp. 59-60). It is possible to show further [37 , pp. 4-5] ; see also for some comments the address [46] that the transformation $\mathcal{A}$ $(\psi, \vartheta) \rightarrow (\psi', \vartheta')$ defined through the non linear problem in ψ', ϑ' :

$$\Delta\Delta\psi' = T\frac{\partial\vartheta}{\partial\xi} + \frac{\partial(\psi, \Delta\psi)}{\partial(\xi, \zeta)},$$

$$\Delta\vartheta' = \frac{\partial\psi}{\partial\xi} + \frac{\partial(\psi, \vartheta)}{\partial(\xi, \zeta)},$$

and by requiring that (ψ, ϑ) be in B, and (ψ', ϑ') be in $\mathcal{Y}$ is a complete functional transformation of the space B into itself. Its fixed points are the solutions of our problem.

Similary the transformation $\mathcal{B}$: $(\psi, \vartheta) \rightarrow (\psi', \vartheta')$ defined through the linear problem

$$\Delta\Delta\psi' = T\frac{\partial\vartheta}{\partial\xi},$$

$$\Delta\vartheta' = \frac{\partial\psi}{\partial\xi},$$

and by requiring that (ψ, ϑ) be in B, and (ψ', ϑ') be in $\mathcal{Y}$ is also a complete linear transformation of B into itself. The fixed points are the eigenfunctions of the problem of sect. 8.

Now it can be proved that $\mathcal{B}$ is the Frèchet differential of the transformation $\mathcal{A}$ at the point (0,0) of space B.

All we have said remains true if we substitute the space B with the subspace B_1 of the vectors (ψ, ϑ) of B such that ψ is odd and ϑ is even in ξ. The advantage of considering our problem in B_1 is this, that (as remarked in Sect. 8) there exists choices of q such that to the associated eigenvalue $\hat{T}_1$ there correspons only one eigenvector:

$\hat{T}_1$ has multiplicity 1.

Then, for a theorem of Leray-Shauder, $\hat{T}_1$ is a branching point for the solutions of the problem (33), (22): a non-trivial solution of our problem must exist if the value of T is chosen within a sufficiently small interval ($\hat{T}_1, \hat{T}_1 + d, d>0$).

10. Numerical study of the non-linear elliptic problem

The analytical developments of Sects 8, 9 assure us of the existence of non-trivial steady state solutions of the non-linear perturbation equations (22), (32) for $T > \tilde{T}_1$, hence of a branching of the fundamental solution. They allow us also to calculate approximately the value of the Taylor number that characterizes the transition. In practice one would like to know the amplitude of the perturbation as a function of T beyond the critical value $\hat{T}_1$. Knowledge of that amplitude leads for instance to an evaluation of the couples acting on the rotating cylinders, more precisely of the excess of those couples beyond the value that would be predicted for Couette flow. For such an evaluation a recourse to numerical methods is essential. One can pursue either the numerical integration of the steady state equations (22), (23) through a process of discretization and successive approximations [40] or a numerical integration of the diffusion equations (22), (23) until a state is reached sufficiently near the steady state [22].

We give here first of all some details of the first process of discretization in a special case considering the finite-difference problem, which derives from (22), (33) for the choice q = 1, when the net points are chosen to have coordinates $\xi = mh$, $\zeta = \pm (p - \frac{1}{2}) h$ with $h = (n+1)^{-1}$, $m = 1, 2, \ldots n$, $p = 1, 2, \ldots n+1$. The boundary conditions for ψ at $\xi = 0$, $\xi = 1$ call for the use of fictitious external points, whereas the conditions at $\zeta = \pm 1$ imply properties of the operating matrices.

Precisely, the finite difference problem can be written as follows

$$\mathcal{U}_1 \underline{\psi} = \frac{h^3}{2} T \; B_1 \underline{\nu} + h M_1 (\underline{\psi}),$$
$$\mathcal{U}_2 \underline{\nu} = \frac{h}{2} \; B_1 \underline{\psi} + M_2 (\underline{\psi}, \underline{\nu}), \tag{34}$$

where $\underline{\psi}$, $\underline{\nu}$ are two vectors with $2n^2 + 2n$ components, each of which gives the approximate value of the functions ψ, ν over the mesh points ordered from left to right and from top to bottom ; $\mathcal{U}_1$, $\mathcal{U}_2$, B_1 are block-circulant and block-symmetric matrices of order 2n + 2 in submatrices of order n :

$$\mathcal{U}_1 = \{A, B, I, O, \ldots, O, I, B\},$$
$$\mathcal{U}_2 = \{C, I, O, \ldots, O, I\},$$
$$B_1 = \{O, -I, O, \ldots, O, I,\},$$

with O null matrix of order n ; I identity matrix of order n ; A, B, C symmetric matrices of order n, of which the first is pentadiagonal and the other two tridiagonal :

$$a_{1,1} = a_{n,n} = 21 \quad ; \; a_{i,i} = 20 \quad , \; i = 2, \ldots , n-1 \; ,$$

$$a_{i-1,i} = a_{i,i-1} = -3 \qquad , \; i = 2, \ldots , n \; ,$$

$$a_{i-2;i} = a_{i,i-2} = 1 \qquad , \; i = 3, \ldots , n \; ;$$

$$b_{i,i} = -3 \qquad , \; i = 1, \ldots , n \; ,$$

$$b_{i-1,i} = b_{i,i-1} = 2 \qquad , \; i = 2, \ldots , n \; ;$$

$$C = \frac{1}{2} B \quad .$$

M_1, M_2 are non-linear operators acting the first on the vector $\underline{\psi}$ and the second on the compound vector ($\underline{\psi}$ | $\underline{v}$).

As in the case of the differential problem, it is possible to show that the values of T , for which the associated linear homogeneous system

$$\left(\begin{array}{c|c} \mathcal{U}_1 & -h^3 T B_1 \\ \hline -hB_1 & \mathcal{U}_2 \end{array} \right) \left(\begin{array}{c} \underline{\psi} \\ \hline \underline{v} \end{array} \right) = 0 \; , \qquad (35)$$

has non-trivial solutions, i.e. the values of T which are roots of the algebraic equation of degree $2n^2 + 2n$

$$\left| \begin{array}{c|c} \mathcal{U}_1 & -h^3 T B_1 \\ \hline -hB_1 & \mathcal{U}_2 \end{array} \right| = 0 \; ,$$

has solutions, are positive (we will refer to these values as the eigenvalues of the problem). In fact, if $\hat{\underline{\psi}}$, $\hat{\underline{v}}$ is a solution of (35) corresponding to the eigenvalue T , then

$$\underline{\hat{\psi}}^T \, \mathcal{U}_1 \, \underline{\hat{\psi}} = \frac{h^3}{2} \, \hat{T} \quad \underline{\hat{\psi}}^T \, B_1 \underline{\hat{v}} \quad ,$$

$$\underline{\hat{v}}^T \, \mathcal{U}_2 \, \underline{\hat{v}} = \frac{h}{2} \, \underline{\hat{\psi}}^T \, B_1 \underline{\hat{v}} \quad ;$$

hence

$$\hat{T} = \frac{\underline{\hat{\psi}}^T \, \mathcal{U}_1 \, \underline{\hat{\psi}}}{h^2 \, \underline{\hat{v}}^T \, \mathcal{U}_2 \, \underline{\hat{v}}} \quad ;$$

but $\underline{x}^T \, \mathcal{U}_1 \, \underline{x}$, $-(\underline{x}^T \, \mathcal{U}_2 \, \underline{x})$ are positive definite quadratic forms in the components of $\underline{x}$. It follows that $\hat{T}$ must be positive; it follows also that $\underline{\hat{\psi}}$, $\underline{\hat{v}}$ can be taken to have real components.

Again as in the case of the differential problem it is found that non trivial solutions of the non- linear problem (34) may exist only for values of T greater than the lowest eigenvalue $\hat{T}_c$ of (35). To reach a proof of actual existence of a solution under the condition $T > \hat{T}_c$, some preliminary results are required.

Firstly we remark that eigensolutions of (35) can be written as follows

$$\underline{\psi} = \begin{pmatrix} k_1 \underline{\varphi} \\ \cdot \\ \cdot \\ \cdot \\ k_{n+1} \underline{\varphi} \\ \hline D \begin{vmatrix} k_1 \underline{\varphi} \\ \cdot \\ \cdot \\ \cdot \\ k_{n+1} \underline{\varphi} \end{vmatrix} \end{pmatrix} , \qquad \underline{v} = \begin{pmatrix} h_1 \underline{w} \\ \cdot \\ \cdot \\ \cdot \\ h_{n+1} \underline{w} \\ \hline -D \begin{vmatrix} h_1 \underline{w} \\ \cdot \\ \cdot \\ \cdot \\ h_{n+1} \underline{w} \end{vmatrix} \end{pmatrix}$$

where

$$k_i = \sin \left[(2i-1) \, r \alpha \right] , \qquad \alpha = \frac{\pi}{2n+2} , \quad 1 \leqslant r \leqslant n \quad ;$$

$$h_i = \cos \left[(2i-1) \, r \alpha \right] ,$$

$$D = \begin{pmatrix} 0 & \cdots & 0 & -I \\ 0 & \cdots & -I & 0 \\ \cdot & \cdots & \cdot & \cdot \\ -I & \cdots & 0 & 0 \end{pmatrix}.$$

and $\underline{\varphi}$, $\underline{w}$ are n-vectors which satisfy the equations

$$\begin{aligned} &(A + 2\cos 2r\alpha\, B + 2\cos 4r\alpha\, I)\,\underline{\varphi} = h^3 T \sin 2r\alpha\, \underline{w} \\ &(C + 2\cos 2r\alpha\, I)\,\underline{w} = -h \sin 2r\alpha\, \underline{\varphi}. \end{aligned} \tag{37}$$

Eliminating $\underline{w}$ one obtains the equation in $\underline{\varphi}$

$$\underline{\varphi} = h^4\, T\, \sin^2 2r\alpha\; C_r^{-1} \underline{\varphi}, \tag{38}$$

with

$$C_r = -(C + 2\cos r\alpha\, I)\,(A + 2\cos 2r\alpha\, B + 2\cos 4r\alpha\, I).$$

Hence solutions of our problem (35) exist provided that $h^4 T \sin^2(2r\alpha)$ coincides with one of the eigenvalues of the matrix C_r ($r = 1, \ldots n$). Some elementary developments show [47] that the matrices $(A + 2\cos 2r\alpha\, B + 2\cos 4r\alpha I)^{-1}$ ($r = 1, \ldots n$) are positive; on the other hand the matrices $-(C + 2\cos 2r\alpha\, I)$ ($r = 1, \ldots n$) are irreducible, diagonally dominant matrices with positive diagonal elements and non-positive off-diagonal element; so that the matrices $-(C + 2\cos 2r\alpha\, I)^{-1}$ are positive.

Hence the C_r are positive; more precisely

$$C_n < C_{n-1} < \ldots < C_1. \tag{39}$$

If we chose for T a value $\hat{T}$ so that $h^4 \hat{T} \sin^2 2r\alpha$ is an eigenvalue of C_r, we can construct through (38) the corresponding eigenvector $\underline{\varphi}$ and, successively, through the second eqn (37) and formulae (36) the eigenvector $(\underline{\psi}, \underline{v})$ of our original problem (35).

Equivalently we could say that the system

$$\underline{\psi} = \frac{h^4}{4} \hat{T} \, \mathcal{V}_1^{-1} B_1 \mathcal{V}_2^{-1} B \, \underline{\psi} \qquad (40)$$

admits of solutions.
Thus, we can determine n^2 eigenvalues of $M = \mathcal{V}_1^{-1} B_1 \mathcal{V}_2^{-1} B$ (if each is counted with the appropriate multiplicity). Actually it can be checked that to each eigenvalue so determined there correspond two eigenvectors ($\underline{\psi}$, $\underline{v}$) ; the first is of the form (36) the second has a similar structure but the rôle of the trigonometric functions in the defining formulae is reversed.
We have accounted so far for $2n^2$ eigenvalues of M ; the remaining $2n$ are zero; in fact M is a singular matrix.

Let us consider now the spectral radius $\rho(C_r)$ of C_r; a theorem of Perron-Frobenius and the inequalities (39) assure us that $\rho(C_r) < \rho(C_{n-1}) < \ldots < \rho(C_1)$, and that $\rho(C_1)$ is a simple eigenvalue of C_1 so that $\rho(C_1)$ is not an eigenvalue of $C_2, \ldots C_n$.
In conclusion

$$\hat{T}_c = \frac{4}{h^4 \rho(C_1)}$$

is the minimum value of T for which the problem (35) has solution. To this value of T there corresponds a unique solution (apart from a constant factor) of the type

$$\underline{\psi} = \left(\frac{\underline{\psi}_1}{D\underline{\psi}_1} \right) \quad , \quad \underline{v} = \left(\frac{\underline{v}_1}{-D\underline{v}_1} \right) \qquad (41)$$

where $\underline{v}_1$, $\underline{\psi}_1$ are n(n+1)-vectors.

Consider now the vector space B of the 4n(n+1)-vectors ($\underline{\psi} \,|\, \underline{v}$) with $\underline{\psi}$, $\underline{v}$ of the type (41); let B be normed ℓ_∞. If we choose any vector ($\underline{\psi} \,/\, \underline{v}$) in B and calculate the vectors

$$\left(\frac{\frac{h^3}{2}\, T\, \mathcal{V}_1^{-1}\, B_1 \underline{\nu}}{\frac{h}{2}\, \mathcal{V}_2^{-1}\, B_1 \underline{\psi}} \right), \quad \left(\frac{\frac{h^3}{2}\, T\, \mathcal{V}_1^{-1} B_1 \underline{\nu} + h\, \mathcal{V}_1^{-1}\, M_1\, (\underline{\psi})}{\frac{h}{2} \mathcal{V}_2^{-1}\, B_1 \underline{\psi} \; + \mathcal{V}_2^{-1}\, M_2\, \left(\frac{\underline{\psi}}{\underline{\nu}}\right)} \right),$$

these belong also to B . Hence we can consider the eigenvectors of (35) and the non-trivial solutions of (34) respectively as fixed points of the following compact mappings of B into itself

$$\mathcal{T} \ldots \begin{cases} \underline{\psi}^1 = \frac{h^3}{2}\, T\, \mathcal{V}_1^{-1}\, B_1 \underline{\nu}\,, \\ \underline{\nu}^1 = \frac{h}{2} \mathcal{V}_2^{-1}\, B_1 \underline{\psi}\,, \end{cases}$$

and

$$\mathcal{T}' \ldots \begin{cases} \underline{\psi}^1 = \frac{h^3}{2}\, T\, \mathcal{V}_1^{-1}\, B_1 \underline{\nu} + h\, \mathcal{V}_1^{-1}\, M_1\, (\underline{\psi})\,, \\ \underline{\nu}^1 = \frac{h}{2} \mathcal{V}_2^{-1}\, B_1 \underline{\psi} + \mathcal{V}_2^{-1}\, M_2\, \left(\frac{\underline{\psi}}{\underline{\nu}}\right), \end{cases}$$

Now $\mathcal{T}$ is the Fréchet differential of $\mathcal{T}'$ calculated over the null element of B . If we seek solution of our problems (34), (35) exclusively within B then we find that for $T = \hat{T}_c$ there corresponds a simple eigenvalue of (35) .

A theorem of Leray Schauder assures us, then, of the existence of a non-trivial solution of (34) for each choice of T in an appropriate interval ($\hat{T}_c$, $\hat{T}_c + d$) , $d > 0$; in other words $\hat{T}_c$ is a branching point for the solutions of (34).

11. Notes on the numerical experiments

We have dealt so far with fundamental questions related to the system (34) ; we comment now briefly on problems connected with the planning of actual numerical experiments.

Because of the non-linear nature of system (34), its practical solution calls for an iterative procedure of the type envisaged in eqn (20); for simplicity we make reference here to the scheme

$$\begin{aligned} \underline{\psi}^{(n+1)} &= \frac{h^3}{2} T \mathcal{V}_1^{-1} B_1 \underline{v}^{(n)} + h\mathcal{V}_1^{-1} M_1 (\underline{\psi}^{(n)}) , \\ \underline{v}^{(n+1)} &= \frac{h}{2}\mathcal{V}_2^{-1} B_1\underline{\psi}^{(n)} + \mathcal{V}_2^{-1} M_2 (\underline{\psi}^{(n)}, \underline{v}^{(n)}) , \end{aligned} \tag{42}$$

although the alternative scheme

$$\underline{\psi}^{(n+1)} = \frac{h^3}{2} T\mathcal{V}_1^{-1} B_1 v^{(n)} + h\mathcal{V}_1^{-1} M_1 (\underline{\psi}^{(n)}) ,$$

$$\underline{v}^{(n+1)} = \frac{h}{2} \mathcal{V}_2^{-1} B_1 \underline{\psi}^{(n+1)} + \mathcal{V}_2^{-1} M_2 (\underline{\psi}^{(n+1)}, \underline{v}^{(n)}) ,$$

seems to be faster.

It is found that, if the step h is chosen to be small enough, the vector $(\underline{\psi}^{(n)}, \underline{v}^{(n)})$ tends with increasing n to the null vector when $T < \hat{T}_c$, whereas it converges towards the non-trivial solution of (34) when $T > \hat{T}_c$.

Notice that this happens although for $T > \hat{T}_c$, eqn (34) admits always a trivial solution. A precise analysis of this behaviour is not available; we can add here only a heuristic argument which indicates a bound on h for the stability of the process (42). This bound was verified closely in practice; it is of the type mentioned in Sect. 4.

Let us call $\underline{\varepsilon}_{\psi}^{(n)}$, $\underline{\varepsilon}_{\vartheta}^{(n)}$ the errors in $\underline{\psi}^{(n)}$ and $\underline{\vartheta}^{(n)}$ respectively . Then, from (42) we get

$$\underline{\varepsilon}_{\psi}^{(n+1)} = \frac{h^3}{2} T \mathcal{V}_1^{-1} B_1 \underline{\varepsilon}_{\vartheta}^{(n)} + h \mathcal{V}_1^{-1} \left\{ M_1(\underline{\psi}^{(n)}) - M_1(\hat{\underline{\psi}}) \right\} ,$$

$$\underline{\varepsilon}_{\vartheta}^{(n+1)} = \frac{h}{2} \mathcal{V}_2^{-1} B_1 \underline{\varepsilon}_{\psi}^{(n)} + \mathcal{V}_2^{-1} \left\{ M_2(\underline{\psi}^{(n)}, \underline{\vartheta}^{(n)}) - M_2(\hat{\underline{\psi}}, \hat{\underline{\vartheta}}) \right\} , \qquad (43)$$

where now $\hat{\psi}$, $\hat{\vartheta}$ stand for the solution of (34) .

We accept the approximate equalities

$$M_1(\underline{\psi}^{(n)}) - M_1(\hat{\underline{\psi}}) \simeq \left(\frac{\partial M_1}{\partial \underline{\psi}}\right)_{\psi=\hat{\psi}} \underline{\varepsilon}_{\psi}^{(n)} ,$$

$$M_2(\underline{\psi}^{(n)}, \underline{\vartheta}^{(n)}) - M_2(\hat{\underline{\psi}}, \hat{\underline{\vartheta}}) \simeq \left(\frac{\partial M_2}{\partial \underline{\psi}}\right)_{\substack{\psi=\hat{\psi} \\ \vartheta=\hat{\vartheta}}} \underline{\varepsilon}_{\psi}^{(n)} + \left(\frac{\partial M_2}{\partial \underline{\vartheta}}\right)_{\substack{\psi=\hat{\psi} \\ \vartheta=\hat{\vartheta}}} \underline{\varepsilon}_{\vartheta}^{(n)} ,$$

so that it follows from system (43) that

$$\begin{pmatrix} \underline{\varepsilon}_{\psi}^{(n)} \\ \hline \underline{\varepsilon}_{\vartheta}^{(n)} \end{pmatrix} = \left(\begin{array}{c|c} h \mathcal{V}_1^{-1} \left(\widehat{\frac{\partial M}{\partial \underline{\psi}}}\right) & \frac{h^3}{2} T \mathcal{V}_1^{-1} B_1 \\ \hline \frac{h}{2} \mathcal{V}_2^{-1} B_1 + \mathcal{V}_2^{-1} \left(\widehat{\frac{\partial M}{\partial \underline{\psi}}}\right) & \mathcal{V}_2^{-1} \left(\widehat{\frac{\partial M_2}{\partial \underline{\vartheta}}}\right) \end{array}\right)^n \begin{pmatrix} \underline{\varepsilon}_{\psi}^{(o)} \\ \hline \underline{\varepsilon}_{\vartheta}^{(o)} \end{pmatrix} .$$

This equality implies that the error decreases only if the spectral radius $\rho^{(\mathfrak{M})}$ of the matrix

$$\mathfrak{m} = \left(\begin{array}{c|c} h\,\mathcal{V}_1 \left(\frac{\partial \hat{M}_1}{\partial \underline{\psi}}\right) & \frac{h^3}{2}\, T\, \mathcal{V}_1^{-1}\, B_1 \\ \hline \frac{h}{2}\mathcal{V}_2^{-1}\, B_1 + \mathcal{V}_2^{-1} \left(\frac{\partial \hat{M}_2}{\partial \underline{\psi}}\right) & \mathcal{V}_2^{-1} \left(\frac{\partial \hat{M}_2}{\partial \underline{\vartheta}}\right) \end{array} \right)$$

does not exceed unity. $\rho(\mathfrak{m})$ depends on h , T and also on $\hat{\underline{\psi}}$, $\hat{\underline{\vartheta}}$; but these two last vectors are unknown to start with : a reasonable guess for the solution is required in practice for an evaluation of the conditions of convergence; such conditions will put then restrictions on h depending on the value of T. However the calculation of the spectral radius of $\mathfrak{m}$ is not an easy matter; as a consequence one is forced to rely on rougher estimates, such as the following one.

Assume that in the vector $\underline{E}^{(k-1)} = \left(\frac{\underline{\varepsilon}_{\psi}^{(k-1)}}{\varepsilon_{\vartheta}^{(k-1)}}\right)$ all but one component vanish, for instance the error component relative to the value of $\underline{v}$ over a certain mesh point P. Then we may take, as an approximation, that only a few components of $\underline{E}^{(k)}$ are different from zero, precisely those relative to values of $\underline{v}$ over meshpoints adjoining P.

If P is sufficiently far from the boundary the components taken to be non-null are those of order m - n, m - 1, m + 1 , m + n .

During the next iteration, leading to $\underline{E}^{(k+1)}$, there is a "backfire" effect of the spread error over the m-th component $\delta_m^{(k+1)}$. As a rough estimate of the condition of stability it is required that

$$\left|\delta_m^{(k+1)}\right| < \left|\delta_m^{(k-1)}\right| . \qquad (44)$$

Notice incidentally that, if this criterion is adapted to apply to the heat-transfer equation $\frac{\partial u}{\partial t} = \sigma \frac{\partial^2 u}{\partial x^2}$ in the finite difference form $u_j^{n+1} - u_j^n = [\sigma \Delta t/(\Delta x)^2]\,[u_{j+1}^n - 2u_j^n + u_{j-1}^n]$,

it leads to the stability rule $[\sigma \Delta t/(\Delta x)^2] \leq 2/3$, as can be easily checked; here the notation is obvious. As is well known a more appropriate analysis in this case suggests the upper limit 1/2 rather 2/3 for the ratio $[\sigma \Delta t/(\Delta x)^2]$, at least in the case of simple boundary conditions.

Similary, if the criterion is adapted to apply to the wave equation $\frac{\partial^2 u}{\partial t^2} = c^2 \frac{\partial^2 u}{\partial x^2}$ in the finite-difference form

$$u_j^{n+1} - 2u_j^n + u_j^{n-1} = (c\,\Delta t/\Delta x)^2\,(u_{j+1}^n - 2u_j^n + u_{j-1}^n)$$

it leads to the rule $(c\,\Delta t/\Delta x) \leq 1$.

Returning now to our problem, we are interested in the solution of a linear system extracted from the system

$$\mathcal{V}_2\, \underline{\varepsilon}_{\nu}^{(k)} = \left(\frac{\partial \hat{M}_2}{\partial \underline{\nu}}\right) \varepsilon_{\nu}^{(k-1)}$$

in fact we have supposed that ε_{ψ} be null.

Because all components of $\varepsilon_{\nu}^{(k-1)}$ but one are also null, we intend to examine the approximation where all but five components of $\varepsilon_{\nu}^{(k)}$ vanish. These components satisfy the following reduced system

$$\begin{pmatrix} -4 & 0 & 1 & 0 & 0 \\ 0 & -4 & 1 & 0 & 0 \\ 1 & 1 & -4 & 1 & 1 \\ 0 & 0 & 1 & -4 & 0 \\ 0 & 0 & 1 & 0 & -4 \end{pmatrix} \begin{pmatrix} \delta^{(k)}_{m-n} \\ \delta^{(k)}_{m-1} \\ \delta^{(k)}_{m} \\ \delta^{(k)}_{m+1} \\ \delta^{(k)}_{m+n} \end{pmatrix} = \frac{1}{4}\delta^{(k-1)}_{m} \begin{pmatrix} \psi_{m+1} - \psi_{m-1} \\ \psi_{m-n} - \psi_{m+n} \\ 0 \\ \psi_{m+n} - \psi_{m-n} \\ \psi_{m-1} - \psi_{m+1} \end{pmatrix}$$

which can be easily solved. In the second step leading to $\delta^{(k+1)}$ a linear system with 13 unknowns is involved; we leave out details to quote the result

$$\delta^{(k+1)}_{m} = \frac{5}{448}\left\{(\psi_{m+1} - \psi_{m-1})^2 + (\psi_{m+n} - \psi_{m-n})^2\right\}\delta^{(k-1)}_{m} \qquad (45)$$

From a numerical point of view inequality (44) may be interpreted now as a constraint imposed upon the change of ψ over two mesh steps; in fact (44) implies, in view of (45),

$$(\psi_{m+1} - \psi_{m-1})^2 + (\psi_{m+n} - \psi_{m-n})^2 < \frac{448}{5} .$$

An alternative, physically significant, interpretation of (45) is possible; consider the modulus S of the projection in the (r, z)-plane of the velocity of the fluid and indicate with R_N the Reynolds number based on S, the physical size $(r_2 - r_1)\ h$ of the mesh and the viscosity ν. Then (45) can be written

$$R_N < \sqrt{\frac{112}{5}} \simeq 4.7$$

These criteria of stability, though rough, have proved to be very useful in the preparation of computer programs. For samples of results of numerical work we make reference, for instance, to paper [23] .

REFERENCES

[1] L. F. Richardson, Weather prediction by numerical process. Cambridge Univ. Press., London 1922.

[2] R. Barker, Intégration des équations du mouvement d'un fluide visqueux incompressible. Encyclopedia of Physics, vol.8/2, Springer (1966).

[3] R. Finn, Stationary solutions of the Navier-Stokes equations. Proc. Symposia Appl. Math., 17 (1965), 121-153.

[4] D.N. de Allen, Relaxation Methods in Engineering and Science, Mc Graw-Hill, 1954.

[5] A. Thom, C.J. Apelt, Field Computations in Engineering and Physics. Van Nostrand, 1961.

[6] J. von Neumann, Collected Works, vol. 5. Pergamon Press, 1963.

[6 A] id., vol. 6

[7] B. Alder, S.Fernbach, M. Rotenberg, ed. , Methods in computational physics. Academic Press. 1, Statistical Physics (1962); 2, Quantum Mechanics (1963); 3, Fundamental Methods in Hydrodynamics (1964); 4, Applications in Hydrodynamics (1965); 5, Nuclear Particle Kinematics (1966); 6, Nuclear Physics.

[8] F.H. Harlow, The particle -in -cell methods for numerical solution of problems in fluid dynamics. Proc. Symposia Appl. Math., 15 (1963), 269-288.

[9] J.R. Pasta, S. Ulam, Heuristic numerical work in some problems of hydrodynamics. Math. Tables Other Aids Comp., 13 (1959), 1-12.

[10] A. Blair, N. Metropolis, J. von Neumann, A.H. Taub, M. Tsingou, A study of a numerical solution to a two-dimensional hydrodynamical problem. Math. Tables Other Aids Comp., 13 (1959), 145-184.

[11] J.E. Welch, F.H. Harlow, J.P. Shannon, B.J. Daly, The MAC method, a computing technique for solving viscous, incompressible, transient fluid-flow problems involving free surfaces. Los Alamos Scient. Lab., LA - 3425.

[12] F.H. Harlow, J.E. Welch, Numerical calculation of time-dependent viscous incompressible flow of fluid with free surface. Phys. Fluids 8 (1965), 2182-2189.

[13] F.H. Harlow, J.E. Welch, Numerical study of large-amplitude free-surface motions. Phys. Fluids, 9 (1966), 842-851.

[14] J.D. Hellums, S.W. Churchill, Computation of natural convection by finite difference methods. Proc. Int. Conference on Heat Transfer, Inst. Mech. Eng., London (1961).

[15] M.R. Abbott, A numerical method for solving the equations of natural convection in a narrow concentric cylindrical annulus with a horizontal axis. Quat. Journ. Mech. Appl. Math., 17 (1964), 471-481.

[16] C.E. Leith, Numerical simulation of the earth's atmosphere in [1] , vol. 4, 1-28.

[17] A. Kasahara, E. Isaacson, J.J. Stoker, Numerical studies of frontal motion in the atmosphere. Tellus, 17 (1965), 1.

[18] E. Isaacson, Fluid dynamical calculations in Numerical Solution of Partial Differential Equations, J.H. Bramble, ed.,Academic Press. New York (1966), 35-49.

[19] J.E. Fromm,F.H. Harlow, Numerical solution of the problem of vortex street development. Phys. Fluids, 6 (1963), 975-982.

[20] F. H. Harlow, J.E. Fromm, Dynamics and heat transfer in the von Kármán wake of a rectangular cylinder. Phys. Fluids, 7 (1964), 1147-1156.

[21] B.J. Daly, A numerical study of two fluid Rayleigh-Taylor instability. The Physics of Fluids, 10 (1967), 297.

[22] A. L. Krilov, E.K. Proizvolova, Numerical analysis of the fluid flow between two rotating cylinders. Proceedings (CБОРНИК РАБОТ) Computing Centre Moscow Univ., 2 (1963), 174-181.

[23] G. Capriz, G.Ghelardoni, G.Lombardi, Numerical study of the stability problem for Couette flow. Phys. Flyids, 9 (1966), 1934-1936.

[24] F.H. Harlow, J.E. Fromm, Computer experiments in fluid dynamics. Scientific American, 212 (1965), 104-110.

[25] F.H. Harlow, J.P. Shannon, J.E. Welch, Liquid waves by computer. Science, 149 (1965), 1092-1093.

[26] J.E. Welch, Computer simulation of water waves, Datamation 12 (1966), 41.

[27] F.H. Harlow, J.P. Shannon, J.E. Welch, Un calculateur qui fait des vagues. Sciences, 7 (1966), 14.

[28] D. Greenspan, P.C. Jain, R. Manohar, B. Noble, A. Saburai, Numerical studies of the Navier-Stokes equations. Math. Res. Center, Techn. Summary Rept. 482 (1964).

[29] P.C. Jain, Numerical study of the Navier-Stokes equations for the production of small eddies from large ones. Math. Res. Center. Techn. Summary Rept. 491 (1964).

[30] E. De Luca, Numerical studies of point perturbations in laminar plane Poiseuille motion. Army Material Res. Agency, Tech. Rept. AMRA TR 63-10.

[31] M. Capovani, G.Capriz, G.Lombardi, Studio numerico della stabilità del moto di un fluido viscoso in un canale. Calcolo 2, Suppl. 1 (1965), 33-49.

[32] E. Bellomo, A numerical program for dealing with finite-amplitude disturbance in plane parallel laminar flows. Meccanica, 2 (1967), 95-108.

[33] D.F. De Santo, H.B. Keller, Numerical studies of transition from laminar to turbulent flow over a flat plate. J. Soc. Ind. Appl. Math. , 10 (1962), 569-595.

[34] J.A.T. Bye, Obtaining solutions of the Navier-Stokes equation by relaxation processes. Comp. J., 8 (1965-66), 53-56.

[35] L.H. Thomas, The stability of plane Poiseuille flow. Phys. Rev. 91 (1953), 780-783.

[36] W. Velte, Stabilitätsverhalten und Verzweigung stationärer Lösungen der Navier-Stokesschen Gleichungen. Arch. Rat. Mech. Anal., 16 (1964), 97-125.

[37] W. Velte, Stabilität und Verzweigung Stationärer Lösungen der Navier-Stokesschen Gleichungen beim Taylor Problem. Arch. Rat. Mech. Anal., 22 (1966), 1-14.

[38] K. Kirchgässner, Die Instabilität der Strömung zwischen zwei rotierenden Zylindern gegenüber Taylor-Wirbeln fur beliebige Spaltbreiten. Z.A.M.P. 12 (1961), 14-30.

[39] G.Ghelardoni, Questioni connesse coll'impostazione analitica e la soluzione numerica di un problema di idrodinamica. Calcolo, 2, Suppl. 1 (1965), 51-66.

[40] G.Ghelardoni, G. Lombardi, Soluzione numerica di un problema di stabilità idrodinamica.
Calcolo, 2, Suppl. 1 (1965), 67-80.

[41] G.I. Taylor, Stability of a viscous liquid contained between two rotating cylinders. Phil. Trans. Roy. Soc. (London) A 223 (1923), 289-343.

[42] R.J. Donnelly, D. Fultz, Experiments on the stability of spiral flow between rotating cylinders.
Proc. Nat. Acad. Sci., 46 (1960), 1150-1154.

[43] D.Coles, Transition in circular Couette flow.
Navy Dept. Rept., Cambridge, Mass.

[44] S. Chandrasekhar, Hydrodynamic and hydromagnetic stability.
Oxford Univ. Press (1961).

[45] H. Witting, Über den Einfluss der Strömlinien-Krümmung auf die Stabilität laminärer Strömungen.
Arch. Rat. Mech. Anal., 2 (1958), 243-283.

[46] G.Prodi, Problemi di diramazione per equazioni funzionali.
Atti VIII Congresso U.M.I., Trieste (1967), Under press.

[47] G.Ghelardoni, Considerazioni numeriche su un problema di idrodinamica.
Calcolo, Under press.

CENTRO INTERNAZIONALE MATEMATICO ESTIVO

(C.I.M.E.)

A. DOU

ENERGY INEQUALITIES IN AN ELASTIC CYLINDER

Corso tenuto ad Ispra dal 3-7 luglio 1967

ENERGY INEQUALITIES IN AN ELASTIC CYLINDER

by

A. DOU

(University of Madrid)

1 The principle of Saint - Venant.

More than a centrary ago, in 1855 and 1856, B. de Saint-Venant [1] stated a principle that has been applied steadily in the calculus of beams, but has not yet been proved. From the point of view of Numerical Analysis the importance of the Principle of Saint-Venant is obvious and great.

We consider an homogeneous, isotropic and perfectly elastic cylinder Ω_ℓ,

(1) $$\Omega_\ell \equiv \Omega = \{(x, y, z) \mid (x, y) \in Q, |z| < \ell\} \subset R^3 .$$

For simplicity we assume also

$$Q = \{(x, y) \mid |x| < \pi, |y| < \pi\} \subset R^2 .$$

We assume throughout the paper that there are no body forces and no lateral surface forces; and also that the cylinder is in statical equilibrium.

We are given the surface forces $T(x, y)$ and $T^-(x, y)$ acting upon the bases of the cylinder $z = \ell$ and $z = -\ell$ respectively, and we may assume without loss of generality that they are either odd and write $T^{(1)}$,

$$\text{for } z = \ell : T^{(1)} = \{T_1^{(1)}(x, y), T_2^{(1)}(x, y), T_3^{(1)}(x, y)\}$$

$$\text{for } z = -\ell : T^{-(1)} = \{-T_1^{-(1)}(x, y), -T_2^{(1)}(x, y), T_3^{(1)}(x, y)\}, \quad x, y \in Q ,$$

or that they are even and write $T^{(2)}$,

$$\text{for } z = \ell : \quad T^{(2)} = \left\{ T_1^{(2)}(x,y),\ T_2^{(2)}(x,y),\ T_3^{(2)}(x,y) \right\},$$
$$\text{for } z = -\ell : \quad T^{-(2)} = \left\{ T_1^{(2)}(x,y),\ T_2^{(2)}(x,y),\ -T_3^{(2)}(x,y) \right\}, \quad x, y \in \bar{Q}.$$

When there is no need to specify, we shall write $T^{(g)}$, $g = 1, 2$, or simply T if no confusion is possible. The specifications refer to the component $\tau_{33}^{(g)}$ of the corresponding stress tensor with respect to the independent variable z . These corresponding stress tensor will be called, respectively, odd and even ; obviously all their components are odd or even functions of z.

We shall consider only those surface forces

$$(2) \qquad T^{(g)} \equiv T = \left\{ T_1(x,y),\ T_2(x,y),\ T_3(x,y) \right\}, \qquad x, y \in \bar{Q},$$

which satisfy the following conditions :

$$(3a) \qquad T_i \in C(\bar{Q}), \quad T_{\alpha,\alpha} \in C(\bar{Q}), \qquad i = 1, 2, 3, \quad \alpha = 1, 2,$$

$$(3b1) \qquad \iint_Q T_i \, dx\, dy = \iint_Q x\, T_3 \, dx\, dy = \iint_Q y\, T_3 \, dx\, dy = 0,$$

$$(3b2) \qquad \iint_Q (x\, T_2 - y T_1)\, dx\, dy = 0,$$

$$(3c) \qquad T_1(\pi, y) = T_1(-\pi, y) = T_2(x, \pi) = T_2(x, -\pi) = 0, \quad |x| \leq \pi, |y| \leq \pi.$$

In (3a) and throughout the paper the subindex k after the comma means partial derivative with respect to x_k, $(x_1, x_2, x_3) \equiv (x, y, z)$. The conditions (3b) express that the surface forces are self-equilibrated at each base, and (3c) insures that the corresponding stress tensor be continuous in $\bar{\Omega}$.

Surface forces satisfying conditions (3) shall be called permissible surface forces (psf) , and the corresponding boundary conditions shall be called permissible boundary conditions (pbc) . The set of all psf is obvious by a vector space over the reals .

Let $\|S\|_2$ be the $L_2(Q)$ norm of $S(x,y)$ and let $\|\tau(\bar{x})\|$ be the euclidean norm of the stress tensor $\tau(\bar{x})$, $\bar{x} \equiv (x,y,z) \in \bar{\Omega}$. Then the principle of Saint-Venant can be stated thus :

Except in the neighborhood of the bases of the cylinder Ω_ℓ, $\|\tau(\bar{x})\|$ is small compared with $\max_{i=1,2,3} \|T_i\|_2$. In practical applications "neighborhood" means, for instance, that the point $x \in \bar{\Omega}$, is such that $z \geqslant \frac{4\ell}{5}$, $\ell > 10.\pi$; and "small" means that $\tau(\bar{x})$ can be neglected altogether, say $\|\tau(\bar{x})\| \leqslant (1/10) \max_{i=1,2,3} \|T_i\|_2$, for $|z| < 4\ell/5$, $\ell > 10\pi$. Today the best justification of this principle is its successfuly steady application for more than hundred years. Mathematically speaking I do not think that this principle, as stated, is true, although there must be some very general conditions, under which it will be true.

The application of the principle to the calculus of beams is indeed both, obvious and useful . In general beams are bouded with forces $F(x,y)$, which satisfy conditions (3a) , (3c) and the first of (3 b 1) for $i = 1, 2$; but one has

$$\iint_Q F_3(x,y)\, dx\, dy = R,$$

$$\iint_Q F_3\, y\, dx\, dy = M_1, \quad \iint_Q F_3\, x\, dx\, dy = M_2,$$

$$\iint_Q (x F_2 - y F_1)\, dx\, dy = M_3 .$$

The principle allows to dispense with the functions F_1, F_2, F_3 and take into account only the resultant R, the bending moments M_1, M_2 and the torsion moment M_3. To these quantities corresponds in Ω a unique elementary solution $\tau(x, y, z)$ of Saint-Venant's type, i.e. independent of z, and therefore is a linear combination of uniform traction or compression, pure bending and pure torsion.

We remark the following interesting corollary :

The principle of Saint-Venant implies that the only bounded solutions for the stress tensor $\tau(x,y,z)$ in the infinite cylinder Ω_∞ are those of Saint Venant's type, i.e. independent of z.

In the remainder of this seminar I shall give three inequalities that bear on the principle of Saint-Venant and proved by R.A. Toupin [2] , J.J. Roseman [3] and myself, [4,5] and outline the proof of the third. Finally I shall comment on related questions.

2 Energy inequalities.

The first two inequalities due to Toupin and Roseman assume that one end, $z = \ell$, is loaded with psf $T(x,y)$ and the other and, $z = -\ell$, is free of forces. This is achieved in our presentation setting $T^{(1)} = T^{(2)} = (1/_2)\, T\,(x,y)$ and loading the cylinder with psf $T^{(1)} + T^{(2)}$.

The result of Toupin asserts that the total elastic energy of that part of the cylinder beyond a distance s from the loaded end, $U(s)$, satisfies the inequality

$$U(s) < U(0)\ .\ \exp\left\{-\frac{s-h}{c(h)}\right\},$$

where $0 < h < s$ and $c(h)$ is a characteristic decay length depending on elastic constants and on the smallest characteristic frequency of free vibration of Ω_h . Substantially it says that , if we move away from the loaded end, then the total elastic energy in the remaining part of the cylinder decreases exponentially.

The result of Roseman asserts that, if $P(s)$ is a point of the cylinder whose distance to the loaded end is s, then

$$\|\tau(P)\|^2 < \frac{K}{\theta^3 a^3}(U(s-a) - U(s+a)),$$

where the first member and U have already been defined, K is a constant depending on elastic constants of the body and θ and a are positive constants depending on the geometry of the cross-section. This result is similar to the previous pointwise estimates of J. H. Bramble and L. E. Payne [6], but this one is good up to the boundary.

The third inequality is contained in the following theorem :

Assume that τ is the stress tensor in Ω_ℓ corresponding to any psf T = = (T_1, T_2, T_3), let $\mathcal{E}_\tau(\Omega)$ be the total elastic potential energy of τ in Ω_ℓ and let $0 < \ell_o \leq \ell$. Then there is a constant K depending only on ℓ_o such that

(4) $$\mathcal{E}_\tau(\Omega) < \frac{K}{\mu}(\|T_1\|_2^2 + \|T_2\|_2^2 + \|T_3\|_2^2 + \|T_{1,1}\|_2^2 + \|T_{2,2}\|_2^2) .$$

Now I shall outline the proof of this theorem given in [5]. We need the constituent equations relating the stresses with the displacements in an homogeneous, isotropic and perfectly elastic body, the equations of Elasticity governing the dispalcements and the theorem of Castigliano.

The stress tensor $\tau = ((\tau_{ij}))$, i, j = 1, 2, 3, analytic in Ω and continuous in $\bar{\Omega}$, corresponding to the psf $T^{(g)} = (T_1^{(g)}, T_2^{(g)}, T_3^{(g)})$, g = 1, 2, is given in terms of the first derivatives of the displacements $u(x, y, z) = (u_1, u_2, u_3)$, $(x,y,z) \in \Omega$, by

(5) $$\tau_{ij} = \lambda\delta_{ij}u_{h,h} + \mu(u_{i,j} + u_{j,i}) , \qquad u_{h,h} = \operatorname{div} u ,$$

$$\lambda > 0 \quad , \quad \mu > 0 \quad 0 < \sigma = \frac{\lambda}{2(\lambda+\mu)} < \frac{1}{2} , \ \delta_{ij} \text{ Kronecker symbol.}$$

The displacements u are given by the unique solution of the Navier equations of elasticity

(6) $$\mu\Delta u_i + (\lambda+\mu) u_{h,hi} = 0 ,$$

$$u_{h,hi} = \frac{\partial u_{h,h}}{\partial x_i} , \qquad (x_1, x_2, x_3) = (x, y, z) ,$$

satisfying the following permissible boundary conditions (pbc) :

(7)
$$\tau_{1i}(\pi, y, z) = \tau_{1i}(-\pi, y, z) = 0 , \quad |y| \leq \pi, \ |z| \leq \ell ,$$
$$\tau_{2i}(x, \pi, z) = \tau_{2i}(x, -\pi, z) = 0 , \quad |x| \leq \pi , \quad i = 1, 2, 3,$$

which express that no lateral surface forces are present, and

(8)
$$\tau_{3i}(x, y, \ell) = T_i^{(g)}(x, y) , \quad \tau_{3i}(x, y, -\ell) = -T_i^{-(g)}(x, y) ,$$
$$x, y \in Q, \ i = 1, 2, 3 ,$$

corresponding to the psf in the bases of the cylinder.

A tensor $\theta(x, y, z) \equiv ((\theta_{ij}))$, i, j = 1, 2, 3, continuous in $\bar{\Omega}$ is called a virtual stress tensor for τ in Ω , if it satisfies the same pbc (7) and (8) as τ and moreover is a solution of the equilibrium equations

(9)
$$\theta_{ih, h} = 0, \quad i = 1, 2, 3, \quad x, y, z \in \Omega .$$

Let

(10)
$$W_\rho(x, y, z) = \frac{1}{2E}\left[(1+\sigma)\sum_{i, j=1}^{3} \rho_{ij}^2 - (\rho_{11} + \rho_{22} + \rho_{33})^2\right] ,$$
$$x, y, z \in \Omega, \quad \sigma \text{ as in (5)} , \quad E = 2\mu(1+\sigma) ,$$

be the positive definite quadratic form of the elastic energy density due to the tensor $\rho = ((\rho_{ij}))$. Let τ be a stress tensor in Ω and θ a virtual stress tensor for τ. Integrating W_τ over Ω, one obtains the total elastic energy $\mathcal{E}_\tau(\Omega)$ due to the stress tensor τ . Then , the theorem of Castiglano, or the minimum strain energy theorem says that

(11)
$$\mathcal{E}_\tau(\Omega) \leq \iiint_\Omega W_\theta(x, y, z)\, dx\, dy\, dz .$$

The main idea of the proof is to construct for any psf, or at least for some types of psf, a virtual stress tensor such, that its virtual energy or energy integral is bounded , even when

$\ell \to \infty$.

Now, the equilibrium equations (9) are automatically satisfied by any tensor $\theta(x, y, z)$ of the form

$$\theta_{11} = \alpha_{11} M_{,22}(x, y) \, . \, N''(z)$$

$$\theta_{12} = \alpha_{12} M_{,21} \, . \, N'' \, , \quad \theta_{22} = \alpha_{22} M_{,11} \, . \, N'' \, ,$$

$$\theta_{13} = \alpha_{13} M_{,122} \, . \, N' \, , \quad \theta_{23} = \alpha_{23} M_{,112} \, . \, N'$$

$$\theta_{33} = \alpha_{33} M_{,1122} \, . \, N \, ,$$

provided that the real numbers α_{ij} satisfy

$$\alpha_{11} + \alpha_{12} + \alpha_{13} = 0$$

$$\alpha_{12} + \alpha_{22} + \alpha_{23} = 0$$

$$\alpha_{13} + \alpha_{23} + \alpha_{33} = 0 \, .$$

It is easy to conjecture and verify, that the function $N(z)$ must increase exponentially, and in the present proof the functions $\sinh \beta z$ for $g = 1$ and $\cosh \beta z$ for $g = 2$ have been taken. Although the boundary conditions (7), (8) must also be satisfied, there is also abundant room to try to take care of every condition.

Now we may outline the proof in three steps. First to get sufficient types of psf such that meet the two following requirements:

1) every set of psf can be decomposed in a linear combination of these types;

2) it will be possible for each type to construct a virtual stress tensor in Ω_ℓ. It turns out that, for a cross-section like Q, already five types of psf are sufficient, each type depending on one arbitrary function and possibly some real constants. The strongest condition that each type of psf $T^{(g)} = (\gamma_1 T_1, \gamma_2 T_2, \gamma_3 T_3)$, where the

$\gamma_1, \gamma_2, \gamma_3$ are the three arbitrary constants, must satisfy, is that

(13) $$T_{1,1} = T_{2,2} = T_3 \quad ,$$

as can be conjectured from (12).

In the second step we consider each type of psf $T^{(g)}$. For each $T^{(g)}$ there is a corresponding stress tensor $\tau^{(g)}$ in Ω_ℓ and for each $\tau^{(g)}$ we construct a virtual stress tensor $\theta^{(g)}$ such, that its virtual energy satisfies for all ℓ the following inequality

(14) $$\mathcal{E}_\tau^{(g)}(\Omega_\ell) \leq \frac{C}{\mu}\|\gamma\|^2 \cdot \|T_3\|_2^2 \quad ,$$

where C is a constant depending on a lower bound ℓ_o of ℓ, $\ell_o < \ell$, if and only if $g = 1$, that is only if $T^{(g)}$ is odd; $\|\gamma\|$ is any norm of $\gamma = (\gamma_1, \gamma_2, \gamma_3) \in R^3$; μ is one Lamé constant; and T_3 (x, y) is the function of the third component of the psf $T^{(g)} = (\gamma_1 T_1, \gamma_2 T_2, \gamma_3 T_3)$.

In the third and last step of the proof we decompose any given psf T in a linear combination of at most sixteen summands, each one of them being of the five types defined in the first step. Because of the linearity of the Navier and of equilibrium equations, and of conditions (3) and (7), the stress tensor τ corresponding to the psf T may be decomposed also the same linear combination of at-most sixteen corresponding stress tensors.

Then we apply the second step of the proof to each summand of the decomposition and therefore we get un upper bound of the total elastic energy of the cylinder Ω_ℓ in terms of the squares of the norms of the third component of each summand. It turns out, that the square of the norm of the third component can be always estimated by means of a linear combination of the squares of the norms that appear in the inequality (4) . Therefore, the inequality is proved.

3. Related questions

a) When we stated the inequality (14), we said, that, if and only if the psf T are even , that is of the form $T^{(2)}$, then the constant C of (14) does not depend on a lower bound of ℓ. On the contrary, if the given psf T are odd in (4) , then C in (14) or K in (4) does depend on a lower bound ℓ_0 of ℓ . (See the last formula of page 91 in [5]) .

I call the attention on the physical interpretation of this result and how intuitive it looks.

b) From inequality (4) it follows that if τ is a stress tensor in the infinite cylinder Ω_∞ , without external forces, and if

$$(15)\quad \sum_{i=1}^{3} \left| \tau_{3i}(x,y,z) \right| + \sum_{\alpha=1}^{2} \left| \tau_{3\alpha,\alpha}(x,y,z) \right|$$

is bounded in Ω_∞ , then τ is an elementary solution of Saint-Venant's type. This corollary of (4) may be called a weak principle of Saint-Venant. I recall that this result is close to the above mentioned corollary of the principle of Saint-Venant.

It would be interesting to know if there exists in Ω_∞ a counterexample of the principle , allowing $\sum_{\alpha=1}^{2} \left| {}_{3\alpha,\alpha}(x,y,z) \right|$ to diverge when $|z| \to \infty$, in such a way that at the same time $\sum_{i=1}^{3} \left| \tau_{3i}(x,y,z) \right|$ remains bounded.

c) The mentioned results of Toupin and Roseman, and also the proof of (4) , imply that in the cylinder Ω_ℓ the stresses decay exponentially .This assertion about the "decaying" is not necessarily true in the neighborhood of the bases of the cylinder, no matter how long the cylinder may be.

However, it seems, that for any given psf T it is possible to aproximate T as much as desired by psf T^* in such a way, that the corresponding stress tensor τ^* does take the maximum value in the bases of the cylinder Ω_ℓ, provided that Ω_ℓ is sufficiently long.

d) Dividing the Navier equations (6) by μ, they take the form

$$\Delta u + \frac{1}{1-2\sigma} \text{ grad div } u = 0 . \tag{16}$$

All the stated results are valid for the Poisson's ratio σ, such that $-1 < \sigma < 1/2$, although I do not know of any practical elastic material with σ negative.

One consequence of the proved weak principle of Saint-Venant is that in Ω_∞ there are only elementary solutions of Saint-Venant's type for σ such that $-1 < \sigma < 1/2$.

But it has been proved by the author, [7], that if $\sigma = -1 - \varepsilon$, ε positive and small, then there are, in the infinite cylinder with a circle for cross-section and without external forces, periodic in z solutions of (16), so that the principle of Saint-Venant does not hold for $\sigma < -1$, in spite of the fact that the Navier equations remain elliptic.

REFERENCES

[1] de Saint-Venant, B., Mémoires de l'Académie des Sciences des savants étrangers, 14(1855), 233-560 and Mémoire sur la flexion des prismes, Journal de Liouville, Ser. 2, (1856), 89-189.

[2] Toupin, R.A., Saint-Venant's principle, Arch. Rational Mech. Anal. 18(1965), 83-96

[3] Roseman, J.J., A pointwise estimate for the stress in a cylinder and its application to Saint-Venant's principle, Arch. Rational Mech. Anal. 2 1(1965), 23-48.

[4] Dou, A., On the Principle of Saint-Venant, MRC Thecnical Summary Report # 472, May 1964.

[5] Dou, A., Upper Estimate of the Potential Elastic Energy of a Cylinder, Comm. Pure Appl. Math. 19(1966), 83-93.

[6] Bramble, J.H., and Payne, L.E., Pointwise bounds in the first biharmonic boundary value problem, J. Math. and Phys. 42 (1963), 278-286.

[7] Dou, A., Bounded solutions of the Elasticity equations in the in the unbounded cylinder, Proceedings of the Inter. Congress of Math., Moscow, 1966. Abstracts of short communications.

CENTRO INTERNAZIONALE MATEMATICO ESTIVO

(C. I. M. E.)

TODD DUPONT

ON THE EXISTENCE OF AN ITERATIVE METHOD FOR THE SOLUTION OF ELLIPTIC DIFFERENCE EQUATIONS WITH AN IMPROVED WORK ESTIMATE

Corso tenuto ad Ispra 3-11 luglio 1967

On the Existence of an Iterative Method for the Solution of Elliptic Difference Equations with an Improved Work Estimate

Todd Dupont

Rice University, Houston

1. Introduction

The iterative solution of the difference equations associated with $\nabla\cdot(a(x)\nabla u) = f$ is a problem which has received a great deal of attention in the literature. In this paper it will be shown that for rather general domains in the plane there exists an iterative method with work estimates which are better than those now available. This is an abstract existence theorem, not a complete specification of an improved iterative procedure.

On a rectangle with a(x) constant the Peaceman-Rachford [5] procedure gives a work estimate of $O(h^{-2}\log h^{-1}\log \varepsilon^{-1})$ for reduction of the L^2-norm of the error by a factor ε. Still on a rectangle but for more general equations the procedure of Gunn [3] gives the same work estimate for the reduction of the analogue of the Dirichlet integral of the error by a factor ε. Gunn's procedure also gives an estimate of $O(h^{-2}(\log h^{-1})^2\log \varepsilon^{-1})$ for reduction of the uniform norm of the error by a factor ε again on a rectangle. We shall show that, if the domain is the union of rectangles with sides parallel to the coordinate axes and if a(x) is twice continuously differentiable in the closure of the domain, a work estimate of the form $O(h^2(\log h^{-1})^2\log \varepsilon^{-1})$ can be obtained for the reduction of the uniform norm of the error by a factor ε.

In sections 2 and 3 we define the problem and the iteration and give an analysis of the work required to produce the solution. This analysis is done assuming the main lemma of the paper, Lemma 1, which is proved in sections 4 and 5.

2. Definition of the Problem

We will work with a domain $D = R_1 \cup R_2$, where each R_i is a rectangle in R^2 with sides parallel to the coordinate axis. The choice of two rectangles is made for simplicity, though it is clear that what follows holds for any finite union of such rectangles. Assume that we have a square grid of mesh size h covering R_1 and R_2 and that the sides of R_1 and R_2 lie on grid lines. We also assume that $\partial R_1 \cap \partial R_2 \subset \partial D$. Let D_h denote the grid points in the interior of D, ∂D_h the grid points on ∂D and $\overline{D}_h = D_h \cup \partial D_h$. Similarly, define $R_{i,h}$, $\partial R_{i,h}$ and $\overline{R}_{i,h}$.

We are attempting to solve

$$\begin{aligned} L_h u_h &= f \quad \text{on} \quad D_h , \\ u_h &= g \quad \text{on} \quad \partial D_h , \end{aligned} \tag{2.1}$$

for u_h, a function defined on $\overline{D}_h$, where $L_h u_h$ is any one of the five point approximations (4.4) to $\nabla\cdot(a\nabla u)$ generated by (4.1), (4.2), and (4.3) and f and g are given functions. We shall assume a to be a positive C^2 function in $\overline{D}$. In what follows we are interested only in finding the solution to the algebraic equations (although we shall use the fact that this is an approximation to the solution of the differential problem).

3. Definition and Analysis of the Iteration

The iteration that will be examined here is a nested one. The outer iteration consists of a Schwarz alternation, and the inner iteration consists of a Gunn iteration. The outer iteration is implicit and the equations involved in it will be approximately solved by the Gunn procedure. The outer iteration is defined as follows. Let $u_{i,h}$ be the solution of

$$
\begin{aligned}
L_{j,h} u_{i,h} &= f \quad \text{on} \quad R_{j,h} , \\
u_{i,h} &= g \quad \text{on} \quad \partial R_{j,h} \cap \partial D , \qquad (3.1) \\
u_{i,h} &= u_{i-1,h} \quad \text{on} \quad \partial R_{j,h} \backslash \partial D ,
\end{aligned}
$$

where $j = 1$ if i is odd and $j = 2$ if i is even. $L_{j,h}$ is the restriction to $R_{j,h}$ of L_h. As we intend to solve for $u_{i,h}$ by a Gunn iteration, we need to specify an initial guess on $R_{j,h}$ which we take to be $u_{i-2,h}$ if $i \geq 2$ and $u_{o,h}$ if $i = 1$.

We shall assume the following lemma for the present and proceed with the analysis of the iteration.

Lemma 1. There exists q depending only on R_1, R_2 and a such that, if

$$
L_{j,h} w_h = 0 \quad \text{in} \quad R_{j,h}
$$

$$
w_h = \begin{cases} 0 & \text{on } \partial R_{j,h} \cap \partial D \\ 1 & \text{on } \partial R_{j,h} \backslash D , \end{cases}
$$

then $\max_{\partial R_{j',h} \cap R_{j,h}} w_h < q_h \leq q < 1$, where $j' \neq j$.

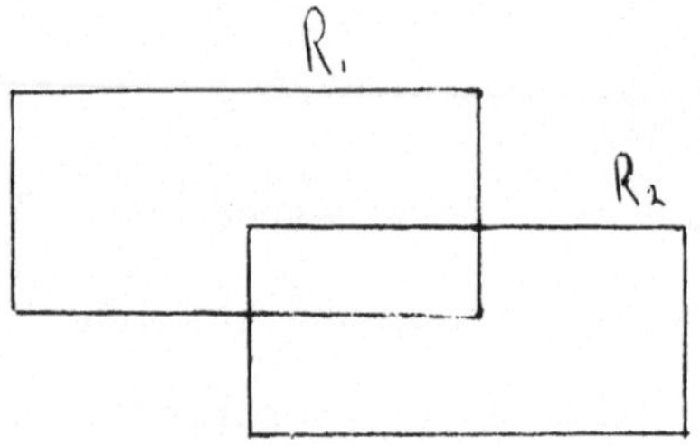

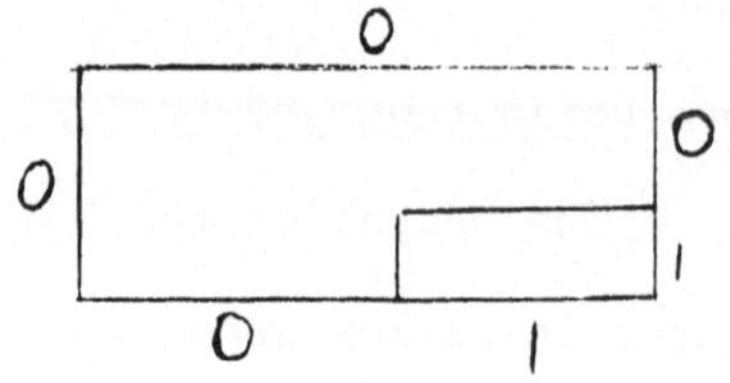

The Domain The Boundary Values

For example, if D is as shown in the figure and w_h is defined on $\bar{R}_{1,h}$, $L_{1,h}w_h = 0$ on $R_{1,h}$, and the boundary values of w_h are as shown, then $w_h \leq q < 1$ on that part of $\partial R_{2,h}$ inside R_1.

Let $w_{i,h}$ be the calculated approximation to $u_{i,h}$. Then

(3.2)
$$L_{j,h}(w_{i,h} + r_{i,h}) = f \quad \text{on } R_{j,h} ,$$
$$w_{i,h} = \begin{cases} g & \text{on } \partial R_{j,h} \cap \partial D, \\ w_{i-1,h} & \text{on } \partial R_{j,h} \setminus \partial D . \end{cases}$$

Letting u_h be the solution of (2.1) and $z_{i,h} = w_{i,h} - u_h$, we obtain

(3.3)
$$L_{j,h}(z_{i,h} + r_{i,h}) = 0 \quad \text{in } R_{j,h} ,$$
$$z_{i,h} = \begin{cases} 0 & \text{on } \partial R_{j,h} \cap \partial D , \\ z_{i-1,h} & \text{on } \partial R_{j,h} \setminus \partial D . \end{cases}$$

It should be noted that $r_{i,h}$, being the difference between the true solution and the approximation to it, is equal to zero on the boundary. Let us adopt the following notation:

$$(3.4)\qquad \begin{aligned} z_i &= \max_{R_{j,h}\setminus R_{j',h}} |z_{i,h}| , \\ x_i &= \max_{R_{j,h}} |z_{i,h}| , \\ \eta_i &= \max_{R_{j,h}\setminus R_{j',h}} |\eta_{i,h}| , \\ \nu_i &= \max_{R_{j,h}} |\eta_{i,h}| , \end{aligned}$$

where $j = 1$ if i is odd and $j = 2$ if i is even and $j \neq j'$. By the maximum principle and Lemma 1 it follows that

$$(3.5)\qquad \begin{aligned} \max_{R_{j,h}\setminus R_{j',h}} |z_{i,h} + \eta_{i,h}| &\leq \max_{\partial R_{j',h}\cap R_{j,h}} |z_{i,h} + \eta_{i,h}| \\ &\leq q \max_{\partial R_{j,h}\cap R_{j',h}} |z_{i,h}| \leq q x_{i-1} . \end{aligned}$$

Therefore,

$$(3.6)\qquad z_i \leq q x_{i-1} + \eta_i \leq q x_{i-1} + \nu_i .$$

Now if we neglect round-off error, let σ denote the norm (with respect to the uniform norm) of the error propagator in the inner iteration, and let $w_{i,h}^{true}$ denote the exact solution of (3.1), we obtain

$$(3.7)\qquad \nu_i \leq \sigma \max_{R_{j,h}} |w_{i,h}^{true} - w_{i-2,h}| .$$

Thus,

$$(3.8)\qquad \begin{aligned} \nu_i &\leq \sigma \max_{R_{j,h}} |w_{i,h} + \eta_{i,h} - u_h + u_h - w_{i-2,h}| \\ &\leq \sigma(x_i + x_{i-2} + \nu_i) . \end{aligned}$$

Therefore,

(3.9) $$\nu_i \le \frac{\sigma}{1-\sigma}(x_i + x_{i-2}) \; .$$

Since

$$\max_{R_{j,h}} |z_{i,h} + \eta_{i,h}| \le \max_{(\partial R_{j,h})\cap R_j} |z_{i-1,h}| \le z_{i-1} \; ,$$

it follows that

(3.10) $$x_i \le z_{i-1} + \nu_i \; .$$

Now, (3.6) and (3.9) imply that

(3.11) $$x_i \le z_{i-1}+\nu_i \le qx_{i-2}+\nu_{i-1}+\nu_i \le qx_{i-2}+\frac{\sigma}{1-\sigma}(x_i+x_{i-1}+x_{i-2}+x_{i-3}).$$

Therefore,

(3.12) $$x_i \le \frac{q(1-\sigma)+\sigma}{1-2\sigma}x_{i-2} + \frac{\sigma}{1-2\sigma}(x_{i-1} + x_{i-3}) \; .$$

This implies that x_i converges to zero if all the roots of

$$x^3 - \frac{\sigma}{1-2\sigma}\, x^2 - \frac{q(1-\sigma)+\sigma}{1-2\sigma}x - \frac{\sigma}{1-2\sigma} = 0$$

are less than one in absolute value. This is the case if, say, $0 \le \sigma < (1-q)/5$.

Thus, if σ is sufficiently small (independently of h), there exists ν such that $0 \le \nu < 1$ and $x_n < C\nu^{n-4}[x_0+x_1+x_2+x_3]$. Hence, it requires $O(\log \varepsilon^{-1})$ outer iterations to reduce the uniform norm of the error by a factor ε.

Thus, it remains only to analyze the work required to reduce the norm of the error by a factor σ at each step. In Gunn's paper it is shown that the work to reduce the A-norm of the error on a rectangle by a factor δ is $O(h^{-2}(\log h^{-1})\log \delta^{-1})$, where the A-norm is $\|u\|_A = (-h^2 \sum_{R_h} u\Delta_h u)^{\frac{1}{2}}$, Δ_h being the standard five point approximation to the Laplace operator. It is easy to see that there exists C_0 and C_1, independent of h, such that

$$C_0 h \max_{\overline{R}_h} |u_h| \leq \|u_h\|_A \leq C_1 h^{-1} \max_{\overline{R}_h} |u_h| .$$

Thus, it suffices to take $\delta = Ch^{-2}\sigma$. Hence, the work at each step is

$$O(h^{-2}(\log h^{-1})^2 \log \sigma^{-1}) = O(h^{-2}(\log h^{-1})^2).$$

Consequently, the estimate for the total work is $O(h^{-2}(\log h^{-1})^2 \log \varepsilon^{-1})$.

Theorem. Let R_1 and R_2 be two-dimensional rectangles with boundaries ∂R_i consisting of segments parallel to the coordinate axes and let $D = R_1 \cup R_2$. Let there exist a sequence (h_m), $0 < h_m \to 0$, such that ∂D falls on a square grid of mesh h_m. Let $\partial R_1 \cap \partial R_2 \subset \partial D$. Let $a(x) \in C^2(\overline{D})$, $a(x) \geq a_0 > 0$. Then, there exists an iterative procedure for obtaining the solution of (2.1) such that the number of arithmetic operations required to reduce the uniform norm of the error in the approximate solution by a factor ε is no more than $O(h_m^{-2}(\log h_m)^2 \log \varepsilon^{-1})$.

4. Proof of Lemma 1

In this section we shall prove a special case of Lemma 1 which is sufficient to imply Lemma 1 when used with the maximum principle. Let $R = \{(x_1,x_2): 0<x_i<d_i\}$. Let $0 \leq c < d_1$ and suppose d_1, d_2 and c are all integral multiples of $h_0 > 0$. Let $h_m = h_0 2^{-m}$. If $\Omega \subset \mathbf{R}^2$, let $\overline{\Omega}_h = \overline{\Omega} \cap \{(ph,qh)$: p,q integers$\}$. Let $\partial\Omega_h$ be the points (x_1,x_2) of $\overline{\Omega}_h$ such that $(x_1 \pm h, x_2)$ or $(x_1, x_2 \pm h)$ is not in $\overline{\Omega}_h$. Let $\Omega_h = \overline{\Omega}_h \setminus \partial\Omega_h$. Let $\nabla_{x_{i,h}} u(x) = h^{-1}(u(x+e_{i,h}) - u(x))$ and $\nabla_{\overline{x}_{i,h}} u(x) = h^{-1}(u(x)-u(x-e_{i,h}))$ for i = 1,2, where $e_1 = (1,0)$, $e_2 = (o,1)$. Let $\Delta_h u(x) = \sum_{i=1}^{2} \nabla_{x_{i,h}} \nabla_{\overline{x}_{i,h}} u(x)$.

for i = 1,2, define $L^{(1)}_{i,h}$, $L^{(2)}_{i,h}$, $L^{(3)}_{i,h}$ as follows:

$$(4.1)\quad L^{(1)}_{i,h}u(x) = h^{-1}\{\frac{a(x+e_{i,h})+a(x)}{2}\nabla_{x_{i,h}}u(x) - \frac{a(x-e_{i,h})+a(x)}{2}\nabla_{\bar{x}_{i,h}}u(x)\}$$

$$(4.2)\quad L^{(2)}_{i,h}u(x) = h^{-1}\{(a(x+e_{i,h})a(x))^{\frac{1}{2}}\nabla_{x_{i,h}}u(x)-(a(x-e_{i,h})a(x))^{\frac{1}{2}}\nabla_{\bar{x}_{i,h}}u(x)\}$$

$$(4.3)\quad L^{(3)}_{i,h}u(x) = h^{-1}\{a(x+\tfrac{1}{2}e_{i,h})\nabla_{x_{i,h}}u(x) - a(x-\tfrac{1}{2}e_{i,h})\nabla_{\bar{x}_{i,h}}u(x)\} .$$

$$(4.4)\quad \text{Let } L^{(j)}_h = \sum_{i=1}^{2} L^{(j)}_{i,h} , \quad j = 1,2,3.$$

Thus, $L^{(1)}_h$, $L^{(2)}_h$, and $L^{(3)}_h$ are the three most common five point approximations to $\nabla\cdot(a\nabla u)$.

<u>Lemma 1'.</u> If $a \in C^2(\bar{R})$ there exists q such that $0 < q < 1$ and such that, if u_h is the solution of

$$L_h u_h = 0 \quad \text{in} \quad R_h ,$$

$$u_h = \begin{cases} 1 & \text{on } \partial R_h \cap \{(x_1,x_2):\ 0 \le x_1 < c\} , \\ 0 & \text{on } \partial R_h \cap \{(x_1,x_2):\ x_1 \ge c\} , \end{cases}$$

for $L_h = L^{(j)}_h$, j = 1,2, or 3, and $h = h_m$ some m, then $\max u_h(c,x_2) < q$.

The proof of Lemma 1' goes as follows.

1) For sufficiently small ε and for all $h = h_m$ there exists q such that $u_h(c,y) < q < 1$ for $y = kh$ and $y \in [0,\varepsilon] \cup [d_2-\varepsilon,d_2]$.

2) For each $\varepsilon > 0$ there exists $q < 1$ such that for h sufficiently small $u_h(c,y) \le q$ for $y = kh \in (\varepsilon,d_2-\varepsilon)$.

3) Take ε for 1) and $h < \delta$ for 2). Note that there are only a finite number of h_n's $\ge \delta$ and that we can apply the strong maximum principle to each of these to get a $q < 1$ for each one. Then take q to be the largest one of the finite set given by 1), 2) and the h_n's $\ge \delta$.

In carrying out the proof of 1) and 2) we use the following:

Lemma 2. If $\Delta_h v_h = -c$ in R_h, $c \geq 0$ and $v_h = f$ and ∂R_h, where $f \in C(\partial R)$ and is twice continuously differentiable on the closure of each edge of R, then

$$|\nabla_{x_j,h} v_h| \leq K \ , \quad i = 1,2,$$

where

$$K = [\max f - \min f]/ \min_{i=1,2} d_j + [c + \max_{\partial R}\{|D^2_{x_1} f|, |D^2_{x_2} f|\}] \max_{i=1,2} d_j \ .$$

(Naturally $|D^2_{x_1} f|$ is used on sides of ∂R which are parallel to the x_1 axis and similarly for $D^2_{x_2} f$.)

Proof: The proof will be carried out for $\nabla_{x_1,h} v_h$. First extend v_h to R'_h and R''_h, where $R' = \{(x_1,x_2): -d_1 \leq x_1 \leq 0 \leq x_2 \leq d_2\}$ and $R'' = \{(x_1,x_2): d_1 \leq x_1 \leq 2d_1, 0 \leq x_2 \leq d_2\}$. This extension is made subject to the constraint $\Delta_h v_h = -c$. This extension is clearly not unique, but take any such extension and note that $\Delta_h(\nabla_{x_1,h}\nabla_{x_1,h} v_h) = 0$ in R_h . Thus, $\max_{\overline{R}_h} |\nabla_{\overline{x}_1,h}\nabla_{x_1,h} v_h| = \max_{\partial R_h} |\nabla_{\overline{x}_1,h}\nabla_{x_1,h} v_h|$. On the sides of R_h given by $x_2 = 0$ and $x_2 = d_2$, $|\nabla_{\overline{x}_1,h}\nabla_{x_1,h} v_h| \leq \max |D^2_{x_1} f|$.

On the sides $x_1 = 0$ and $x_1 = d_1$, $\nabla_{\overline{x}_1,h}\nabla_{x_1,h} v_h = -c - \nabla_{\overline{x}_2,h}\nabla_{x_2,h} v_h$.

Thus, $|\nabla_{\overline{x}_1,h}\nabla_{x_1,h} v_h| \leq c + \max |D^2_{x_2} f|$. Hence, the total change in $|\nabla_{x_1,h} v_h|$ is no more than $[c + \max\{|D^2_{x_1} f|, |D^2_{x_2} f|\}]d_1$. For each x_2 there exist x_1 and x'_1 such that $\nabla_{x_1,h} v_h(x_1,x_2) \leq d_1^{-1}(\max f - \min f)$ and $\nabla_{x_1,h} v_h(x'_1,x_2) \geq -d_1^{-1}(\max f - \min f)$. So, the conclusion of Lemma 2 holds for $\nabla_{x_1,h} v_h$.

The proof of 1) goes as follows. It is clearly sufficient to show 1) for $y < \varepsilon$ only. This is done by first making the change of variables $v_h(x,y) = a(x,y)^{\frac{1}{2}} u_h(x,y)$. This gives $\Delta_h v_h = f_h$, where $|f_h| \leq C$. C is independent of h, since $f_h = (\Delta_h a^{\frac{1}{2}})u_h + a^{-\frac{1}{2}}(L_h^{(2)} - L_h)u_h$. Also,

$$v_h = \begin{cases} a^{\frac{1}{2}} & \text{for } x < c \quad \text{on } \partial R_h \,, \\ 0 & \text{for } x \geq c \quad \text{on } \partial R_h \,. \end{cases}$$

Now, $v_h \leq v_{1,h} + v_{2,h} + v_{3,h}$, where

$$\Delta_h v_{1,h} = 0 \text{ in } R_h \,, \quad v_{1,h} = \begin{cases} a(c,0)^{\frac{1}{2}} \,, & x < c \quad \text{on } \partial R_h \,, \\ 0 \,, & x \geq c \quad \text{on } \partial R_h \,, \end{cases}$$

$$\Delta_h v_{2,h} = 0 \text{ in } R_h \,, \quad v_{2,h} = \tilde{g} \quad \text{on } \partial R_h \,,$$

$$\Delta_h v_{3,h} = -C \text{ in } R_h \,, \quad v_{3,h} = 0 \quad \text{on } \partial R_h \,,$$

where $\tilde{g}$ is continuous on ∂R, $\tilde{g}(c,0) = 0$, and for $x < c$, $\tilde{g}(x,y) \geq a(x,y)^{\frac{1}{2}} - a(c,0)^{\frac{1}{2}}$, and for $x \geq c$, $\tilde{g}(x,y) \geq 0$.

It was shown by Miller [4] that $|v_{1,h}(c,y)| \leq a(c,0)^{\frac{1}{2}} \max\{\frac{1}{2}, cd_1^{-1}\}$ for all y. We know by Lemma 2 that $|v_{3,h}(c,y)| \leq Ky$. Finally, by Walsh and Young [7] $v_{2,h}$ tends uniformly to v_2 as h tends to zero, where v_2 is the solution to $\Delta v = 0$ in R, $v = \tilde{g}$ on ∂R. Thus for h small and y small $|v_{2,h}(c,y)|$ is small. Hence, given $\delta > 0$, we can take y and h sufficiently small such that $u_h(c,y) \leq \max(\frac{1}{2}, cd_1^{-1}) + \delta$. If we insist that the bound on y be smaller than the above bound on h, we get this relation for all h. The proof of 1) is complete.

The proof of 2) is carried out as follows. The boundary data are increased to a C^∞ function g which is still not greater than one and which is still zero for $x_1 = d$. Now if we let $\tilde{u}_h$ denote the solution to the new problem, then $u_h \le \tilde{u}_h$ and $\tilde{u}_h$ tends uniformly to $\tilde{u}$, where $\tilde{u}$ is the solution to the problem $\nabla\cdot(a\nabla u) = 0$ in R and $u = g$ on ∂R, by the results of section 5. We know by the strong maximum principle that $\max_{y\in[\varepsilon,d_2-\varepsilon]} \tilde{u}(c,y) < 1$. Thus $\exists\, q < 1$ such that for h sufficiently small $\max_{y\in[\varepsilon,d_2-\varepsilon]} u_h(c,y) \le q < 1$.

The only thing left to prove is the convergence result used above for $\tilde{u}_h$. This is very similar to many such convergence results but will be included for completeness.

5. Bound on the Discretization Error

Although this is properly part of the proof of Lemma 1′, we shall state it separately as it has some small amount of independent interest.

Lemma 3. Let $R = \{(x_1,x_2)\colon 0 < x_1 < d_1,\ 0 < x_2 < d_2\}$. Let $a \in C^2(\bar{R}$, $a > 0$, and let g be a continuous function on ∂R. Let u be the solution of $\nabla\cdot(a\nabla u) = 0$ in R such that $u = g$ on ∂R. Let u_h be defined on $\bar{R}_h$ by $L_h u_h = 0$ in R_h and $u_h = g$ on ∂R_h where $L_h = L_h^{(1)}, L_h^{(2)}$, or $L_h^{(3)}$. Then $\max_{R_h} |u_h - u| \to 0$ as $h \to 0$. (Note no rate can be given without more assumptions; see Walsh and Young [7]. If g is assumed Lipschitz continuous, then an error bound of $O(h^{2/7})$ could be obtained.)

This proof is modeled on that of Pucci [6] which is quite similar to that of Bers [1].

Proof of Lemma 3:

1) Note that if $L_h v_h \ge 0$ (or ≤ 0) in R_h, then $\max_{\bar{R}_h} v_h = \max_{\partial R_h} v_h$ (or $\min_{\bar{R}_h} v_h = \min_{\partial R_h} v_h$).

2) If $M \geq \max\{\max_R(|a_{x_1}| + |a_{x_2}|), \max_R a^{\frac{1}{2}}(\max_R(|a_{x_1}|^{\frac{1}{2}} + |a_{x_2}|^{\frac{1}{2}}))\}$ and $\beta \geq M + 2 + (\min_R a)^{-1}$, then there exists $\tilde{h}_0$ such that $0 < h < \tilde{h}_0$ implies

$$L_h(e^{\beta d_1} - e^{\beta x_1}) = L_{1,h}(e^{\beta d_1} - e^{\beta x_1}) \leq -1 \text{ for } L_h = L_h^{(1)}, L_h^{(2)}, \text{ or } L_h^{(3)}.$$

Proof: $L_{1,h}^{1}(e^{\beta x_1})(x_1,x_2) = a(x_1,x_2)\nabla_{x_{1,h}}\nabla_{\overline{x}_{1,h}} e^{\beta x_1} + \frac{h}{2}[a_{x_1}(\xi_1,x_2)\nabla_{x_{1,h}} e^{\beta x_1}$

$$+ a_{x_1}(\xi_2,x_2)\nabla_{x_{1,h}} e^{\beta x_1}]$$

$$\geq (\min_R a)[e^{\beta(x_1-h)}(\beta^2 - \tfrac{h}{2}\beta M e^{2\beta h})] \geq 1$$

for h small. $L_{1,h}^{(2)}$ and $L_{1,h}^{(3)}$ can be treated similarly.

3) There exists K such that, if $R_h^* \subset R_h$, $h < \tilde{h}_0$, and $L_h v_h = f$ in (R_h^*), then $\max_{R_h^*} |v_h| \leq \max_{\partial R_h^*} |v_h| + K \max_{(R_h^*)} |f|$.

Proof: Let $w_h = \max_{\partial R_h^*} |v_h| + (e^{\beta d_1} - e^{\beta x_1}) \max_{(R_h^*)} |f|$. Then $L_h(w_h \pm v_h) \leq 0$ in (R_h^*) and $w_h \pm v_h \geq 0$ on ∂R_h^*. Therefore, $|v_h| \leq |w_h|$. Thus, 3) holds if we take $K = e^{\beta d_1} - 1$.

4) Let $R_h^\delta = R_h \cap R^\delta$, $R^\delta = \{(x_1,x_2) \in R\colon \operatorname{dist}((x_1,x_2),\partial R) \geq \delta\}$. Let $R_h^* = R_h^{h^{1/5}}$. Then $L_h(u-u_h) = O(h^{1/5})$ in R_h^* and the constant is independent of h.

Proof: Schauder interior estimates (see [2]) imply that there exists K such that $|u_{x_1}(x_1,x_2)| + |u_{x_2}(x_1,x_2)| \leq Kh^{-1/5}$ and

$$|u_{x_1x_1}(x_1,x_2) - u_{x_1x_1}(x_1',x_2')| + |u_{x_2x_2}(x_1,x_2) - u_{x_2x_2}(x_1',x_2')| \leq Kh^{1/5}$$

if $|(x_1,x_2) - (x_1',x_2')| \leq 2h$ and $(x_1,x_2),(x_1'x_2') \in R_h^*$. Now letting

$a(x_1 \pm h,x_2) = a(x_1,x_2) \pm ha_{x_1}(x_1,x_2) + \frac{h^2}{2}a_{x_1x_1}(\xi^{\pm},x_2)$, we get

$$L^{(1)}_{1,h}\, u(x_1,x_2) = a(x_1,x_2)\nabla_{x_{1,h}}\nabla_{\bar{x}_{1,h}} u(x_1,x_2) + \frac{a_{x_1}(x_1,x_2)}{2h}[u(x_1+h,x_2) - u(x_1-h,x_2)]$$

$$+ \frac{h}{4}[a_{x_1x_1}(\xi^+,x_2)\nabla_{x_{1,h}} u(x_1,x_2) + a_{x_1x_1}(\xi^-,x_2)\nabla_{x_{2,h}} u(x_1,x_2)]$$

$$= (I) + (II) + (III) .$$

(I): Using $u(x_1 \pm h,x_2) = u(x_1,x_2) \pm hu_{x_1}(x_1,x_2) + \frac{1}{2}h^2\, u_{x_1x_1}(\xi'_{\pm},x_2)$,

we see that $\nabla_{x_{1,h}}\nabla_{\bar{x}_{1,h}} u(x_1,x_2) = u_{x_1x_1}(x_1,x_2) + O(h^{1/5})$ in R^*_h .

(II): Using the above expression for $u(x_1 \pm h,x_2)$, we see that $u(x_1+h,x_2) - u(x_1-h,x_2) = 2h\, u_x(x_1,x_2) + O(h^{2+1/5})$ in R^*_h .

(III): These terms are $O(h^{4/5})$ in R^*_h .

Thus, $L^{(1)}_{1,h}\, u(x_1,x_2) = a(x_1,x_2)u_{x_1x_1}(x_1,x_2) + a_{x_1}(x_1,x_2)u_{x_1}(x_1,x_2) + O(h^{1/5})$. Hence $L^{(1)}_h\, u(x_1,x_2) = (L^{(1)}_{1,h} + L^{(1)}_{2,h})u(x_1,x_2) = O(h^{1/5}) + \nabla\cdot(a\nabla u) = O(h^{1/5})$ in R^*_h. Now, $(a(x_1,x_2)^{\frac{1}{2}} - a(x_1+h_1,x_2)^{\frac{1}{2}})^2 = a(x_1,x_2) - 2(a(x_1,x_2)a(x_1+h,x_2))^{\frac{1}{2}} + a(x_1+n,x)$. Thus, $0 < \frac{a(x_1,x_2) + a(x_1+h,x_2)}{2} - (a(x_1,x_2)a(x_1+h,x_2))^{\frac{1}{2}} \le Ch^2$. Using this relation and the Schauder estimates, we get $L^{(2)}_h u = L^{(1)}_h u + O(h^{4/5}) = O(h^{1/5})$ in R^*_h . Next using $a(x_1 + \frac{h}{2},x_2) = \frac{a(x_1,x_2 + a(x_1+h,x_2)}{2} + O(h^2)$ exactly as above, we get $L^{(3)}_h u = L^{(1)}_h u + O(h^{4/5}) = O(h^{1/5})$ in R^*_h.

5) $\max_{R_h \setminus R^*_h} |u-u_h| \to 0$ as $h \to 0$.

Proof: For $(x_1,x_2) \in R_h \backslash R_h^*$ let (x_1^o,x_2^o) denote the point on ∂R which is closest to (x_1,x_2). Note that $(x_1^o,x_2^o) \in \partial R_h$ and also $|u(x_1,x_2) - u_h(x_1,x_2)| \le |u(x_1,x_2) - u(x_1^o,x_2^o)| + |u(x_1^o,x_2^o) - u_h(x_1,x_2)|$ and $u(x_1^o,x_2^o) = u_h(x_1^o,x_2^o)$. Thus, $|u(x_1,x_2) - u_h(x_1,x_2)|$ $\le |u(x_1,x_2) - u(x_1^o,x_2^o)| + |u_h(x_1^o,x_2^o) - u_h(x_1,x_2)|$. To estimate the first term notice that $|(x_1,x_2) - (x_1^o,x_2^o)| \le h + h^{1/5} \le 2h^{1/5}$, $h < 1$, and therefore $|u(x_1,x_2) - u(x_1^o,x_2^o)| \le \omega(2h^{1/5})$ where ω is the modulus of continuity of u. In order to estimate $|u_h(x_1^o,x_2^o) - u_h(x_1,x_2)|$ we again make the change of variable $v_h(x_1,x_2) = a(x_1,x_2)^{\frac{1}{2}}u_h(x_1,x_2)$. Then, as before, $\Delta_h v_h = f_h$, where $f_h = (\Delta_h a^{\frac{1}{2}})u_h - (a^{-\frac{1}{2}})(L_h^{(i)} - L_h^{(2)})u_h$. Now, estimating crudely, we get $|f_h| \le C$, C independent of h. Next, $v_h = v_{1,h} + v_{2,h}$, where

$$\Delta_h v_{1,h} = 0\ , \quad v_{1,h} = v_h \text{ on } \partial R_h\ ,$$

$$\Delta_h v_{2,h} = f\ , \quad v_{2,h} = 0 \text{ on } \partial R_h\ .$$

Thus, $|u_h(x_1,x_2) - u_h(x_1^o,x_2^o)| \le (\min a^{\frac{1}{2}})^{-1}[|v_{1,h}(x_1,x_2) - v_{1,h}(x_1^o,x_2^o)|$ $+ |v_{2,h}(x_1,x_2) - v_{2,h}(x_1^o,x_2^o)|]$. By Lemma 2 the second term is bounded by $Ch^{1/5}$. Thus, it remains only to estimate $|v_{1,h}(x_1,x_2) - v_{1,h}(x_1^o,x_2^o)|$. This is done by noting that $v_{1,h}$ converges uniformly to the solution v_1 of $\Delta v = 0$ in R^o and $v = a^{\frac{1}{2}}g$ on ∂R which is continuous in R. Hence, we get $|v_{1,h}(x_1,x_2) - v_{1,h}(x_1^o,x_2^o)| \to 0$. This proves 5).

6) Combining 3), 4) and 5) we get $\max_{R_h} |u-u_h| \to 0$ as $h \to 0$. This completes the proof of Lemma 3.

6. Remarks

1) It seems quite probable that the arguments used here can be extended to n dimensions and somewhat more general equations with a work estimate of $O(h^{-n}(\log h^{-1})^2 \log \varepsilon^{-1})$.

2) In the proof of Lemma 1 we were only able to prove the existence of the q and not estimate it. It is, of course, this fact which makes this procedure only an abstract existence theorem. It is sufficient, by arguments in Miller's paper, to find q for $c = \frac{1}{2}d_1$, and in this case we can get a good estimate for q depending only on the constant of ellipticity for what seems to be the worse possible case. That is, of course, the case where $a(x_1,x_2) = \max a$ for $x_1 < c$, $a(x_1,x_2) = \min a$ for $x \geq c$. In this case $q = (\max a + \min a)^{-1} \max a$. However, I have been unable to show that this is the worst case and it seems to be a quite interesting problem.

3) I gratefully acknowledge the encouragement and assistance of Jim Douglas, Jr.

References

1. L. Bers, On mildly nonlinear partial difference equations of elliptic type, J. Res. Nat. Bur. St., 51(1953), 229-236.

2. A. Friedman, Partial Differential Equations of Parabolic Type, Prentice Hall, New York, 1964. (Chapter 3, Section 8).

3. J. E. Gunn, The solution of elliptic difference equations by semi-explicit techniques, J. Soc. Indust. Appl. Math., Numerical Analysis, Ser. B, 2(1964), 24-45.

4. K. Miller, Numerical analogues to the Schwarz alternating procedure, Numerische Mathematik, 7(1965), 91-103.

5. D. W. Peaceman and H. H. Rachford, The numerical solution of parabolic and elliptic differential equations, Journal of the Society of Industrial and Applied Mathematics, 3(1955), 28-41.

6. C. Pucci, Some Topics in Parabolic and Elliptic Equations, University of Maryland Lecture Series, No. 36, 1958.

7. J. L. Walsh and D. Young, On the degree of convergence of solutions to difference equations to the solution of the Dirichlet problem, J. of Math. and Phys., 33(1954), 80-93.

CENTRO INTERNAZIONALE MATEMATICO ESTIVO

(C. I. M. E.)

J. R. CANNON and JIM DOUGLAS

THE APPROXIMATION OF HARMONIC AND PARABOLIC FUNCTIONS ON HALF-SPACES FROM INTERIOR DATA

Corso tenuto ad Ispra dal 3-11 Luglio 1967

The Approximation of Harmonic and Parabolic Functions on Half-Spaces from Interior Data

J. R. Cannon (Minneapolis) and Jim Douglas, Jr., (Chicago)

1. Introduction. The object of this paper is to discuss the numerical approximation of functions either harmonic or parabolic in a half-space given the approximate values of the functions on some subset of the open half-space. Two things are immediately obvious. One is that, without some knowledge of the global behavior of the functions, nothing can be said about the functions in the whole half-space. The second is that, if there exists one function extending the data to the half-space, in general there are infinitely many. Throughout the paper the notation is based on approximating any one of the solutions; it is clear that the error estimates apply equally well to every solution.

The global constraint imposed on harmonic functions is a convenient generalization of boundedness (with a known bound) on the half-space; in the parabolic case the solutions are assumed non-negative. These different assumptions lead to minor differences in the arguments and results; however, each assumption could be employed in either case. Two cases are treated for the measurement of the data. First, it is assumed that the function is measured with a known accuracy on an entire hyperplane parallel to the boundary of the half-space. Later the data are measured with a prescribed accuracy on a rectangle on the same hyperplane.

This research was supported in part by the National Science Foundation and the Air Force Office of Scientific Research.

When the data are given on all of the hyperplane, it is proved in both the harmonic and parabolic cases that an approximate analytic continuation can be found in a practical manner and that the accuracy of the approximation on any specified rectangle in the open half-space depends in a Hölder-fashion on the accuracy of the measurement and the amount of computation involved in the discrete procedure. When the data are given on a compact subset of the hyperplane parallel to the boundary of the half-space, the situation is somewhat different. In the harmonic case the dependence of the approximate solution on the data is very weak; in fact, the a priori estimate of the error derived herein is of such a nature as to preclude practical calculation. We do not know whether the estimate can be improved; however, we do show that, if the data surface is rotated ever so trivially, no a priori estimate is feasible. Indeed, even uniqueness fails. In the parabolic case for data on a compact subset of the hyperplane, our a priori estimate of the error is again weaker than that for data on all the hyperplane, but the estimate does indicate that the computing problem is practical. No example is known to us to indicate that trivial rotation of the data surface should lead to failure; the authors disagree as to the likelihood of such a failure for parabolic functions

Harmonic functions in two variables are treated at considerable length in section 2 and the results for two variables are extended rather easily to several variables in section 3. Solutions of the heat equation are treated analogously in section 4.

A number of the results in this paper have been presented [2,3,4] in several earlier lectures by one of the authors; however, no proofs have been given and the theorems here stated are stronger than those given earlier. In particular, the results for data given on a compact set are much more precise here. Some related numerical experiments were reported in [1].

2. Harmonic Functions in Two Variables. Let u be a harmonic function in the half-plane $\{y>0\}$ having the representation

$$u(x,y) = \int_{-\infty}^{\infty} P(x-\xi,y)d\mu(\xi), \quad y>0, -\infty<x<\infty, \tag{2.1}$$

where

$$P(x,y) = \pi^{-1}\, y[x^2+y^2]^{-1} \tag{2.2}$$

and μ is a signed measure such that

$$|\mu|([n,n+1)) = \mu^+([n,n+1))+\mu^-([n,n+1))\le M, n = 0,\pm 1,\pm 2,\ldots, \tag{2.3}$$

where $\mu = \mu^+ - \mu^-$ is the decomposition of μ into the difference of nonnegative measures. The constraint (2.3) is the global restriction imposed upon u to force continuous dependence of u on the data to be specified later.

Assume first that u is known approximately on the line $\{y = Y>0\}$:

$$|u(x,Y) - g(x)|\le \epsilon, \ -\infty<x<\infty, \tag{2.4}$$

where $g(x)$ is continuous. Later this condition will be modified. It is clear that

$$|u(x,y) - \int_{-\infty}^{\infty} P(x-\xi,y-Y)g(\xi)d\xi| \le \epsilon, \ y>Y; \tag{2.5}$$

thus, only the values of u on the strip $\{0<y<Y\}$ are of interest. The first question to be discussed will be the approximation of u on the strip $\{\delta\le y<Y\}$, $\delta>0$.

Note that, if $m\le x+X\le m+1$,

$$\Big|\int_{|\xi-x|>X} P(x-\xi,Y)d\mu(\xi)\Big| \le \Big|\int_{x+X}^{m+1} P(x-\xi,Y)d\mu(\xi)+\sum_{k=m+1}^{\infty}\int_{k}^{k+1} P(x-\xi,Y)d\mu(\xi)\Big|$$
$$+\Big|\int_{-m-1}^{-x-X} P(x-\xi,Y)d\mu(\xi)+\sum_{k=m+1}^{\infty}\int_{-k-1}^{-k} P(x-\xi,Y)d\mu(\xi)\Big|$$

(2.6)

$$\le 2P(X,Y)M + 2M \sum_{k=m+1}^{\infty} P(k-x,Y)$$

$$\le 2M[P(X,Y) + \int_{[X]}^{\infty} P(\zeta,Y)d\zeta]$$

$$< 3\pi^{-1}YM[X]^{-1},\ X\ge 2,$$

by (2.3). Thus,

$$|u(x,Y) - \int_{x-X}^{x+X} P(x-\xi,Y)d\mu(\xi)| \le CM[X]^{-1}, \quad (2.7)$$

where C denotes a constant depending on Y. Let $x_i = \xi_i = ih$, $i=0, \pm 1, \pm 2, \ldots$, where $h = n^{-1}$, n an integer. If m and q are integers such that $m \le x_i < m+1$ and $q \le x_i + X < q+1$, then

$$\int_{x_i}^{x_i+X} P(x_i-\xi,Y)d\mu(\xi) = \int_{x_i}^{m+1} P(x_i-\xi,Y)d\mu(\xi) + \sum_{k=m+1}^{q-1} \int_{k}^{k+1} P(x_i-\xi,Y)d\mu(\xi)$$

(2.8)

$$+ \int_{q}^{x_i+X} P(x_i-\xi,Y)d\mu(\xi).$$

Then, from (2.3) it follows that

$$|\int_{x_i}^{m+1} P(x_i-\xi,Y)d\mu(\xi) - \sum_{x_i \le \xi_j < m+1} P(x_i-\xi_j,Y)\mu((\xi_j,\xi_{j+1}))|$$

(2.9)

$$\le Mh \max_{x_i \le \xi \le m+1} |\frac{\partial P}{\partial \xi}(x_i-\xi,Y)|.$$

Similarly,

$$|\int_{k}^{k+1} P(x_i-\xi,Y)d\mu(\xi) - \sum_{k \le \xi_j < k+1} P(x_i-\xi_j,Y)\mu([\xi_j,\xi_{j+1}))|$$

(2.10)

$$\le Mh \max_{k \le \xi \le k+1} |\frac{\partial P}{\partial \xi}(x_i-\xi,Y)|.$$

Since $P_x(x,y) = -2\pi^{-1}xy(x^2+y^2)^{-2}$,

$$(2.11)\qquad \left|\frac{\partial P}{\partial \xi}(x_i-\xi,Y)\right| \le \begin{cases} 2\pi^{-1}Y^{-2}, & x_i \le \xi \le m+1, \\ 2\pi^{-1}Y(k+1-x_i)(k^2+Y^2)^{-2}, & k \le \xi \le k+1. \end{cases}$$

Hence, it follows from (2.7)-(2.11) that

$$(2.12)\qquad \left|u(x_i,Y) - \sum_{x_i-X\le\xi_j<x_i+X} P(x_i-\xi_j,Y)\mu([\xi_j,\xi_{j+1}))\right| \le CM([X]^{-1}+h),$$

since

$$(2.13)\qquad \max_{x_i\le\xi\le m+1}\left|\frac{\partial P}{\partial \xi}(x_i-\xi,Y)\right| + \sum_{k=m+1}^{\infty}\max_{k\le\xi\le k+1}\left|\frac{\partial P}{\partial \xi}(x_i-\xi,Y)\right| \le C.$$

Let A denote the class of all sequences $\{(a_i,b_i)\} = \alpha$ such that

$$(2.14)\qquad 0 \le a_j, b_j, \qquad j = 0, \pm1, \pm2, \ldots,$$

$$\sum_{n\le\xi_j<n+1}(a_j+b_j) \le M, \qquad n = 0, \pm1, \pm2, \ldots.$$

For $\alpha \in A$ set

$$(2.15)\qquad u(x,y;\alpha) = \sum_{|x-\xi_j|\le X} P(x-\xi_j,y)(a_j-b_j), \quad -\infty<x<\infty,\ y>0.$$

The sequence $\{a_i\}$ represents the discretization of the unknown measure μ^+ and $\{b_i\}$ that of μ ; note that, as a consequence of (2.7), the Poisson integral can be truncated with a predictable effect on the solution. Now, fit $u(x,y;\alpha)$ to the data on $\{y = Y\}$ as follows.

Set

(2.16) $$e(\alpha) = \sup_{-\infty<i<\infty} |u(x_i,Y;\alpha) - g(x_i)|, \alpha \in A.$$

For $\alpha \in A$ this supremum is finite, since both $u(x,Y;\alpha)$ and $g(x)$ are bounded as a result of (2.3) and its discretization (2.14). Let

(2.17) $$\zeta = \inf_{\alpha \in A} e(\alpha).$$

Since $\{(\mu^+([\xi_j,\xi_{j+1})), \mu^-([\xi_j,\xi_{j+1})))\} \in A$, (2.4) and (2.12) imply that

(2.18) $$\zeta \leq \epsilon + CM([X]^{-1} + h).$$

What is wanted is a solution of the minimization problem given by the equation $e(\alpha) = \zeta$ for some $\alpha \in A$; but, since this is an infinite-dimensional linear programming problem, it is not at all clear that a solution exists. However, (2.12) implies the existence of $\alpha' \in A$ such that

(2.19) $$e(\alpha') \leq \epsilon + CM([X]^{-1}+h),$$

although it is not obvious how one would obtain such an α'. For the moment assume that an α' has been found and set

(2.20) $$v(x,y) = u(x,y;\alpha').$$

The function v is then the approximation to u in the strip $\{0<y<Y\}$; consider now the question of how good an approximation it is to u. Note that since terms appear and disappear from v as x moves from $-\infty$ to ∞, consequently, v is neither continuous nor harmonic. Let

$$(2.21) \qquad z(x,y) = \sum_{-\infty<\xi_j<\infty} P(x-\xi_j,y)(a_j-b_j), \{(a_j,b_j)\} = \alpha';$$

z is both continuous and harmonic in $\{y>0\}$. Moreover, it follows from (2.6) that

$$(2.22) \qquad |z(x_i,Y)-v(x_i,Y)| \le CM[X]^{-1}, \; i = 0,\pm 1, \pm 2,\ldots.$$

From (2.4), (2.19), and (2.22),

$$(2.23) \qquad |z(x_i,Y)- u(x_i,Y)| \le 2\epsilon + CM([X]^{-1}+h).$$

The global constraints (2.3) and (2.14) imply that u_x and z_x are bounded on $\{y = Y\}$ by a multiple of M depending only on Y. Hence,

$$(2.24) \qquad |z(x,Y)-u(x,Y)| \le 2\epsilon + CM([X]^{-1}+h), -\infty<x<\infty.$$

The bounding of z-u in the strip $\{0<y<Y\}$ is facilitated by the following lemma.

Lemma 1: Let $w(x,y)$ be harmonic for $y>0$ and let w have the representation (2.1). Let $|\mu|([n,n+1)) \le M$, $n = 0,\pm 1, \pm 2,\ldots.$ If $|w(x,Y)| \le \epsilon$ and $\eta>0$, there exist $\epsilon_0>0$ and a constant C such that

$$|w(x,y)| \le CM^{1-\frac{y}{Y}+\eta}\,\epsilon^{\frac{y}{Y}-\eta}, \; -\infty<x<\infty, \; \eta\le y\le Y,$$

for $0< \epsilon < \epsilon_0$.

Proof: It is easy to see that it is sufficient to bound $w(0,y)$. Let $0<\delta<Y$. Since

$$(2.25) \qquad w(x,y) = \frac{y-Y}{\pi} \int_{-\infty}^{\infty} \frac{w(\xi,Y)}{(x-\xi)^2+(y-Y)^2}\, d\xi,$$

it follows that

(2.26) $$|w_y(x,Y+\delta)| \le \frac{\varepsilon}{\pi}\int_{-\infty}^{\infty} \frac{|\xi^2-\delta^2|}{(\xi^2+\delta^2)^2}\, d\xi \le \frac{4\varepsilon}{\pi\delta}\ , \quad -\infty<x<\infty.$$

The maximum principle implies that $|w(x,Y+\delta)| \le \varepsilon$. Let $w^*(x,y)$ be the harmonic conjugate of w in $\{y>0\}$ vanishing at $(0,Y+\delta)$. Then,

(2.27) $$|w^*(x,Y+\delta)| \le \int_0^{|x|} |w_y(\xi,Y+\delta)|\, d\xi \le \frac{4|x|\varepsilon}{\pi\delta}\ .$$

The function w is dominated for fixed y by the value at $(0,y)$ of the harmonic function generated by point masses $\mu(\{0\}) = 2M$ and $\mu(\{n\}) = M$, $n = \pm 1, \pm 2, \ldots$. Thus,

(2.28) $$\begin{aligned} |w(x,y)| &\le \pi^{-1}M[2\delta^{-1}+2\sum_{k=1}^{\infty}\delta(k^2+\delta^2)^{-1}] \\ &\le 2(\pi^{-1}\delta^{-1}+6^{-1}\pi\,\delta)M\ , \quad y\ge\delta. \end{aligned}$$

It is also easily seen that

(2.29) $$|w_x(x,y)| \le 4\pi^{-1}[3^{3/2}4^{-2}\delta^{-2}+\delta\sum_{k=1}^{\infty}k^{-3}]M,\quad \delta\le y,\quad \delta\le 3^{1/2}.$$

If $f(x+iy) = w(x,y) + iw^*(x,y)$, $y>0$, then it is clear that for $\delta<1$

(2.30) $$|f(x+iy)| \le C_1(\delta)M + C_2(\delta)|x|\ \varepsilon,\quad C_i(\delta) = O(\delta^{-3+i}),\ y\ge\delta.$$

Also,

(2.31) $$|f(x+i(Y+\delta))| \le (1+4\pi^{-1}\delta^{-1}|x|)\ \varepsilon\ , \quad -\infty<x<\infty.$$

Set

(2.32) $$K = C_1(\delta)M+C_2(\delta)\beta\varepsilon$$

and

(2.33) $$-g(x,y) = \log K^{-1}|f(x+iy)|.$$

Then, g is a harmonic function that is nonnegative on the segments $\{(x,y): -\beta\le x\le\beta, y=\delta\}$ and $\{(x,y): x=\pm\beta, \delta\le y\le Y+\delta\}$, and

(2.34) $$g(x,Y+\delta)\ge -\log\frac{(1+4\pi^{-1}\delta^{-1}\beta)\,\varepsilon}{C_1(\delta)M+C_2(\delta)\beta\varepsilon} = \gamma,\quad |x|\le\beta.$$

A harmonic minorant for g is given by the harmonic function vanishing on the three lower sides of the rectangle and equal to $\gamma\cos\pi x/2\beta$ on the upper side. Thus,

(2.35) $$g(x,y)\ge \quad - \quad \frac{\gamma\cos\frac{\pi x}{2\beta}\,\sinh\frac{\pi(y-\delta)}{2\beta}}{\sinh\frac{\pi Y}{2\beta}} \quad ,$$

and

(2.36) $$|f(iy)|\le \quad K\quad \left[\frac{(1+4\pi^{-1}\delta^{-1}\beta)\varepsilon}{K}\right]^{\frac{\sinh\pi(y-\delta)/2\beta}{\sinh\pi Y/2\beta}}$$

Since $\sinh qx/\sinh qy = xy^{-1}[1+6^{-1}q^2(x^2-y^2)+\cdots]$,

(2.37) $$\frac{\sinh\pi(y-\delta)/2\beta}{\sinh\pi Y/2\beta} = \frac{y-\delta}{Y}\,[1+O(\beta^{-2}((y-\delta)^2-Y^2))]$$

$$\ge \frac{y}{Y}-\eta,$$

provided that

(2.38) $$0<\delta\le\tfrac12 Y\eta,\qquad O(\beta^{-2}Y^2)\le\tfrac12\eta,\qquad y\ge\delta.$$

Choose β so that the second inequality holds. Then choose $\delta\le\min(\eta,\frac12 Y\eta)$. Finally choose ε_0 so that $\beta\varepsilon_0\le M$. Then, $K\le \mathrm{const}\cdot M$ and it follows that

(2.39) $$|w(0,y)|\le|f(iy)|\le CM^{1-\frac{y}{Y}+\eta}\,\varepsilon^{\frac{y}{Y}-\eta},\quad \eta\le y\le Y,$$

and the lemma has been proved.

The restriction to small ε is not important, but it is only in that case that the result is useful. The trivial example

$u = e^{-y} \sin x$ shows that the lemma is close to best possible.

The application of Lemma 1 to the function z-u is immediate, since

$$(2.40) \qquad (z-u)(x,y) = \int_{-\infty}^{\infty} P(x,\xi,y)\,d\nu(\xi), \quad |\nu|([n,n+1)) \le 2M.$$

It follows from (2.24) and (2.40) that

$$(2.41) \qquad |(z-u)(x,y)| \le C_1 M^{1-\frac{y}{Y}+\eta} [\varepsilon + CM(\lceil X\rceil^{-1}+h)]^{\frac{y}{Y}-\eta}, \quad -\infty<x<\infty, \eta \le y \le Y.$$

Recall that z differs from $v(x,y) = u(x,y;\alpha')$, the approximate harmonic continuation, by not more than $CM\lceil X\rceil^{-1}$ along the line $\{y=Y\}$. It is easily seen that

$$(2.42) \qquad \begin{aligned} |(z-v)(x,y)| &\le \frac{2M}{\pi} \sum_{k=\lceil X\rceil}^{\infty} \frac{y}{k^2+y^2} \le 2y\,M\,\lceil X\rceil^{-1} \\ &\le CM^{1-y/Y+\eta}(\varepsilon + M\lceil X\rceil^{-1}+Mh)^{\frac{y}{Y}-\eta}, \quad X \ge 2. \end{aligned}$$

Thus, the following theorem has been proved.

Theorem 1. Let $\alpha' \in A$ be any solution of (2.29) and define $u(x,y;\alpha')$ by (2.15). If $\eta > 0$ and if ε, X^{-1}, and h are sufficiently small, there exists a constant C such that

$$|u(x,y)-u(x,y;\alpha')| \le CM^{1-\frac{y}{Y}+\eta}(\varepsilon+M\lceil X\rceil^{-1}+Mh)^{\frac{y}{Y}-\eta} \quad -\infty<x<\infty, \eta\le y\le Y,$$

provided that u has the representation (2.1) and (2.3) and (2.4) hold.

Note that the error estimate is in terms of the global constraint M, the measurement error ε, and the parameters h and X of the approximation. Obviously, the data need have been known only at the points (x_i, Y), but this is a trivial reduction.

As was remarked on earlier, the infinite-dimensional linear programming problem (2.17) imposes serious practical limitations. If an approximation to the solution u is desired on the rectangle $R = \{|x| \le X_1, \eta \le y \le Y\}$, then it seems intuitively clear that the values of a_i and b_i assigned at points $(x_i, 0)$ at a great distance from R should have a negligible effect on the approximation in R and, consequently, could be set equal to zero. A finite linear programming problem would result. The intuition is correct, but the proof seems nontrivial. Two arguments, both complicated, will be presented. The first method will give a better error estimate, but the harmonic conjugate will be introduced, limiting the argument to two-dimensional problems. The second argument also is based strongly on complex analysis, but the analytic function arises from extending an independent variable to the complex domain. This method of attack can be applied to harmonic functions in several variables and to solutions of the heat equation. One pays for the generality by obtaining weaker error estimates.

During the argument to be given below quantities X_2, X_3, X_4, and X_5, in addition to the X_1 of the definition of R, will be introduced. $X_i, i = 2,\ldots,5$, will tend to infinity in obtaining the estimates, and the following relations will hold: $X_1 < X_5 < X_4 < X_3 = X_4 + X_2$. More precise requirements for these terms will appear later.

Retain the representation (2.1), (2.3) and the measurement (2.4). The estimate (2.12) carries over in the form

$$|u(x_i, Y) - \sum_{x_i - X_2 \le \xi_j < x_i + X_2} P(x_i - \xi_j, Y)\mu([\xi_j, \xi_{j+1}))| \le CM([X_2]^{-1} + h), \tag{2.43}$$

where $X_2 > 0$. Let $X_3 = X_2 + X_4$, where $X_4 > X_1$. Let A now represent the sequences $\alpha = \{(a_i, b_i)\}$, where

$$a_i, b_i \geq 0 \ , \ i = 0, \pm 1, \pm 2, \ldots,$$

(2.44) $$a_i = b_i = 0, \ |x_i| > X_3,$$

$$\sum_{n \leq \xi_j < n+1} (a_j + b_j) \leq M, \ n = 0, \pm 1, \pm 2, \ldots.$$

Set

(2.45) $$u(x,y;\alpha) = \sum_{|x-\xi_j| \leq X_2} P(x-\xi_j, y)(a_j - b_j), -\infty < x < \infty, y > 0,$$

and let

(2.46) $$e(\alpha) = \max_{|x_i| \leq X_4} |u(x_i, Y; \alpha) - g(x_i)|, \ \alpha \in A.$$

If

(2.47) $$\zeta = \inf_{\alpha \in A} e(\alpha),$$

the determmation of an $\alpha' \in A$ such that $e(\alpha') = \zeta$ is a standard finite-dimensional linear programming problem. It follows again that

(2.48) $$\zeta \leq \varepsilon + CM([X_2]^{-1} + h),$$

since the error is treated only for $|x_i| \leq X_4$ and the weights are not set to zero until $|x_i| > X_4 + X_2$. Again set $v(x,y) = u(x,y;\alpha')$ for some solution α' of the mimimization problem and set

$$(2.49) \qquad z(x,y) = \sum_{|\xi_j|\le X_3} P(x-\xi_j,y)(a_j-b_j),\ \{(a_j,b_j)\} = \alpha'.$$

As before, if $w = u-z$,

$$(2.50) \qquad |w(x_i,Y)| \le 2\epsilon + CM([X_2]^{-1}+h) \equiv \gamma_1,\ |x_i|\le X_4.$$

Clearly, w is harmonic in $\{y>0\}$ and

$$w(x,y) = \int P(x-\xi,y)d\nu(\xi),$$

$$(2.51) \qquad |\nu|([n,n+1]) \le 2M.$$

Since $|w_x(x,Y)| \le CM$, (2.50) holds for all $x \in [-X_4,X_4]$ with a different C.

The remainder of the analysis of w is similar to that given in Lemma 1, but the limitation of the bound in (2.50) to $\{|x|\le X_4\}$ leads to certain additional features. First, it is clear that

$$(2.52) \qquad |w(x,Y)| \le CM,\ |x| \ge X_4.$$

Now let $X_1 < X_5 < X_4$. Then, if $\delta>0$,

$$|w(x,Y+\delta)| \le \frac{\delta}{\pi}\int_{-X_4}^{X_4} \frac{\gamma_1}{(x-\xi)^2+\delta^2}\,d\xi + \frac{\delta}{\pi}\int_{|\xi|>X_4} \frac{CM}{(x-\xi)^2+\delta^2}\,d\xi$$

$$(2.53) \qquad \le \gamma_1 + \frac{2CM\,\delta}{X_4-X_5} \equiv \gamma_2 \quad ,\ |x| \le X_5 \quad .$$

Similarly,

$$(2.54) \qquad |w_y(x,Y+\delta)| \le 4\delta^{-1}\gamma_1 + \frac{2CM}{X_4-X_5} \equiv \gamma_3,\ |x|\le X_5 \ .$$

Clearly, $|w_x(x,y)| \le C(\delta)M$, $y \ge \delta > 0$. The estimation of w is reduced to exactly the same problem as was treated in Lemma 1, except that it is not sufficient to look at $w(0,y)$. The rectangle used to estimate the harmonic function g introduced in (2.33) should be centered in x at an arbitrary $x_o \in [-X_4, X_4]$, the harmonic conjugate w^* should vanish at $(x_o, Y+\delta)$, and β should be replaced by $X_5 - X_1$. The constant K of (2.32) should be replaced by

(2.55) $$K = C(\delta)M + C(\delta)(X_5 - X_1)\gamma_3 + \gamma_2.$$

Choose $\delta \le \min(\eta, \frac{1}{2}Y\eta)$ and X_5 such that

(2.56) $$\frac{\sinh \pi(y-\delta)/2(X_5-X_1)}{\sinh \pi Y/2(X_5-X_1)} \ge \frac{y}{Y} - \eta, \quad y \ge \delta.$$

Then, choose X_2' and X_4' sufficiently large and h' sufficiently small that

(2.57) $$(X_5 - X_1)\gamma_3 \le M, \quad \gamma_2 \le M.$$

For $X_2 \ge X_2'$, $X_4 \ge X_4'$, and $h \le h'$, it follows that

(2.58) $$|w(x,y)| \le C M^{1-\frac{y}{Y}+\eta} \gamma_3^{\frac{y}{Y}-\eta}, \quad y \ge \eta, \ |x| \le X_1.$$

It is easy to see that

(2.59) $$\gamma_3 \le C(\eta)\,\epsilon + C(\eta)\, M\, \{h + [X_2]^{-1} + (X_4 - X_5)^{-1}\}.$$

Theorem 2. Let $\alpha' \in A$ be a solution of the linear programming problem (2.47) and define $u(x,y;\alpha')$ by (2.45). Let $\eta > 0$. Then there exist $X_5 > X_1$ and $X_2' > 0$, $X_4' > X_5$ and $h' > 0$ such that

$$|u(x,y) - u(x,y;\alpha')| \le C M^{1-\frac{y}{Y}+\eta} \gamma_3^{\frac{y}{Y}-\eta}, \quad (x,y) \in R,$$

if $X_2 \ge X_2'$, $X_4 \ge X_4'$, and $h \le h'$. Moreover, γ_3 can be bounded as in (2.59). In particular, the error in the approximation depends in a

Hölder-continuous fashion on the measurement error ϵ and the amount of calculation performed, as indicated by h, X_2, X_4, and X_5.

A second deviation of an error estimate can be based on the extension of the variable y to the complex domain,

(2.60) $$\sigma = y + iy^*.$$

This analysis begins with the relations (2.50), (2.51), and (2.52). Note first that

(2.61)
$$|w(x,y)| \le \gamma_1 + 2\frac{y-Y}{\pi}\int_{X_4-X_1}^{\infty}\frac{CMd\xi}{\xi^2+(y-Y)^2}$$
$$\le \gamma_1 + CM(y-Y)(X_4-X_1)^{-1},\ |x|\le X_1,$$
$$\le \gamma_1 + CM(Y_1-Y)(X_4-X_1)^{-1} \equiv \gamma_4,\ |x|\le X_1,\ Y\le y\le Y_1.$$

Consider the integral

(2.62) $$w(x,\sigma) = \int_{-\infty}^{\infty} P(x-\xi,\sigma)\, d\nu(\xi),\ \mathrm{Re}\ \sigma > 0.$$

Since

$$|(x-\xi)^2+\sigma^2|^2 = (x-\xi)^4 + (y^2+y^{*2})^2 + 2(x-\xi)^2(y^2-y^{*2})$$
$$\ge 4y^2\, y^{*2}$$

and $|(x-\xi)^2 + y^2|^2 \ge y^4$, the denominator of $P(x-\xi,\sigma)$ is bounded away from zero for any σ such that Re $\sigma > 0$. Thus, the integral converges for any x and any σ such that Re $\sigma > 0$, and $w(x,\sigma)$ is, for each x, a holomorphic function of σ in the half-plane $\{\mathrm{Re}\ \sigma > 0\}$.

Moreover, for $y^* > 0$,

$$|w(x,\sigma)| \le \frac{1}{\pi y} \int_{\{|x-\xi|<y^*+1\}} d|\nu|(\xi) + \frac{1}{\pi} \int_{\{|x-\xi|\ge y^*+1\}} \frac{|\sigma|}{|(x-\xi)^2+\sigma^2|} d|\nu|(\xi)$$

(2.63)

$$\le \frac{2M(y^*+2)}{\pi y} + CM(y^*+Y_1)$$

$$\le CM(y^*+Y_1) \quad , \delta \le y \le Y_1 .$$

Hold x fixed in the interval $[-X_1, X_1]$ and consider the rectangle $Q = \{Y \le y \le Y_1, 0 \le y^* \le Y_1^*\}$ in the σ-plane. Then, it follows from (2.61) and (2.63) that

$$(2.64) \qquad |w(x,\sigma)| \le \begin{cases} \gamma_4, \ y^* = 0, \ \sigma \in \partial Q \\ CM, \ \sigma \in \partial Q \ , \ C = C(Y,Y_1,Y_1^*). \end{cases}$$

It then follows from (2.31)-(2.36) that

$$(2.65) \qquad |w(x,\tfrac{1}{2}(Y+Y_1)+iy^*)| \le (CM)^{1-\alpha(y^*)} \gamma_4^{\alpha(y^*)}, \ |x| \le X_1,$$

where

$$(2.66) \qquad \alpha(y^*) = \frac{\sinh \pi(Y_1^*-y^*)/(Y_1-Y)}{\sinh \pi Y_1^*/(Y_1-Y)}, \ 0 \le y^* < Y_1^* .$$

Now, consider the rectangle $Q' = \{\delta \le y \le \frac{1}{2}(Y+Y_1), |y^*| \le Y_2^*\}$, where $0 < \delta < \eta$ and $0 < Y_2^* < Y_1^*$ and x remains fixed. The same result holds for $y^* < 0$. Then,

$$(2.67) \qquad |w(x,\sigma)| \le \begin{cases} (CM)^{1-\alpha}\gamma_4^{\alpha} \ , \ y = \frac{1}{2}(Y+Y_1), \ \sigma \in \partial Q', \alpha = \alpha(Y_2^*), \\ C(\delta, \frac{1}{2}(Y+Y_1), Y_2^*)M, \ \sigma \in \partial Q' . \end{cases}$$

Hence,

$$(2.68) \qquad |w(x,y)| \le (CM)^{1-\beta(y)} [(CM)^{1-\alpha}\gamma_4^{\alpha}]^{\beta(y)}, |x| \le X_1 \ , \ \delta \le y < \tfrac{1}{2}(Y+Y_1) \ ,$$

where

$$(2.69) \qquad \beta(y) = \frac{\sinh \pi(y-\delta)/2Y_2^*}{\sinh \pi(\frac{1}{2}(Y+Y_1)-\delta)/2Y_2^*} .$$

Since $R = \{|x| \le X_1, \eta \le y \le Y\}$, there exist a constant C and a positive number β such that

$$(2.70) \qquad \max_{(x,y)\in R} |w(x,y)| \le CM^{1-\beta} Y_4^{\beta} .$$

The remainder of the argument leading to a theorem of the same type as Theorem 2 is the same as before, although it is clear that the exponen β is not so large as before. However, the advantage of this approach will be seen immediately below, as well as later in several other applications.

So far it has been assumed that the data g(x) has been known for all x. Clearly, it is not practical to measure g for all x; consequently, it is both practically useful and mathematically interesting to limit the data to the following:

$$(2.71) \qquad |u(x,Y)-g(x)| \le \epsilon , \quad |x| \le X_4,$$

the choice of the subscript being made to coincide in the argument below with the previous X_4. Again let us approximate u on the rectangle R, but it is not assumed that $X_1 < X_4$. If the approximation is made as before, relations (2.43)-(2.52) remain valid, although X_4 is fixed now and not a parameter to be chosen.

The estimation of the error w will involve first the extension of the x-variable to the complex domain and then the y-variable. Let $\chi = x+ix^*$ and consider the integral

$$(2.72) \qquad w(\chi,y) = \int_{-\infty}^{\infty} P(\chi-\xi,y)d\nu(\xi) = \frac{1}{\pi}\int_{-\infty}^{\infty} \frac{y\,d\nu(\xi)}{(x-\xi+ix^*)^2+y^2} ,$$

which can easily be seen to converge and to be holomorphic in χ for $|x^*| < y$. In fact, if $|x^*| \le \rho Y$, $0 < \rho < 1$, then

(2.73) $$|P(\chi-\xi,Y)| \le \frac{1}{\pi}\ \frac{Y^2}{(1-\rho^2)Y^2+(x-\xi)^2} = (1-\rho^2)^{-\frac{1}{2}}\ P(x-\xi,(1-\rho^2)^{\frac{1}{2}}Y),$$

and

(2.74) $$|w(\chi,Y)| \le C(\rho)M, \quad |\mathrm{Im}\ \chi| \le \rho\ Y.$$

Thus, the holomorphic function $w(\chi,Y)$ satisfies the following bounds:

$$|w(x,Y)| \le \gamma_1 \quad , \ |x| \le X_4 \ ,$$

(2.75) $$|w(\pm X_4+ix^*,Y)| \le C(\rho)M, \quad |x^*| \le \rho\ Y,$$

$$|w(x+i\rho Y,Y)| \le C(\rho)M, \quad |x| \le X_4 \ .$$

By the argument leading to (2.36),

$$|w(ix^*,Y)| \le (C(\rho)M)^{1-\alpha(x^*)}\ \gamma_1^{\alpha(x^*)},$$

(2.76) $$\alpha(y) = \sinh \frac{\pi(\rho Y-y)}{2X_4} \Big/ \sinh \frac{\pi\rho Y}{2X_4} \quad , \ 0 \le y \le \rho\ Y.$$

By symmetry the above result holds with y replaced by $|y|$ for $|y| \le \rho\ Y$. Thus, $w(\chi,Y)$ satisfies the following constraints on the boundary of the half-strip $\{|\mathrm{Im}\chi| \le y_0 < \rho Y, \mathrm{Re}\chi \ge 0\}$:

$$|w(ix^*,Y)| \le \gamma_2 = (C(\rho)M)^{1-\alpha(y_0)}\ \gamma_1^{\alpha(y_0)}, \quad |x^*| \le y_0 \ ,$$

(2.77)

$$|w(x\pm iy_0,Y)| \le C(\frac{y_0}{Y})M \ , \quad x \ge 0 \ .$$

Let x be fixed so that $x > X_4$, and let $x_1 > x$. Then, by the same argument as above,

$$(2.78)\qquad \begin{aligned} |w(x,Y)| &\le \left(C\left(\frac{y_o}{Y}\right)M\right)^{1-\beta(x,x_1)} \gamma_2^{\beta(x,x_1)} , \quad 0 \le x < x_1 , \\ \beta(x,x_1) &= \sinh \frac{\pi(x_1-x)}{2y_o} \Big/ \sinh \frac{\pi x_1}{2y_o} . \end{aligned}$$

The ratio $\beta(x,x_1)$ is an increasing function of x_1 and has the limit

$$(2.79)\qquad \beta(x) = e^{-\pi x/2y_o}$$

as x_1 tends to infinity. Hence.

$$(2.80)\qquad |w(x,Y)| \le C(y_o,M)\gamma_2^{\beta(x)} = C(\rho,y_o,M)\gamma_1^{\alpha(y_o)\beta(x)} , \quad x > X_4 .$$

The argument leading from (2.50) to Theorem 2 can be repeated with the symbol X_4^* replacing the parameter X_4 in the earlier argument. Let X_5 be chosen so that (2.56) is valid and let $X_1<X_5<X_4^*$. It follows that the quantities γ_1, γ_2, and γ_3 defined in (2.50), (2.53), and (2.54), respectively, can be replaced as follows:

$$(2.81)\qquad \begin{aligned} \gamma_1' &= C(\rho,y_o,M)\gamma_1^{\alpha(y_o)\beta(X_4^*)} , \quad \gamma_1 = 2\varepsilon + CMX_2^{-1} + Ch , \\ \gamma_2' &= \gamma_1' + C(M,\delta)(X_4^* - X_5)^{-1} , \\ \gamma_3' &= C(\delta)\gamma_1' + C(M)(X_4^* - X_5)^{-1} . \end{aligned}$$

The relation (2.58) becomes

$$(2.82)\qquad |w(x,y)| \le C(M,\eta)\gamma_3'^{\frac{y-\eta}{Y-\eta}} , \quad y \ge \eta, \ |x| \le X_1 ;$$

hence,

$$|w(x,y)| \le C(M,\rho,\delta,y_0,\eta)\{[\varepsilon + X_2^{-1} + h]^{\alpha(y_0)\beta(X_4^*)} + (X_4^* - X_5)^{-1}\}^{\frac{y}{Y}-\eta}, \quad y \ge \eta, \ |x| \le X_1 \tag{2.83}$$

Note that the estimate (2.83) contains in particular two parameters, X_4^* and X_5, that were introduced into the analysis for convenience but which are not involved in the calculation. Obviously, these parameters should be selected so as to minimize the bound. The minimum is closely approximated by equating the two additive terms in the bracket:

$$[\varepsilon + X_2^{-1} + h]^{\alpha(y_0)\beta(X_4^*)} = (X_4^* - X_5)^{-1} . \tag{2.84}$$

As $\varepsilon + X_2^{-1} + h$ tends to zero,

$$\frac{\pi X_4^*}{2y_0} = \frac{\log\log(\varepsilon + X_2^{-1} + h)^{\alpha(y_0)}}{1 - \frac{2y_0}{\pi X_4^*}\log\log(X_4^* - X_5)} \sim \log\log(\varepsilon + X_2^{-1} + h)^{\alpha(y_0)} . \tag{2.85}$$

Thus,

$$(\varepsilon + X_2 + h^{-1})^{\alpha(y_0)\beta(X_4^*)} + \frac{1}{X_4^* - X_5} = \frac{2}{X_4^* - X_5} \sim \frac{2}{X_4^*} \sim \frac{\pi}{y_0 \log\log(\varepsilon + X_2^{-1} + h)^{-\alpha(y_0)}} . \tag{2.86}$$

Consequently,

$$|w(x,y)| \le C[\log\log(\varepsilon + X_2^{-1} + h)^{-\alpha(y_0)}]^{-\frac{y}{Y}+\eta}, \quad y \ge \eta , \ |x| \le X_1 . \tag{2.87}$$

The constant C depends on M, ρ, δ, y_0, and η. The inequality (2.87) implies the following result.

<u>Theorem 3</u>. Let $\alpha' \in A$ be a solution of the minimization problem (2.44) - (2.47), where X_4 is defined by (2.71). Let $u(x,y;\alpha')$ be defined by (2.45). Then, there exist positive constants C, γ, and β depending on the rectangle R, X_4, and M such that

$$\max_R |u(x,y) - u(x,y;\alpha')| \le C[\log \log(\epsilon + X_2^{-1} + h)^{-\gamma}]^{-\beta} ,$$

provided that u is a harmonic function in the upper half-plane having the representation (2.1) - (2.3).

The error estimate above is not adequate for practical usage, although it does establish a quite weak continuous dependence relationship. It is perhaps of some theoretical interest to consider the asymptotic form of the estimate as X_4 tends to zero. As X_4 tends to zero,

$$\alpha(y_0) \sim e^{-\pi y_0/2X_4} , \tag{2.88}$$

and becomes small quite rapidly. It should also be noted that the method of analysis used to derive (2.87) could be extended to give a unique continuation theorem for harmonic functions in a half-plane.

One consequence of Theorem 3 is a uniqueness theorem for harmonic functions. Let Γ_θ be a line segment of length $2X_4$ centered at $(0,Y)$ and making an angle θ with the x-axis. Theorem 3 implies that a harmonic function that is bounded on $\{y>0\}$ and is constant on Γ_0 (i.e., $\alpha=0$) is identically constant. This is false for $\theta \neq 0$. For example, let L_θ be the line determined by Γ_θ and let x_θ be its intersection with the x-axis. Let

$$\text{(2.89)}\qquad \begin{aligned} &\Delta u_\theta = 0 \,, \quad y > 0 \,, \\ &u_\theta(x,0) = \begin{cases} 1 \,, & x < x_\theta \,, \\ 0 \,, & x > x_\theta \,. \end{cases} \end{aligned}$$

L_θ is a level curve of u_θ. This indicates that the weak continuity resulting from Theorem 3 is to be expected. It also implies that no a priori estimate can be obtained from data given on Γ_θ .

3. Harmonic Functions in More than Two Variables. The results of the last section can rather easily be generalized to harmonic functions in more than two variables. Let $x = (x^1, \ldots, x^n) \in \mathbf{R}_n$ and $y \in R_1$. Let $P(x,y) = c_n y(|x|^2+y^2)^{-\frac{n}{2}-1}$ be the Poisson kernel for the half-space $\{y>0\}$. Let u be a harmonic function in the half-space $\{y>0\}$ having the representation

$$u(x,y) = \int P(x-\xi,y)d\mu(\xi) , \quad y > 0 . \tag{3.1}$$

Denote the cube $\{s^i \le x^i < s^i+1\}$, s^i an integer, by S. Assume that

$$|\mu|(S) \le M , \quad \text{all } S . \tag{3.2}$$

Also, assume that, for some $Y > 0$,

$$|u(x,Y) - g(x)| < \epsilon , \quad x \in \mathbf{R}_n . \tag{3.3}$$

Clearly, (3.1)-(3.3) are the direct analogues of (2.1)-(2.4). Again let

$$R = \{|x^j| \le X_1 , \quad \delta \le y \le Y\} \tag{3.4}$$

be the rectangle on which it is desired to approximate u. The approximation will be accomplished in precisely the same fashion as before and the method of proof will be a straight-forward generalization of the earlier proof based on extending y to the complex variable $\sigma = y + iy^*$.

Let $I = (i_1, \ldots, i_n)$ be a multi-index of integers, $x_i^j = ih$, and $x_I = (x_{i_1}^1, \ldots, x_{i_n}^n)$. Set $Z_I = \{x_{i_j}^j \le x^j < x_{i_j+1}^j\}$. Then it is easy to see that

$$(3.5) \qquad \left| u(x,Y) - \sum_{|\xi^j_{i_j} - x^j| \le X_2} P(x-\xi_I, Y)\mu(Z_I) \right| \le CM(X_2^{-1}+h) .$$

Let A denote the sequences $\alpha = \{(a_I, b_I)\}$ such that

$$(3.6) \qquad \begin{aligned} & a_I, b_I \ge 0 , \quad \text{all } I , \\ & a_I = b_I = 0 , \quad \text{if } \max|\xi^j_{i_j}| > X_3 = X_2 + X_4 , \\ & \sum_{Z_I \subset S} (a_I + b_I) \le M , \quad \text{all } S . \end{aligned}$$

For $\alpha \in A$, set

$$(3.7) \qquad u(x,y;\alpha) = \sum_{|\xi^j_{i_j} - x^j| \le X_2} P(x-\xi_I, y)(a_I - b_I) , \quad y > 0 ,$$

and let

$$(3.8) \qquad e(\alpha) = \max_{|x^j_{i_j}| \le X_4} |u(x_I, Y;\alpha) - g(x_I)| .$$

Then,

$$(3.9) \qquad \zeta = \inf_{\alpha \in A} e(\alpha) \le \varepsilon + CM(X_2^{-1}+h) ,$$

and again a solution $\alpha' \in A$ of the minimization problem can be obtained by linear programming. Set $v(x,y) = u(x,y;\alpha')$ for some solution α' and set

$$(3.10) \qquad z(x,y) = \sum_{|\xi^j_{i_j}| \le X_3} P(x-\xi_I, y)(a'_I - b'_I) .$$

Then, v is the approximation to the harmonic function u. As before, v is not harmonic, but z is and differs from v on R by CMX_2^{-1}. Thus, it is sufficient to estimate $w(x,y) = u(x,y) - z(x,y)$ on R.

The harmonic function w satisfies the following relations:

$$(3.11)\quad \begin{aligned} & w(x,y) = \int P(x-\xi,y)d\nu(\xi), \quad y>0, \\ & |\nu(S)| \le 2M, \text{ all } S, \\ & |w(x,Y)| \le 2\epsilon + CM(X_2^{-1}+h) \equiv \gamma_1, \quad |x^j| \le X_4, \\ & |w(x,Y)| \le CM \qquad\qquad, \quad \text{all } x. \end{aligned}$$

Repeating the argument beginning with (2.60) with σ as before and $Y \le y \le Y_1$, we find that

$$(3.12)\quad |w(x,y)| \le \gamma_1 + CM(X_4-X_1)^{-1} \equiv \gamma_4, \quad |x^j| \le X_4, \quad Y \le y \le Y_1.$$

The integral

$$(3.13)\quad w(x,\sigma) = \int P(x-\xi,\sigma)d\nu(\xi), \quad \operatorname{Re}\sigma > 0,$$

converges and is holomorphic for $\operatorname{Re}\sigma > 0$, and $|w(x,\sigma)| \le CM$ if $\delta \le y \le Y_1$. Hold x fixed and set $Q = \{Y \le y \le Y_1, \; 0 \le y^* \le Y_1^*\}$. The relations (2.64) and (2.65) again hold with $\alpha(y^*)$ still given by (2.66). Define Q′ as follows (2.66). The remainder of the argument is unaltered and

$$(3.14)\quad \max_R |w| \le CM^{1-\beta}\gamma_4^{\beta},$$

where $\beta > 0$. Clearly, X_4 can be chosen so that

$$(3.15)\quad \max_R |w| \le C(M,R)(\epsilon + X_2^{-1} + h)^{\beta}.$$

In any event, the estimates in the two variable case carry over essentially unaltered to the higher dimensional case.

The argument given in the two variable case for the data known only on an interval can be carried over, but the dependence on the data and parameters remains weak.

4. Parabolic Functions. The approximation of solutions of the heat equation from interior data can be accomplished in a fashion very similar to the method applied to harmonic functions. It is sufficient to consider the case of a single space variable; the extension to several space variables is as immediate as in the harmonic case.

The most natural global constraint to apply to a solution u of the heat equation is that it be nonnegative, since temperatures are naturally bounded on one side along with all the other common physical quantities satisfying parabolic equations. In addition, let us assume that u is bounded on each half-space $\{t \geq t_0 > 0\}$ by $M(t_0)$, where $M(T) = 1$ and $M(t)$ is otherwise unknown:

$$(4.1)\qquad \begin{aligned} &u_t = u_{xx}\,, \quad -\infty < x < \infty\,, \quad 0 < t < \infty\,, \\ &0 \leq u(x,T) \leq 1\,, \quad -\infty < x < \infty\,. \end{aligned}$$

Assume also that

$$(4.2)\qquad |u(x,T) - g(x)| \leq \varepsilon\,, \quad -\infty < x < \infty\,.$$

It follows from a theorem of Widder [5] that

$$(4.3)\qquad u(x,t) = \int_{-\infty}^{\infty} K(x-\xi,t)\,d\mu(\xi)\,, \quad t > 0\,,$$

where

$$(4.4)\qquad K(x,t) = (4\pi t)^{-\frac{1}{2}} e^{-x^2/4t}$$

and $\mu = \mu^+$. Now,

$$1 \geq u(n+\tfrac{1}{2},T) \geq \int_n^{n+1} K(n+\tfrac{1}{2}-\xi,T)d\mu(\xi) \tag{4.5}$$
$$\geq K(\tfrac{1}{2},T)\mu([n,n+1]) .$$

Hence,

$$\mu([n,n+1)) \leq K(\tfrac{1}{2},T)^{-1} \equiv M . \tag{4.6}$$

Thus, the constant M assumed in the harmonic problem has been calculated under the natural assumptions above. (If the harmonic function had been assumed nonnegative, the same calculation would have been valid.)

It is easy to see that

$$0 \leq u(x,T) - \int_{x-X_2}^{x+X_2} K(x-\xi,T)d\mu(\xi) \leq CMe^{-X_2^2/4T} . \tag{4.7}$$

Also,

$$\left|\int_{x_i-X_2}^{x_i+X_2} K(x_i-\xi,T)d\mu(\xi) - \sum_{|x_i-\xi_j|\leq X_2} K(x_i-\xi_j,T)\mu([\xi_j,\xi_{j+1}))\right| \tag{4.8}$$
$$\leq CMh.$$

Hence,

$$\left|g(x_i) - \sum_{|x_i-\xi_j|\leq X_2} K(x_i-\xi_j,T)\mu([\xi_j,\xi_{j+1}))\right| \leq \epsilon + CM(h+e^{-X_2^2/4T}) \tag{4.9}$$

Let A denote the set of sequences $\alpha = \{a_j\}$ such that

$$a_j \geq 0 , \quad \text{all } j \tag{4.10}$$
$$a_j = 0 , \quad |x_j| > X_3 = X_2 + X_4 ,$$

and set

$$u(x,t;\alpha) = \sum_{|x-\xi_j|\leq X_2} K(x_i-\xi_j,t)a_j . \tag{4.11}$$

Note that the constraint (4.6) does not appear in a discretized form in the definition of the admissible sequences A; only the nonnegativity is required. Let

(4.12) $$e(\alpha) = \inf_{|x_i|\le X_4} |g(x_i) - u(x_i,T;\alpha)| .$$

It follows from (4.9) that

(4.13) $$= \inf_{\alpha\in A} e(\alpha) \le \epsilon + CM(h + e^{-X_2^2/4T}) .$$

Let $\alpha' = \{a_j'\}$ be a solution of the optimization problem (4.13) and let $v(x,t) = u(x,t;\alpha')$. In distinction from the previous harmonic cases, $v(x,t)$ is not the desired approximation to u. Again, let us attempt to approximate u on the rectangle $R = \{|x| \le X_1,\ \delta \le t \le T\}$, noting that it is again trivial to estimate u for $t > T$.

It follows from (4.13) that

(4.14) $$\begin{aligned} 0 \le \sum_{|\xi_j - x_i|\le X_2} K(x_i-\xi_j,T)a_j' &\le |g(x_i)| + \epsilon + CM(h+e^{-X_2^2/4T}) \\ &\le 1 + 2\epsilon + CM(h+e^{-X_2^2/4T}),\ |x_i|\le X_4. \end{aligned}$$

Consequently, if $X_2 \ge 1$,

(4.15) $$\sum_{n\le\xi_j<n+1} K(x_i-\xi_j,T)a_j' \le 1 + 2\epsilon + CM(h+e^{-X_2^2/4T}) ,\ |x_i|\le X_4 .$$

Choose x_i so that $|x_i-n-\frac{1}{2}| \le \frac{1}{2}h$; then,

(4.16) $$\sum_{n\le\xi_j<n+1} a_j' \le K(\tfrac{1}{2}+\tfrac{1}{2}h,T)^{-1}[1+2\epsilon+CM(h+e^{-X_2^2/4T})] ,$$

if $[n,n+1) \subset \{|x| \le X_4+\frac{1}{2}-\frac{1}{2}h\}$; (4.16) is the discrete analogue of (4.6) and is a consequence of the procedure for $|x| \le X_4+\frac{1}{2}-\frac{1}{2}h$. This is the reason that some version of (4.6) was not imposed into the definition of A. Since X_2 grows rather slowly to enable the $CM\exp(-X_2^2/4T)$ term to become small, it is not too restrictive to require that

(4.17) $$X_5 = X_4 + \frac{1}{2} - \frac{1}{2}h - X_2 > X_1 .$$

Set

(4.18) $$a_j^* = \begin{cases} a_j' , & |\xi_j| \le X_4+\frac{1}{2}-\frac{1}{2}h , \\ 0 , & \text{otherwise} , \end{cases}$$

and let

(4.19) $$V(x,t) = \sum_{|x-\xi_j|\le X_2} K(x-\xi_j,t)a_j^* .$$

The function V is the approximate solution of the continuation problem. It follows that

(4.20) $$|V(x,T) - v(x,t)| \le CMe^{-X_2^2/4T} , \quad |x| \le X_5 .$$

If

(4.21) $$z(x,t) = \Sigma K(x-\xi_j,t)a_j^*$$

and $w(x,t) = z(x,t) - u(x,t)$, then

$$w_{xx} = w_t , \quad t > 0 ,$$

$$w(x,t) = \int K(x-\xi,t)d\nu(t) ,$$

(4.22) $$0 \le \nu([n,n+1)) \le 3M , \quad X_2^{-1} \text{ and } h \text{ sufficiently small,}$$

$$|w(x,T)| \le \gamma_1 = 2\varepsilon + CM(e^{-X_2^2/4T}+h) , \quad |x| \le X_5 ,$$

$$|w(x,T)| \le CM , \quad \text{all } x .$$

The estimation of w on R will follow the harmonic argument based on extending y to the complex domain. First, note that

$$(4.23)\quad \begin{aligned}|w(x,t)| &\le \gamma_1 + 2\int_{X_5-X_1}^{\infty} K(\xi,t-T)CMd\xi \\ &\le \gamma_1 + CM(T_1-T)^{-\frac{1}{2}}e^{-(X_5-X_1)^2/4(T_1-T)} \equiv \gamma_2 ,\end{aligned}$$

if $T < t \le T_1$, $T_1 - T < \frac{1}{2}(X_5-X_1)^2$, and $|x| \le X_1$. Let $\sigma = t + it^*$. Then,

$$(4.24)\quad |K(x,\sigma)| \le |1+it^*/t|^{\frac{1}{2}}K(x,t(1+t^{-2}t^{*2})),$$

and

$$(4.25)\quad w(x,\sigma) = \int K(x-\xi,\sigma)d\nu(\xi) , \quad \text{Re } \sigma > 0 ,$$

is holomorphic in σ in Re $\sigma > 0$ for each fixed x. It follows from (4.22) and (4.24) that

$$(4.26)\quad |w(x,\sigma)| \le CM , \quad 0 < \tfrac{1}{2}\delta \le t, \ |t^*| \le T_1^* , \ -\infty < x < \infty .$$

Let x be fixed, $|x| \le X_1$, and let $Q = \{T \le t \le T_1, 0 \le t^* \le T_1^*\}$. Then,

$$(4.27)\quad \begin{aligned}&|w(x,\tfrac{1}{2}(T+T_1)+it^*)| \le (CM)^{1-\alpha(t^*)}\gamma_2^{\alpha(t^*)} , \\ &\alpha(t^*) = \frac{\sinh \pi(T_1^*-t^*)/(T_1-T)}{\sinh \pi T_1^*/(T_1-T)} ,\end{aligned}$$

by the proof of Lemma 1. Let $Q' = \{\frac{1}{2}\delta \le t \le \frac{1}{2}(T+T_1), \ |t^*| \le T_2^*\}$, $T_2^* < T_1^*$. Then,

$$(4.28)\quad |w(x,\sigma)| \le \begin{cases} (CM)^{1-\alpha(T_2^*)}\gamma_2^{\alpha(T_2^*)} , & \sigma \in \partial Q' \cap \{t = \frac{1}{2}(T+T_1)\} \\ C(\delta,T_2^*)M , & \sigma \in \partial Q' . \end{cases}$$

Thus, by another application of the proof of Lemma 1,

$$(4.29)\qquad \begin{aligned} |w(x,t)| &\le CM^{1-\beta(t)}\gamma_2^{\alpha(T_2^*)\beta(t)} , \quad |x|\le X_1,\ \tfrac{1}{2}\delta\le t\le\tfrac{1}{2}(T+T_1) , \\ \beta(t) &= \frac{\sinh \pi(t-\frac{1}{2}\delta)/2T_2^*}{\sinh \pi(T+T_1-\delta)/4T_2^*} . \end{aligned}$$

In particular,

$$(4.30)\qquad \begin{aligned} \max_R |w(x,t)| &\le CM^{1-\beta}\gamma_2^{\beta} \\ &\le CM[\varepsilon+M(h+e^{-X_2^2/4T}+e^{-c(X_4-X_1-X_2)^2})^{\alpha}] \end{aligned}$$

for some $\alpha > 0$.

<u>Theorem 4</u>. Let $\alpha' \in A$ be a solution of the minimization problem (4.10),(4.12),(4.13), and let $u(x,t;\alpha')$ be defined by (4.11). Then, there exist positive constants α, c and C depending on the rectangle R and M such that

$$\begin{aligned} \max_R |u(x,t) - u(x,t;\alpha')| &\le C[\varepsilon+h+e^{-X_2^2/4T} \\ &\quad + e^{-c(X_4-X_1-X_2)^2}]^{\alpha} . \end{aligned}$$

The approximation of a parabolic function u from data on a finite interval,

$$(4.31)\qquad |u(x,T) - g(x)| \le \varepsilon , \quad |x| \le X_4 ,$$

can be carried out in a fashion quite similar to that for a harmonic function; however, the dependence of the solution on a rectangle $R = \{\delta\le t\le T,\ |x|\le X_1\}$ on the data is somewhat stronger in the paraboli case. Again let $\chi = x + ix^*$ and set

$$(4.32)\qquad \begin{aligned} w(\chi,t) &= \int K(\chi-\xi,t)d\nu(\xi) , \quad t > 0 , \\ K(\chi,t) &= (4\pi t)^{-\frac{1}{2}}e^{-\chi^2/4t} . \end{aligned}$$

Since $|K(\chi,t)| \le K(x,t)e^{x^{*2}/4t}$, it follows that $w(\chi,t)$ is entire in χ for each $t > 0$ and that $|w(\chi,t)| \le CM$ if $0 < \delta \le t$, $|x^*| \le X_1^*$. The relation analogous to (2.76) is

$$(4.33)\qquad \begin{aligned} |w(ix^*,T)| &\le (CM)^{1-\alpha(x^*)}\gamma_1^{\alpha(x^*)} ,\\ \alpha(x^*) &= \frac{\sinh \pi(X_1^*-x^*)/2X_4}{\sinh \pi X_1^*/2X_4} ,\end{aligned}$$

since (4.22) implies that w is bounded by γ_1 on $\{t=T, |x|\le X_4\}$. If $0 < X_2^* < X_1^*$, it again follows that

$$(4.34)\qquad \begin{aligned} |w(x,T)| &\le C\gamma_1^{\alpha(X_2^*)\beta(x)} ,\\ \beta(x) &= e^{-\pi x/2X_2^*} .\end{aligned}$$

Since X_1^* is not restricted, then the bound (4.34) can be put in the form

$$(4.35)\qquad \begin{aligned} |w(x,T)| &\le C(M,X_2^*,\eta)\gamma_1^{\alpha_1(x)} ,\\ \alpha_1(x) &= \exp\{-\tfrac{\pi}{2}(\frac{X_2^*}{X_4} + \frac{x}{X_2^*}) - \eta\} \end{aligned}$$

for any $\eta > 0$. The argument leading from (4.23) to (4.30) can be repeated with γ_1 replaced by the estimate (4.35). Hence,

$$(4.36)\qquad \max_R |w| = C[\gamma_1^{\alpha_1(x)} + e^{-c(x-X_1-X_2)^2}]^{\alpha} ,$$

$\alpha = \alpha(R) > 0$. If the quantity in the brackets is approximately minimized by equating the two terms, it can be seen that

$$(4.37)\qquad x \sim c_1 \log\log \gamma_1^{-1} ,\quad \gamma_1 \to 0 .$$

Hence,

$$\max_R |w| \le C(M,R,X_4)e^{-c(R,X_4)(\log\log\gamma_1^{-1})^2}, \tag{4.38}$$

where c and C are positive. Note that this implies, in particular, that

$$\max_R |w| \le C(\log\gamma_1^{-1})^{-M}, \quad \text{any } M, \tag{4.39}$$

if γ_1 is sufficiently small.

Theorem 5. If $\alpha' \in A$ is a solution of the minimization problem (4.10),(4.12),(4.13) for fixed $X_4 > 0$ and if $u(x,t;\alpha')$ is defined by (4.11), there exist positive constants c and C depending on R, M, and X_4 such that

$$\max_R |u(x,t) - u(x,t;\alpha')| \le C \exp\{-c[\log\log(\varepsilon + h + e^{-X_2^2/4T})^{-1}]^2\}$$

While the error bound derived in Theorem 5 is not quite as strong as Hölder continuity, it is nevertheless sufficiently strong to be of practical use.

It is clear that the results above extend to the case of several space variables.

References

1. Cannon, J. R., Some numerical results for the solution of the heat equation backwards in time, Numerical Solutions of Nonlinear Differential Equations, editted by D. Greenspan, John Wiley and Sons, Inc., New York, 1966.

2. Douglas, Jr., Jr., Approximate harmonic continuation, Atti del Convegno su le equazioni alle derivate parziale, Nervi, 25-27 febbraio 1965, Edizioni Cremonese, Roma.

3. ________________, Approximate continuation of harmonic and parabolic functions, Numerical Solution of Partial Differential Equations, editted by J. H. Bramble, Academic Press, Inc., New York, 1966.

4. ________________, The approximate solution of an unstable physical problem subject to constraints, Functional Analysis and Optimization, editted by E. R. Caianiello, Academic Press, Inc., New York, 1966.

5. Widder, D. V., Positive temperatures on an infinite rod, Transactions of the American Mathematical Society 55 (1944) 85-95.

CENTRO INTERNAZIONALE MATEMATICO ESTIVO

(C.I.M.E.)

B.E. HUBBARD

"ERROR ESTIMATES IN THE FIXED MEMBRANE PROBLEM"

Corso tenuto ad Ispra dal 3-11 Luglio 1967

ERROR ESTIMATES IN THE FIXED MEMBRANE PROBLEM

by

B.E. HUBBARD [1)]

Estimates of the discretization error in the fixed membrane problem appear to be considerably more difficult to obtain than for the Dirichlet problem. For example, the detailed estimates by Saulev [6] will illustrate this point. I shall give here a particular discrete analogue which is a natural one to consider and yet one whose error analysis is easily related to that for the Dirichlet problem. At the same time we shall considerably reduce the usual regularity assumptions in our error analyses. The results presented here are drawn, for the most part, from a paper [2] with J.H. Bramble.

Let R be a bounded domain in E_n. We shall approximate the problem

$$\Delta u(x) + \lambda u(x) = 0 \qquad x \in R$$
$$u(x) = 0 \qquad x \in \partial R$$

with eigenvalues $0 < \lambda_1 < \lambda_2 < \ldots$ and eigenfunctions $u_1, u_2, \ldots$ by those of the discrete problem

$$\Delta_h U(x) + \mu U(x) = 0 \qquad x \in R_h$$
$$U(x) = 0 \qquad x \in R_h$$

where $R_h = R'_h \cup R_h$ and ∂R_h, are those defined by Bramble [1].

The operators Δ_h is the usual $O(h^2)$ approximation at points $x \in R'_h$.

For $x \in R_h$ we define

$$V_{x^i x^{-i}}(x) \equiv h^{-2} \left[\frac{V(x + \alpha_i h_i)}{i} - \left(\frac{1}{\alpha_i} + \frac{1}{\beta_i}\right) V(x) + \frac{V(x - \beta_i h_i)}{\beta_i} \right]$$

where $h_i = 0, \ldots, h, 0, \ldots, 0$ is the vector of length h in the x^i direction and $0 < \alpha_i, \beta_i < 1$, to admit the possibility that $x + \alpha_i h_i$,

---

1) University of Maryland, supported in part by the Forschungsinstitut fur Mathematik, E.T.H. Zurich.

$x - \beta_i h_i$ are points where the mesh lines cut ∂R. If $v(x)$ is sufficiently smooth then we note that

$$\frac{\partial^2 v(x)}{(\partial x^i)^2} - \frac{2}{\alpha_i + \beta_i} v_{x^i \bar{x}^i}(x) = O(h)$$

so that, in general, the operator Δ_h defined by

$$\Delta_h V \sum_{i=1}^{n} V_{x^i \bar{x}^i}$$

is only an $O(1)$ approximation to the Laplace operator for $x \in R_h$.

For V, W functions defined at mesh points of E_n, having compact support we define

$$(V, W)_h = h^n \sum_x V(x) W(x)$$

$$D_h(V, W) = h^n \sum_x \sum_{i=1}^{n} V_{x^i}(x)\, W_{x^i}(x)$$

where the sums are taken over grid points in E_n.

It is easily seen that μ_1 satisfies the variational principle

$$\mu_1 = \min_{\substack{V(x) = 0 \\ x \notin R_h}} \frac{D_h(v, v) + h^{n-2} \sum_{x \in R_h^*} b(x)\, V(x)^2}{(V, V)_h}$$

where $b(x)$ is defined by

$$b(x) = \sum_{i=1}^{n} \left(\frac{1}{\alpha_i} + \frac{1}{\beta_i}\right) - 2n \geqslant 0 .$$

Therefore we see that the matrix of our discrete problem is symmetric, positive definite and hence, μ, U exist and $0 < \mu_1 \leqslant \mu_2 \leqslant \ldots$.

We assume the usual normalizations

$$\int_R u_k(x)^2 \, dx = (U_k, U_k)_h = 1 \quad .$$

Multiplying the equationfor u_k by U_j and summing we see that

$$(\Delta u_k, U_j)_h = - \lambda_k (u_k, U_j)_h \quad .$$

Likewise, multiplication of the equatio.ı for U_j by u_k and summing yields the second equation

$$-(\Delta_h u_k, U_j)_h = -(u_k, \Delta_h U_j)_h = \mu_j (u_k, U_j)_h \quad ,$$

while the first is the discrete analogue of Green's second identity applied to u_k, U_j vanishing outside of R_h . Adding these equations yields

$$(1) \qquad (\Delta u_k - \Delta_h u_k, U_j)_h = (\mu_j - \lambda_k)(u_k, U_j)_h \quad .$$

If we were to set $j = k$ this equation relates $(\mu_k - \lambda_k)$ to the local error, $(\Delta u_k - \Delta_h u_k)$. If this local error were uniformly $O(h^2)$ then , upon proving that for $h \leqslant h_o(k)$

$$(2) \qquad \left| (u_k, U_k)_h \right| > 1 - \delta_k > 0$$

we would have established that $(\mu_k - \lambda_k) = O(h^2)$ under the usual assumptions that $u \in C^4(R)$. Since the local error at points of R_h^* is only $O(1)$ however, we take the following additional steps. We introduce the Green's function for the operator Δ_h on R_h , which, for parametric values $y \in R_h$, satisfies the discrete equation

$$_{h,x} G_h(x, y) = h^{-n} \, (x, y) \qquad x \in R_h$$

$$G_h(x, y) = \, (x, y) \qquad x \in \partial R_h \, .$$

Many properties of $G_h(x,y)$ are discussed by Bramble [1]. For example, the discrete representation formula applied to U_j is

$$U_j(x) = h^n \sum_{y \in R_h} G_h(x,y) \left[-\Delta_h U_j(y) \right]$$

$$= \mu_j h^n \sum_{y \in R_h} G_h(y,x)\, U_j(y) \quad ,$$

after using the equation for U_j and the symmetry of G_h over the set R_h

Thus

$$(3) \qquad (\Delta u_k - \Delta_h u_k, U_j)_h = \mu_j (\Phi_k, U_j)_h$$

where Φ_k is defined by

$$(4) \qquad \Phi_k(x) = h^n \sum_{y \in R_h} G_h(x,y) \left[\Delta_h u_k(y) - \Delta u_k(y) \right] .$$

But this is the mapping $\Delta_h u_k - \Delta u_k \longrightarrow e$, of the local error to the discretization error, $e(x)$, in the Dirichlet problem. Just as in the Dirichlet problem we obtain the estimate $\Phi_k = O(h^2)$ when $u_k \in C^4(R)$ from the local estimate

$$\left| \Delta_h u_k(x) - \Delta u_k(x) \right| \leqslant \begin{cases} KM_4 h^2 & x \in R'_h \\ KM_2 & x \in R^*_h \end{cases}$$

and from the inequalities

$$h^n \sum_{y \in R_h} G_h(x,y) < K$$

$$h^n \sum_{y \in R^*_h} G_h(x,y) \leqslant K h^2 .$$

The number M_j , is a function of the derivations of u_k of order less than or equal to j .

To prove (2) and thus complete the classical error estimate for the eigenvalues we will make two assumptions

(a) $\mu_j \quad \lambda_j$ as $h \to 0$, $j = 1, 2, \ldots$

(b) λ_k is simple.

Both are made for convenience since, (a) convergence theorems exist for much more general domains than those which shall be considered here and (b) the corresponding error estimates can be obtained for multiple eigenvalues 2 , but with somewhat greater effort.

It follows from (1) and (3) that for given k

$$(5) \qquad (u_k, U_j)_h = \left[\frac{\mu_j}{\mu_j - \lambda_k} \right] (\Phi_k, U_j)_h$$

If $j \neq k$ our assumption that $\mu_j \to \lambda_j$ as $h \to 0$ implies that

$$\left[\frac{\mu_j}{\mu_j - \lambda_k} \right]^2 < \Lambda_k < \infty$$

and consequently

$$(6) \qquad (u_k, U_j)_h^2 < \Lambda_k (\Phi_k, U_j)_h^2 \qquad j \neq k .$$

Since the matrix of our problem is symmetric, and positive definite the eigenvectors $\{U_j\}$ span the space of our discrete problem and we have from Parseval's identity

$$\|u_k\|_h^2 := (u_k, u_k)_h = (u_k, U_k)_h^2 + \sum_{j \neq k} (u_k, U_j)_h^2 .$$

Then (6) implies that

(7) $$\left\|u_k\right\|_h^2 - (u_k, U_k)_h^2 < \Lambda_k \left\|\Phi_k\right\|_h^2 .$$

The inequality (2) follows immediately in the classical case since $(u_k, u_k)_h \to \int_R u_h^2 dx = 1$. This fact can be shown more directly by writing

$$\left| (u_k, U_k)_h^2 - 1 \right| < E_h(u_k) + \Lambda_k \left\|\Phi_k\right\|_h^2$$

where

$$E_h(u_k) := \left| \left\|u_k\right\|_h^2 - \int_R u_k(x)^2 \, dx \right| .$$

so that $\delta_k \to 0$ as $h \to 0$. In fact, if we choose the sign of U_k so that $(u_k, U_k)_h > 0$, then δ_k can be defined by

(8) $$\left| (u_k, U_k)_h - 1 \right| < E_h(u_k) + \Lambda_k \left\|\Phi_k\right\|_h^2 = \delta_k .$$

Setting $j = k$ in (5) leads to the first inequality of the following theorem

Theorem 1 : If λ_k is a simple eigenvalue then

$$\left| \mu_k - \lambda_k \right| < \frac{\mu_k}{1 - \delta_k} \left\|\Phi_k\right\|_h$$

$$\left\|u_k - U_k\right\|_h < \left[\Lambda_k \left\|\Phi_k\right\|_h^2 + \delta_k^2\right]^{\frac{1}{2}}$$

$$\left|u_k - U_k\right|_h < \left|\Phi_k\right|_h + K_k\left\{\left|\lambda_k - \mu_k\right| + \left\|u_k - U_k\right\|_h\right\}$$

where $\left|u_k - U_k\right|_h := \max_{x \in R_h} \left|u_k(x) - U_k(x)\right|$, and K_k is a bounded quantity.

Proof : To prove the second inequality we use (7) and (8) in the equation

$$\|u_k - U_k\|_h^2 = \|u_k\|_h^2 - 2(u_k, U_k)_h + 1$$

$$= \left[\|u_k\|_h^2 - (u_k, U_k)_h^2\right] + \left[(u_k, U_k)_h - 1\right]^2$$

To prove the third inequality we substitute $u_k(x) - U_k(x)$ into the discrete Green's formula,

$$u_k(x) - U_k(x) = h^n \sum_{y \in R_h} G_h(x,y) \left[-\Lambda_h u_k(y) + \Lambda_h U_k(y)\right]$$

$$= h^n \sum_{y \in R_h} G_h(x,y) \left\{ \left[-\Lambda_h u_k(y) + \Lambda_h u_k(y) + \right.\right.$$

$$\left.\left. + \lambda_k \; u_k(y) - U_k(y)\right] + (\lambda_k - \mu_k) U_k(y) \right\}$$

$$= -\Phi_k(x) + (\lambda_k - \mu_k) h^n \sum_{y \in R_h} G_h(x,y) U_k(y) +$$

$$+ \lambda_k h^n \sum_{y \in R_h} G_h(x,y) \;\; u_k(y) - U_k(y) \; .$$

As was mentioned in the lectures by Bramble, $G_h(x, y)$ has the following majorant

$$0 < G_h(x,y) < \begin{cases} \dfrac{1}{\nu_2} \ln \dfrac{\rho(d_o)^2}{\rho(|x-y|)^2} & n = 2 \\[2ex] \dfrac{1}{(n-2)\nu_n} \rho(\; x-y \;)^{2-n} & n \geqslant 3 \; , \end{cases}$$

where $\rho(|x|)^2 = |x|^2 + \alpha h^2$ and ν_n, α, d_o are geometrical constants. Thus if $n = 2, 3$ we see that the result of apllying Schwarz' inequality to the second and third terms of the above identity yields

(9) $$\left| u_k(x) - U_k(x) \right| < \left| \Phi_k(x) \right| + K_k \left[\left| \lambda_k - \mu_k \right| + \left\| u_k - U_k \right\|_h \right]$$

from which the final inequality in the theorem follows. If $n > 4$ the result can still be obtained by an iteration process that begins by noting that

$$h^n \sum_{y \in R_h} G(x,y)\, U_k(y) = (\mu_k)^j h^n \sum_{y_1 \in R_h} \cdots h^n \sum_{y_j \in R_h} G_h(x, y_1) \cdots$$

$$\cdots G_h(y_{j-1}, y_j)\, U_k(y_j)$$

where j is large enough so that Schwarz' inequality will give bounded quantities. Having done this we have an inequality of the form

$$u_k(x) - U_k(x) \leqslant Q + \lambda_k h^n \sum_{y \in R_h} G_h(x,y) \quad u_k(y) - U_k(y)$$

where Q is a bounded quantity, computable in terms of known quantities. By repeatedly substituting this inequality into itself we will obtain terms to which we can apply Schwarz' inequality. Thus the theorem is proved. An analogous theorem holds if λ_k is a multiple eigenvalue.

We shall now give some specific error estimates as corollaries of this theorem . The first of these gives an error estimate of the classical type.

Corollary 1 : If $\Lambda_k < \infty$ and $u \in C^4(R)$ then , for λ_k simple,

$$\left| \lambda_k - \mu_k \right| = O(h^2), \quad \left\| u_k - U_k \right\|_h = O(h^2), \quad \left| u_k - U_k \right|_h = O(h^2).$$

Proof : Just as in the Dirichlet problem it follows from (4) and estimates for the discrete Green's function that $\left| \Phi_k \right|_h = O(h^2)$. It is easily seen that $E_h(u_k) = O(h^2)$ which completes the proof.

In the remaining two corollaries we shall relate the error estimates directly to the smoothness of the boundary.

Corollary 2 : If $\partial R \in H_1(2, A, \gamma)$, i.e. if the functions which give the equation of the boundary in local coordinates have two derivatives and the second derivatives satisfy a Holder condition with exponent $\gamma > 0$ and A then

$$|\lambda_k - \mu_k| + |u_k - U_k|_h < K(k; A, \gamma)\, h^2$$

where K is a computable constant.

Proof : It is known, Gunter 3 that $u_k \in H(2, A', \gamma')$, the class of $u \in C^2(R)$ whose second derivatives satisfy a Holder condition with exponent $\gamma' \in (O, \ )$, and constant A'. Using this information and the mean value theorem for u_k it is easily established that

$$|\Delta_h u_k(y) - \Delta u_k(y)| \leqslant \begin{cases} K(k, A', \gamma') & y \in R \\ K(k, A', \gamma') d(y)^{\gamma' - 2} h^2 & y \in R' \end{cases}$$

where $d(y) = \min\limits_{x \in R} \ x - y$. In view of the inequality

$$h^n \sum_{y \in R_h} G_h(x, y)\, d(y)^{\gamma' - 2} \leqslant K(\gamma') \qquad \gamma' > 0$$

we see that $|\Phi_k| \leqslant Kh^2$. Once again $E_h(u_k) = O(h^2)$ so that the conclusion follows.

Corollary 3 : Let R be a bounded domain in E_2 whose boundary is composed of p analytic arcs. Let $\frac{\pi}{\alpha_i}$, $i = 1, \ldots, p$, the interior angles at the corners, x_i , be such that $\frac{\pi}{\alpha_i} < \frac{\pi}{\alpha} < 2\pi$. Then

$$|\mu_k - \lambda_k| + \|u_k - U_k\|_h \leqslant K(\epsilon) \left[h^{2\alpha-\epsilon} + h^2\right]$$

$$|u_k(x) - U_k(X)| \leqslant K(\epsilon) \left[h^{\alpha-\epsilon} + h^2\right] .$$

<u>Proof</u> : The work of S. Lehman 5 implies that

$$|\Delta_h u_k(x) - \Delta u(x)| \leqslant K \begin{cases} h^2 \sum_{i=1}^{p} |x-x_i|^{\alpha_i - 4} & x \in R'_h \\ \sum_{i=1}^{p} |x-x_i|^{\alpha_i - 2} & x \in R_h \end{cases} ,$$

if we require that x_i are greater than a distance ch from the mesh lines. We can investigate the behaviour of $_k(x)$ by means of the following inequalities

$$h^2 \sum_{y \in R'_h \cap S_\rho(x_i)} G_h(x,y)\, |y-x_i|^{-\beta-2} \leqslant K(\beta)\, |x-x_i|^{-\beta'}$$

$$h^2 \sum_{y \in R^*_h \cap S_\rho(x_i)} G_h(x,y)\, |y-x_i|^{-\beta} < K(\beta)\, |x-x_i|^{-\beta} h^2$$

where $0 < \beta < \beta' < \alpha_i$, and ρ sufficiently small. We note that $K(\beta) \to \infty$ as $\beta \to \alpha_i$. A direct calculation then yields for $x \in S_\rho(x_i)$

$$(10) \quad \Phi_k(x) < \begin{cases} K(\epsilon) \left[h^{2(\alpha_i - \epsilon)} |x-x_i|^{\epsilon - \alpha_i} + h^{2\alpha-\epsilon}\right]; & \frac{1}{2} < \alpha_i \leqslant 1 \\ K h^2 |x-x_i|^{\alpha_i - 2} + K(\epsilon) \left[h^{2(\alpha-\epsilon)} + h^2\right], & 1 < \alpha_i \end{cases} .$$

This estimate is consistent with those obtained by P. Laasonen [4] for the Dirichlet problem and in certain respects gives improved estimates. Clearly, $\|\Phi_k\|_h$ will have the form indicated in the theorem .

Likewise we can show that $E_h(u_k) < K(\epsilon)\left[h^{2(\alpha-\epsilon)} + h^2\right]$ so that the corollary follows. We see from (9) that $|u_k(x) - U_k(x)|$ satisfies a point dependent estimate of the type (10) .

REFERENCES

1 J. Brabmle : (C.I.M.E. Lectures, this volume)

2 J. Bramble and B. Hubbard : Effects of boundary regularity on the discretization error in the fixed membrane eigenvalue problem. (To appear)

3 Gunter : Die Potentialtheorie und ihre Anwendung auf Grundlagen der Mathematischen Physik B.G. B.G. Teubner, Leipzig (1957) .

4 P.Laasonen: On the degree of convergence of discrete approximations for the solutions of the Dirichlet problem. Ann. Acad. S i. Fenn. Series A. 246, 1-19 (1957) .

5 S.Lehman : Developments at an analytic corner of solutions of partial differential equations, J.Math.Mech .8, 727-760 (1959) .

6 V.K. Saulev : On the solution of the problem of eigenvalues by the method of finite differences, Translation in English in A.M.S. Translations (2) Vol. 8, 277-287 (1958) .

CENTRO INTERNAZIONALE MATEMATICO ESTIVO

(C. I. M. E.)

K. JORGENS

CALCULATION OF THE SPECTRUM OF A SCHRODINGER OPERATOR

Corso tenuto ad Ispra dal 3-11 Luglio 1967

CALCULATION OF THE SPECTRUM OF A SCHRODINGER OPERATOR

by

K. Jörgens

Let T denote a selfadjoint linear operator in a complex Hilbert space H, $\sigma(T)$ the spectrum of T, $\sigma_d(T)$ the discrete spectrum of T (that is the set of all isolated eigenvalues of T of finite multiplicity) and $\sigma_e(T) = \sigma(T) - \sigma_d(T)$ the essential spectrum of T. The operator T is said to be semibounded (below), if $\sigma(T)$ has a finite lower bound; in this case we denote by $\nu(T)$ and $\mu(T)$ the greatest lower bound of $\sigma(T)$ and $\sigma_e(T)$ respectively, so that $\nu(T) \leqslant \mu(T)$ holds for every semibounded operator T, and $\nu(T) < \mu(T)$ holds if and only if $\nu(T)$ belongs to $\sigma_d(T)$.

Semibounded operators play an important role in the quantum theory of systems with a finite number of degrees of freedom. The operator T of total energy of a N-particle system in the Schrödinger representation is a selfadjoint semibounded differential operator of second order in the Hilbert space $L_2(R^m)$ (with $m = 3N$). An eigenvalue $\lambda < \mu(T)$ is the total energy corresponding to a stable state of the system; in particular $\nu(T)$ (if this number is smaller than $\mu(T)$) is the enrgy of the ground state. The number $\mu(T)$ is the smallest possible energy corresponding to an unstable state of the system and $\mu(T) - \nu(T)$ is the minimal energy required to lift the system from its ground state to an unstable state.

Considerable effort in numerical analysis has been devoted to the problem of computing $\nu(T)$ and other eigenvalues of T below $\mu(T)$. There are the well-known variational procedures (Rayleigh-Ritz or Galerkin) which give upper bounds to these numbers, and more recently the methods of Weinstein, Aronszajn and Bazley (see [2] and the litera ture quoted there) for obtaining lower bounds, such that it is now possible to compute these eigenvalues up to a given error with a reasonable amount of work. On the other hand little has been done to find numerical approximations for $\mu(T)$ although this number is clearly as important as $\nu(T)$. We discuss here, for Schrödinger operators T, two methods which make it possible to compute $\mu(T)$ in many cases :

<u>The perturbation method.</u> Let T be representable as a sum $T = T_0 + A$, where T_0 is selfadjoint and semibounded and A is T_0-compact (that is the domain $D(A)$ of A contains $D(T_0)$ and every sequence $\{u_n\} \subset D(T_0)$ with $\|u_n\| + \|T_0 u_n\| < C$ contains a subsequence $\{u_{n_j}\}$ such that $\{Au_{n_j}\}$ converges strongly in H). Then it follows that $\sigma_e(T) = \sigma_e(T_0)$, and in particular $\mu(T) = \mu(T_0)$. The method consists in finding a representation of this sort such that $\mu(T_0)$ is known or at least is computable by known methods.

<u>The method of splitting.</u> Let $m = \sum_{j=1}^{N} m_j$, $x = (\bar{x}_1, \bar{x}_2, \ldots, \bar{x}_N)$ with $\bar{x}_j \; R^{m_j}$ for $x \; R^m$. Let

$$T = \sum_{j=1}^{N} T_j + \sum_{j<k} p_{jk} \tag{1}$$

where T_j is a Schrödinger operator in $L_2(R^{m_j})$ acting on functions of the variable $\bar{x}_j$ and p_{jk} the operator of multiplication by a real function $p_{jk}(\bar{x}_j, \bar{x}_k)$. Now consider the operators

$$S_i = \sum_{j \neq i} T_j + \sum_{\substack{j,k \neq i \\ j<k}} p_{jk} \tag{2}$$

acting on functions in $L_2(R^{m-m_i})$ for $i = 1, 2, \ldots, N$. Then under certain hypotheses we have the formula

$$\mu(T) = \min_{(i)} \nu(S_i) + \mu(T_i) \;) \tag{3}$$

which reduces the problem to the (supposedly easier) computation of the numbers $\mu(T_i)$ and to the well-solved problem of computing the numbers (S_i).

Schrödinger operators. We consider operators of the form

$$T_o = \sum_{j=1}^{m} (i\partial_j + a_j)^2 + p$$

in $L_2(R^m)$, where $\partial_j = \frac{\partial}{\partial x_j}$ and a_j and p are functions on R^m

satisfying the following conditions :

(4) $$\begin{cases} a_j \text{ and } p \text{ are real}, \quad (a_j)^2 \in L_{2,loc}(R^m), \\ p \in L_{2,loc}(R^m), \quad \sum_{j=1}^{m} \partial_j a_j = 0 \quad \left(\begin{matrix}\text{in the sense of}\\ \text{distributions}\end{matrix}\right) \end{cases}$$

Then the operator T_o is defined and symmetric on $C_o^\infty(R^m)$ and can also be written as

$$T_o = -\Delta + 2i \sum_{j=1}^{m} a_j \partial_j + \sum_{j=1}^{m} (a_j)^2 + p$$

We now use a uniqueness theorem which is a slight extension of a special case of a theorem of Ikebe and Kato [4].

Theorem 0 (Ikebe-Kato). Let the coefficients a_j and p satisfy (4) and

(5) $$a_j \in C^1(R^m) \quad \text{and}$$

(6) $$\int_{|x-y|<1} |p(y)|^2 \, |x-y|^{4-m-\alpha} dy \leqslant C$$

for some $\alpha > 0$, $C > 0$ and for all $x \in R^m$. Then T_o (defined in $C_o^\infty(R^m)$) has a unique selfadjoint extension (also denoted by T_o) with domain

$$D(T_0) = \left\{ u \;\middle|\; u \in H^2_{loc}(R^m) \cap L_2(R^m),\ T_0 u \in L_2(R^m) \right\}.$$

Furthermore T_0 is semibounded below.

<u>The perturbation method.</u> We consider perturbations of T_0 of the form

$$(7) \qquad T = \sum_{j=1}^{m} (i\partial_j + a_j + b_j)^2 + p + q$$

where the coefficients a_j and p of T_0 are as in theorem 0 and the perturbed coefficients $a_j + b_j$ and $p+q$ satisfy (4) (when substituted for a_j and p respectively). Then T is defined and symmetric in $C_0^\infty(R^m)$ and can be written as $T = T_0 + A$ with

$$A = 2 \sum_{j=1}^{m} b_j(i \;_j + a_j) + \sum_{j=1}^{m} (b_j)^2 + q$$

<u>Theorem 1.</u> Let a_j and p satisfy the assumptions of theorem 0 and assume that

(i) the derivatives $\partial_j a_k$ are uniformly bounded

(ii) the functions $\sum_{j=1}^{m} (b_j)^2$ and q satisfy condition (6)

(when substituted for p) with some $\alpha > 0$ and $C > 0$.

Then T has a unique selfadjoint extension with domain $D(T) = D(T_0)$ which is semibounded. If in addition the function

$$(iii) \qquad M(x) = \int_{|x-y| \leq 1} \left[\left(\sum_{j=1}^{m} b_j^2(y) \right)^2 + q^2(y) \right] dy$$

tends to zero for $|x| \to \infty$, then A is T_0-compact.

This theorem and some similar ones are contained in [5]. The most remarkable feature of these results is that the coefficients of the unperturbed operator T_0 need not be constant or even bounded. For operators T_0 with constant coefficients Birman [3] and Balslev [1] have obtained very general perturbation theorems using the theory of quadratic forms in Hilbert space .

The proof of theorem 1 rests on an identity of the form

$$u = K_\rho(T_o - p)\, u + H_\rho\, u$$

for functions $u \in H^2_{loc}(R^m)$, where K_ρ and H for every $\rho \in (0, 1$ are integral operators with kernels $K_\rho(x,y)$ and $H_\rho(x,y)$ identically zero for $|x-y| > \rho$ and satisfying the inequalities

$$|K_\rho(x,y)| \leq C\,|x-y|^{2-m}$$

$$|(i\partial_j + a_j)_x K_\rho(x,y)| \leq C\,|x-y|^{1-m}$$

$$|H_\rho(x,y)| \leq C\left[{}^{-m} + |x-y|^{2-m}\right]$$

$$|(i\partial_j + a_j)_x H_\rho(x,y)| \leq C\left[\rho^{-m-1} + |x-y|^{1-m}\right]$$

for $|x-y| \leq \rho$ with a constant C independent of x, y and ρ .

Example 1. $T_o = -\Delta$ (that is $a_j = p = 0$). It is well known that $\sigma(T_o) = \sigma_e(T_o) = (0, \infty)$ in this case. By theorem 1 the operator

$$T = \sum_{j=1}^{m} (i\partial_j + b_j)^2 + q$$ has the essential spectrum $\sigma_e(T) = (0, \infty)$ and therefore $\mu(T) = 0$.

Example 2. Let $m = 3, x = (x_1, x_2, x_3)$ and $a_1(x) = -\frac{1}{2}\beta x_2$, $a_2(x) = \frac{1}{2}\beta x_1$, $a_3(x) = 0$ and $p(x) = 0$. The operator T_o with these coefficients describes the motion of a charged particle under the influence of a homogeneous magnetic field of strength β in the x_3-direction. The spectrum of T_o can be found by a separation of variables (see [5]); we get $\sigma(T_o) = \sigma_e(T_o) = (|\beta|, \infty)$. By theorem 1 any perturbation A satisfying the assumptions of the theorem does not change the essential spectrum. In particular we can take $b_j = 0$ for $j = 1, 2, 3$ and

$q(x) = -\gamma|x|^{-1}$; the perturbed operator $T = T_o + q$ now corresponds to the so-called Zeeman-effect in quantum theory. According to theorem 1 we have $\sigma_e(T) = (|\beta|, \infty)$; furthermore it is possible to show (in the case where $\gamma > 0$) by a Ritz-type variational argument, that the discrete spectrum of T consists of infinitely many eigenvalues below $\mu(T) = |\beta|$ (see [5]) .

The method of splitting. We start with the following example :

Example 3. The Schrödinger operator for a system of N particles of equal charge and mass in a central Coulomb field. Here we have $m = 3N$ and $x = (\bar{x}_1, \bar{x}_2, \dots, \bar{x}_N)$ where $\bar{x}_j \in R^3$ is the position of the j-th particle. The operator is

$$T = -\Delta - \beta \sum_{j=1}^{N} |\bar{x}_j|^{-1} + \gamma \sum_{j<k} |\bar{x}_j - \bar{x}_k|^{-1} \tag{8}$$

with positive constants β and γ. Theorem 1 applied to this operator (with $T_o = -\Delta$) only gives that T has a unique selfadjoint and semi-bounded extension. Condition (iii) is not satisfied, and in fact we shall show that $\mu(T)$ can be computed with the help of formula (3) and is found to be smaller than $\mu(T_o) = 0$ for $N \geq 2$.

Theorem 2. Let T_j for $j = 1, 2, \dots, N$ be operators of the form (7) in $L_2(R^{m_j})$ with coefficients satisfying the assumptions of theorem 1. Denote by $\bar{x}_j$ the variables in T_j and put $m = \sum_{j=1}^{N} m_j$ and $x = (\bar{x}_1, \bar{x}_2, \dots, \bar{x}_N)$. Let p_{jk} be non-negative functions defined on $R^{m_j+m_k}$, satisfying (6) with $m = m_j + m_k$ for some positive constants α and C, and such that the functions

$$M_{jk}(\bar{x}_j, \bar{x}_k) = \int_{|\bar{x}_j - \bar{y}_j| \leq 1} \int_{|\bar{x}_k - \bar{y}_k| \leq 1} |p(\bar{y}_j, \bar{y}_k)|^2 \, d\bar{y}_j d\bar{y}_k$$

tend to zero as $|\bar{x}_j| \to \infty$ for fixed $\bar{x}_k$ and as $|\bar{x}_k| \to \infty$ for fixed $\bar{x}_j$. Then the operators (1) and (2) in $C_o^\infty(R^m)$ and $C_o^\infty(R^{m-m_i})$ respectively have unique selfadjoint and semibounded extensions and formula (3) is valid. If $\sigma_e(T_i)=(\mu(T_i),\infty)$ for the index i giving the minimum in (3), then we also have $\sigma_e(T)=(\mu(T),\infty)$.

A theorem of this type has for the first time been proved by M. G. Zislin [8] for an operator somewhat more general than (8). Theorem 2 above is contained in [6]. Related results can be found in the papers of C. van Winter [7]; here the interaction terms p_{jk} are not assumed to be non-negative and consequently formula (3) has to be replaced by a more complicated one. Finally Zislin and Sigalov [9] have refined the theorem for the operator (8) by taking into account its symmetry with respect to permutation of the variables $\bar{x}_j$. They prove a formula similar to (3) for the restriction of T to the invariant subspace of $L_2(R^m)$ corresponding to an irreducible representation of the symmetric group S_N. A summary of this work and of related results has been given by Sigalov in his recent article [10].

Theorem 2 clearly applies to example 3. Denote the operator in question by T^N; then $T_j = T^1 = -\Delta - \beta|x|^{-1}$ in $L_2(R^3)$ and $\sigma_e(T^1) = (0,\infty)$ by theorem 1. T^1 is the well-known operator for a hydrogen-like atom; its lowest eigenvalue is $\nu(T^1) = -\frac{1}{4}\beta^2$. The operators S_i corresponding to T^N are all identical (except for the numbering of variables) and equal to T^{N-1}. Therefore by theorem 2 we have $\sigma_e(T^N) = (\nu(T^{N-1}),\infty)$. It follows that $\mu(T^N) = \nu(T^{N-1}) \leq \mu(T^{N-1}) \leq$ $\leq -\frac{1}{4}\beta^2$ for $N > 2$. If $\beta \geq N\gamma$ Zislin [8] has shown that $\nu(T^N) < \mu(T^N)$; in this case we have $\mu(T^{M+1}) < \mu(T^M)$ for $M = 1, 2, \dots, N$.

REFERENCES

1 Balslev, E. : The singular spectrum of elliptic differential operators in $L^p(R_n)$. Math. Scand. 19 (1966), 193-210.

3 Birman, M. S. : On the spectrum of singular boundary value problems (russian). Math. Sbornik 97 (1961), 125-174.

2 Bazeley, N. W. : Lower bounds for eigenvalues. J. Math. and Mech. 10 (1961), 289-308.

4 Ikebe, T. and T. Kato: Uniqueness of the self-adjoint extension of singular elliptic differential operators. Arch. Rat. Mech. Analysis 9 (1962), 77-92.

5 Jorgens, K. : Zur Spektraltheorie der Schrodinger-Operatoren. Math. Zeitschr. 96 (1967), 355-372.

6 Jorgens, K. : Uber das wesentliche Spektrum elliptischer Differentialoperatoren vom Schrodinger-Typ. Report, Inst. f. Angew. Math. Univ. Heidelberg (1965).

7 van Winter, C. : Theory of finite systems of particles; I the Green function, II scattering theory. Math.-Phys. Publications of the Danish Royal Academy of Sciences, Volume 2 (1965).

8 Zislin, M. G. : On the spectrum of many-particle Schrödinger operators (russian). Trudy Mosk. Mat. Obsc. 9 (1960), 81-120.

9 Zislin, M. G. and A. G. Sigalov : On the spectrum of the energy operator of atoms on subspaces corresponding to irreducible representations of the permutation group (russian). Izvestija Akad. Nauk SSSR 29 (1965), 835-860.

10 Sigalov, A. G. : On an important mathematical problem in the theory of atomic spectra. (russian). Uspechi Mat. Nauk 22 (1967), 3-20.

CENTRO INTERNAZIONALE MATEMATICO ESTIVO

(C.I.M.E.)

A. LASOTA

CONTINGENT EQUATIONS AND BOUNDARY VALUE PROBLEMS

Corso tenuto ad Ispra dal 3-11 Luglio 1967

CONTINTENT EQUATIONS AND BOUNDARY VALUE PROBLEMS

by

A. LASOTA (Krakow)

It is known that the theory of differential equations with multi-valued right-hand sides (sometimes also called contingent equations) is closely related to the optimal control theory. Our aim is to show that the technique of the contingent equations is useful in the theory of continuous and discrete boundary value problems too. In section 1 we start wtih a new topological method which permits to establish the existence of solutions provided a criterion of uniqueness is fulfilled. Section 2 contains a result of A. Pliś concerning contingent equations [5]. In section 3 and 4 existence theorems for continuous and discrete boundary value problems are given. The main result, an approximation theorem, is stated in section 5 .

1. Fixed point theorem. Let E a Banach space and let n(E) denote the family of all nonempty subsets of E. For a set A in n(E), a mapping $H: A \to n(E)$ is called upper semicontinuous if its graph

$$\{ (x, y): \quad y \in h(x) \}$$

is closed in $A \times E$. The map H is said to be compact if, for any bounded subset B of A, the closure of the set

$$\bigcup_{x \in B} H(x)$$

is compact in E. The map H is called completely continuous if it is upper semicontinuous and compact.

For a single valued mapping $h: A \to E$ the upper semicontinuity of the mapping $H: x \to h(x)$ is equivalent to the continuity of h, the compactness of H is equivalent to that of h and the complete continuity of H means the complete continuity of h.

The following theorem is due to A. Lasota and Z. Opial [4].

Theorem 1. Let U be a neighborhood of 0 in the space E and let $H: U \dashrightarrow n(E)$ be a completely continuous mapping such that

$$(1.1) \qquad x \in H(x), \quad x \in U \Rightarrow x = 0,$$

$$(1.2) \qquad H(0) = 0 .$$

Then for any mapping $h: E \to E$, the condition

$$(1.3) \qquad h(x) - h(y) \in H(x-y) \qquad \text{for } x-y \in U$$

implies that the equation

$$(1.4) \qquad x = h(x)$$

has exactly one solution.

The special case of theorem 1 was presented at the second conference on differential equations (Equadiff II) in Prague 1966.

The proof of theorem 1 is based on the old idea of Hadamard to use the monodromy principle.

If h is a mapping of the form $h(x) = Ax + b$ where $b \in E$ and A is a linear completely continuous operator $(A: E \to E)$, theorem 1 yields the essential part of the first theorem of Fredholm. It suffices to set $H(x) = \{Ax\}$ and to observe that condition (1.1) means the uniqueness of solution of the linear homogeneous equation $x = Ax$. Note that in the Fredholm theorem the map A may be non-contractive in general.

2. Plis lemma. Let $\mathcal{J} = [a, b]$ be a compact interval of the real line R and let $cf(R^n)$ denote the family of all nonempty, convex, closed subsets of the n-dimensional real space R^n. The space of all continuous mappings of $\mathcal{J}$ into R^n with the usual norm

$$\|x\| = \max_{\mathcal{J}} |x(t)|$$

($|.|$ stands for the Euclidean norm in R^n) will be denoted C^n. By $\varrho(p, A)$ $(p \in R^n, A \in n(R^n))$ we denote the Euclidean distance of the point p to the set A.

Lemma 1. Let G be a mapping of $\mathcal{J}$ into $cf(R^n)$ and let a sequence of $\{v_k\} \subset C^n$ of absolutely continuous functions satisfy the following conditions

$$\lim_{k \to \infty} v_k(t) = v(t) \quad \text{for } t \in \mathcal{J} \text{ almost everywhere (a.e.)} \tag{2.1}$$

$$|v_k'(t)| \leq h(t) \quad \text{a.e.}, \quad \int_a^b h(t)\, dt < +\infty \tag{2.2}$$

$$\lim_{k \to \infty} \varrho(v_k'(t), G(t)) = 0 \quad \text{a.e.} \tag{2.3}$$

Then the function v is absolutely continuous and

$$v'(t) \in G(t) \quad \text{a.e.} \tag{2.4}$$

3. Continuous boundary problems

Now pass to the differential equations. Consider a map $F: \mathcal{J} \times R^n \to cf(R^n)$, a map $f: \mathcal{J} \times R^n \to R^n$ and $L: C^n \to R^n$ such that:

(i) $F(t, x)$ is homogeneous in x $(F(t, \lambda x) = \lambda F(t, x)$ for real $\lambda)$ and completely continuous, i.e. the set

$$\{(t, x, y) : y \in F(t, x), |x| = 1\}$$

is compact in R^{2n};

(ii) $f(t, x)$ is continuous in (t, x) and satisfies the condition

$$f(t, x) - f(t, y) \in F(t, x - y) \quad (t, x, y) \in \mathcal{J} \times R^{2n}; \tag{3.1}$$

(iii) L is linear and continuous.

Given the mapping F, f, L and a point $r \in R^n$, we consider the contingent equation

$$x'(t) \in F(t, x(t)) \tag{3.2}$$

with the homogeneous linear condition

(3.3) $$Lx = 0$$

and the ordinary differential equation

(3.4) $$x'(t) = f(t, x, (t))$$

with the linear condition

(3.5) $$Lx = r .$$

An absolutely continuous function $x \in C^n$ will be called a solution (in Carathéodory sense)˙ of equation (3.2) if it satisfies condition (3.2) almost everyahere on $\mathcal{J}$.

From the theorem 1 we can derive the following (see [2], [3]).

Theorem 2. If the functions F, f, L satisfy conditions $1^o, 2^o, 3^o$ and if $x = 0$ is the unique solution of equation (3.2) satisfying (3.3) then there exists one and only one solution of the equation (3.4) satisfying (3.5).

Note that the theorem 2 is not true if to mean solution of (3.2) in the usual (non Carathéodory) sense.

In order to prove theorem 2 consider the mapping H of $E = C^n \times R^n$ into $cf(R^n)$ such that for every point (x, p) its image $H(x, p)$ is a set of all pairs (y, q) given by the formulae

$$y(t) = \int_a^t u(s)\, ds + p, \qquad q = p - Lx$$

for $u(s) \in F(s, x))$ and the mapping h of E into itself such that for every point (x, p) its image $h(x, p)$ is a pair (y, q) given by the formulae

$$y(t) = \int_a^t f(s, x(s))\, ds + p , \qquad q = p + r - Lx .$$

It is easy to see that the maps H, h satisfy conditions (1.1), (1.2), (1.3) and that the existence of solutions of problem (3.4), (3.5) is equivalent to the existence of solutions of the functional equation (1.4). Map H is evidently compact. Thus to end the proof it is sufficient to show that H is upper semicontinuous. Suppose that

$$(y^{k'}, q^k) \in H(x^k, p^k), \quad (x^k, p^k) \to (x, p), \quad (y^k, q^k) \to (y, q).$$

Then we have

(3.6) $$y^k(t) = \int_a^t u^k(s)\,ds + p^k, \qquad u^k(t) \in F(t, x^k(t)),$$

(3.7) $$q^k = p^k + r - Lx^k$$

and consequently

(3.8) $$(y^k(t))' \in F(t, x^k(t)).$$

From (3.8) and the upper semicontinuity of F we obtain

$$\lim_{k \to \infty} \rho\left((y^k(t))', F(t, x(t))\right) = 0.$$

Thus from the lemma 1 it follows immediately that

(3.9) $$y'(t) \in F(t, x(t))$$

In addition, upon passing to the limit in (3.6) (for t = 0) and (3.7) we have

(3.10) $$y(a) = p \qquad q = p + r - Lx$$

and consequently

(3.11) $$y(t) = \int_a^t y'(t)\,dt + p.$$

From (3.9), (3.10) and (3.11) it follows that (y, q) H(x, q) and we are done.

4. Discrete boundary problems. In order to obtain a discrete analogue of theorem 2 consider a finite sequence

$$a = t_o < t_1 < \dots < t_n = b \quad .$$

Then we can replace the contingent equation (3.2) by the multivalued difference equation

$$(4.1) \qquad \frac{\Delta x_i}{\Delta t_i} \in F(t_i, x(t_i)) \qquad i = 0, \dots, n-1$$

and the differential equation (3.4) by the difference equation

$$(4.2) \qquad \frac{\Delta x_i}{\Delta t_i} = f(t_i , x(t_i)) \qquad i = 0, \dots, n-1 .$$

where $\Delta x_i = x_{i+1} - x_i$ and $\Delta t_i = t_{i+1} - t_i$.

We assume that solutions of equations (4.1) and (4.2) are continuous in $\mathcal{J}$ and linear in each interval $[t_i , t_{i+1}]$.

Theoreme 3. If the mappings F, f, L satisfy conditions (i) , (ii) , (iii) and if $x = 0$ is the unique solution of equation (4.1) satisfying (3.3) , then there exists one and only one solution of (4.2) satisfying (3 .5)

More exstensive theorems and some applications in the theory of discrete boundary value problems are due to F?H. Szafraniec [6] . The proof of theorem 3 is based on the same idea as the proof of theorem 2. But in this case the lemma 1 is not needed.

5. Approximation theorem. The development of the theory of differential inequalities allows to obtain precise uniqueness theorems for a general class of differential boundary value problems. It is interesting that on the same way we can establish the existence of solutions and the convergence of finite difference approximations. Namely setting $\tau = (t_o, \dots, t_n)$ and $\delta(\tau) = \max \{ t_{i+1} - t_i : i = 0, \dots , n-1 \}$ we have the following

Theorem 4. If the mappings F, f, L satisfy (i) , (ii) , (iii) and if $x = 0$ is the unique solution of the contingent boundary value problem

(3.2), (3.3) then :

1) there exists a unique solution x^o of problem (3.4), (3.5),

2) for $\delta(\tau)$ sufficiently small there exists a unique solution x^{τ} of problem (3.5), (4.2),

3) $\lim_{\delta(\tau)\to 0} \| x^{\tau} - x^o \| = 0$.

The statement 1° is an immediatel consequence of theorem 2. In order to prove 2° it is sufficient to show that for sufficiently small $\delta(\tau)$ problem (4.1), (3.3) has only the trivial solution $x = 0$ and to use theorem 3. To this end suppose that for each integer k there exists a sequence

$$\tau^k = (t_1^k, \ldots, t_{n_k}^k, \qquad \delta(\tau^k) < 1/k$$

and a corresponding nontrivial solution x^k of problem (4.1), (3.3), i.e.

$$(5.1) \qquad \frac{\Delta x_i^k}{\Delta t_i^k} \in F(t_i^k, x^k(t_i^k)) \qquad i = 0, \ldots, n_k - 1$$

$$(5.2) \qquad Lx^k = 0 .$$

Since x^k is linear in each interval (t_i^k, t_{i+1}^k) we can write (5.1) in the form

$$(5.3) \qquad (x^k(t))' \in F(t_i^k, x^k(t_i^k)) \qquad t \in (t_i^k, t_{i+1}^k)$$

and since problem (5.1), (5.2) is homogeneous we can assume that

$$(5.4) \qquad \| x^k \| = 1 .$$

From this and (5.3) it follows that the functions x^k are equiconti- nuous on J . Upon passing to suitable subsequence, if necessary, we may assume that the sequence $\{x^k\}$ convergences to a function $x \in C^n$. Thus passing to the limit in (5.2) , (5.3) and (5.4) we obtain (see lemma 1)

$$x'(t) \in F(t, x(t)) \quad , \quad Lx = 0 \, , \quad \|x\| = 1$$

which is impossible.

To prove statement 3° consider a sequence $\{\tau^k\}$ such that $\delta(\tau^k) \to 0$ and the corresponding sequence $\{x^k\}$ of solutions of (4.2) , (3.5) . We have

$$\frac{\Delta x_i^k}{\Delta t_i^k} = f(t_i^k, x^k(t_i^k)) \, , \qquad L x^k = r$$

and on other hand

$$(x^o(t))' = f(t, x^o(t)) \, , \qquad Lx^o = r \, .$$

From this and (3.1) it follows that

$$(u^k(t)) \in F(t_i^k, u^k(t_i^k)) + \mathcal{E}_k(t) \quad \text{for} \quad t \in (t_i, t_{i+1}) \, , \quad Lu^k = 0$$

where $u^k = x^k - x^o$ and

$$\mathcal{E}_k(t) = f(t_i^k, x^o(t_i^k)) - f(t, x^o(t))$$

Now suppose that $\|u^k\| \not\to 0$. Then upon passing to the suitable subsequence we may assume that $\|u^k\| \to c$ $(c \in (0, +\infty)$) . Setting $v^k = u^k \|u^k\|^{-1}$ we obtain

$$(v^k(t))' \in F(t_i^k, v^k(t_i^k)) + \varepsilon_k(t) \|u^k\|^{-1} \, , Lv^k = 0, \; \|v^k\| = 1.$$

The functions v^k are evidently equicontinuous and we may also assume that the sequence $\{v^k\}$ converges to a function $v \in C^n$. The lemma 1 yields

$$v'(t) \in F(t, v(t))$$

and obviously $Lv = 0$, $\|v\| = 1$ which is impossible. Thus $\|u^k\| \to 0$ and we are done.

In order to illustrate theorem 4 by an example, consider for the equation (3.4) the following aperiodic boundary value condition

(5.5) $$x(a) + \lambda x(b) = r \qquad \lambda > 0 .$$

We start with a lemma concerning differential inequality

(5.6) $$|x'| < \varphi(t) |x|$$

Lemma 2. If the functions $\varphi(t)$ is nonnegative and such that

(5.7) $$\int_a^b \varphi(t)\, dt < \sqrt{\pi^2 + \log^2 \lambda}$$

then $x = 0$ is the unique solution of the inequality (5.6) satisfying the boundary value condition

$$x(a) + \lambda x(b) = 0 .$$

Note that the estimation (5.7) is the best possible of this type. The lemma 2 is due to S. Kasprzyk and J. Myjak [1].

From theorem 4 and lemma 2 it follows immediately

Theorem 5. If the function $f(t,x)$ is continuous and if

(5.8) $$|f(t,x) - f(t,y)| \leqslant \varphi(t) |x - y|$$

where the continuous nonnegative function φ satisfies inequality (5.7) then

1° there exists a unique solution of (3.4), (5.5);

2° for sufficiently small $\delta(\tau)$ there exists a unique solution x of (4.2), (5.5);

3° $$\lim_{\delta(\tau)\to 0} \| x^{\tau} - x^{o} \| = 0 .$$

REFERENCES

1 S. Kasprzyk, J. Myjak, On the existence and uniqueness of solutions of Floquet boundary value problem , Zeszyty Naukowe U. J., Prace Matematyczne (to appear) .

2 A. Lasota , Une generalisation du premier theorem de Fredholm et ses application à la theorie des equations differentielles ordinaires, Ann. Polon. Math. 18(1966) , 65-77.

3 A. Lasota, Z. Opial, On the existence of solutions of linear problems for ordinary differential equations , Bull. Acad. Polon. Sci., Sér. sci. math. , astronom. et phys. , 14(1966) , 371-376.

4 A. Lasota, Z. Opial, On the existence and uniqueness of solutions of nonlinear functional equations, ibidem (to appear) .

5 A. Plis, Mesurable orientor fields, ibidem, 13(1965) , 565-569.

6 F. H. Szafraniec, Existence theorems for discrete boundary problem, Ann. Polon. Math. (to appear) .

CENTRO INTERNAZIONALE MATEMATICO ESTIVO
(C.I.M.E.)

REDUCTION A DES PROBLEMES DU TYPE CAUCHY-KOWALESKA

Corso tenuto ad Ispra dal 3-11 Luglio 1967

REDUCTION A DES PROBLEMES DU TYPE CAUCHY-KOWALESKA

par

J.L. Lions

Université de Paris

Introduction.

Il intervient assez souvent en pratique des systèmes différentiels qui ne sont pas du type de Cauchy-Kowaleska : en hydrodynamique les équations de Navier Stokes , les équations de la magnéto-hydrodynamique, en metéorologie (Marchouk [2]) etc.....

On donne ici deux procèdés , de portée semble-t-il assez génerale, pour approcher de tels problèmes par des problèmes du type Cauchy - Kowaleska qui sont eux mêmes ensuite approchés par des methodes numériques "standard".

Le premier procéde, dit de pénalisation , est une variante d'un procédé introduit par Courant [1] en calcul des variations es d'usage maintenant fréquent en contrôle optimal ; nous donnons au N. 1 la méthode sur l'exemple des équations de Navier-Stokes, suivant Temam [1],[2].

Une idée voisine avait déja été introduite pour les équations linéaires dans Bramble-Payne [1] . Une variante a été utilisée dans Lattès - Lions [1] pour l'approximation numérique de problèmes mal posés .

Le deuxième procédé, dit des dérivées artificielles, est exposé sur deux exemples : au n. 2 sur l'exemple des équations de Navier Stokes et au n. 3 sur un problème linéaire quelque peu "singulier" - mais qui intervient dans les applications (cf. Garipov [1] un un problème relevant de techniques analogues dans Douglas- Peceman-Rachford [1]) .

1. Equations de Navier-Stokes ; Méthode de pénalisation.

1.1 - Position du problème [1]

Soit Ω un ouvert borné de R^n, $n = 2$ ou 3 de frontière Γ assez régulière.

On cherche une fonction $u = \{u_1, \ldots, u_n\}$ et une fonction p definies dans le cylindre $Q = \Omega \times]0, T[$ telles que

$$\frac{\partial u}{\partial t} - \nu \Delta u + \sum_{i=1}^{n} u_i \frac{\partial u}{\partial x_i} + \text{grad } p = f \quad \text{dans } Q, \quad (\nu > 0) \tag{1.1}$$

$$\text{div } u = 0 \quad \text{dans } Q, \tag{1.2}$$

$$u(x,t) = 0 \quad \text{si } x \in \Omega, \; t > 0, \tag{1.3}$$

$$u(x,0) = u_0(x) \quad \text{dans } \Omega \tag{1.4}$$

où $f = \{f_1, \ldots, f_n\}$ et u_0 sont données dans Q et Ω.

Naturellement il faut préciser à quel sens on cherche une solution de (1.1)–(1.4) ; mais il est de toutes facons clair que, à cause de (1.2) (qui ne contient pas de dérivée $\frac{\partial p}{\partial t}$) le problème mixte (1.1) ... (1.4) n'est pas du type de Cauchy -Kowaleska. Cela conduit à des difficultés numériques : lorsque l'on passe par ex. aux differénces finies, les conditions (1.2) introduisant des "contraintes" compliquant sérieusement la mise en oeuvre des calculs.

Notre objet est donc ici d'approcher le système (1.1)...(1.4) par un système de Cauchy-Kowaleska.

Cela peut se faire grosso modo de deux facons [2] :

(i) par "suppression" de la "contrainte" (1.2) ;

(ii) par introduction d'un "terme artificiel" $\varepsilon \frac{\partial p}{\partial t}$.

---

[1] Cf. aussi les conférences de C. Capriz, ce Volume.

[2] A notre connaissance . Il y a tres probablement beaucoup d'autres procédés.

La "méthode" (i) (1) fait l'objet de ce N. la méthode (ii) [1] fait l'objet du n.2 suivant.

1.2. Pénalisation.

L'idée générale est de supprimer (1.2) et d'ajouter dans (1.1) un terme de la forme

$$-\frac{1}{\varepsilon} \text{ grad } (\text{div } u) \quad ,$$

$\varepsilon > 0$ "petit" - ce qui conduit à une fonction u ne vérifiant plus (1.2) mais à la place de (1.2) à une condition

$$\|\text{div } u\|_{L^2(Q)} \sim \sqrt{\varepsilon} \; . \tag{1.5}$$

Precisons cela - Introduisons dans ce but les outils suivants :

$$V = \left\{ v \;\middle|\; v \in (H_0^1(\Omega))^n, \quad \text{div } v = 0 \right\},\ ^{(2)}$$

$$H = \left\{ f \;\middle|\; f \in (L^2(\Omega))^n, \quad \text{div } f = 0 \right\} ;$$

V et H sont des espaces de Hilbert sur R [3], pour les produits scalaires

$$((u, v)) = \sum_{i,j=1}^{n} \int_\Omega \frac{\partial u_i}{\partial x_j} \frac{\partial v_i}{\partial x_j} \, dx \quad ,$$

$$(f, g) = \sum_{i=1}^{n} \int_\Omega f_i \, g_i \, dx$$

dont les normes correspondantes sont notées $\|u\|$, $|f|$.

---

[1] Convenablement précisée ! !

[2] $H_0^1(\Omega)$ est défini dans les conférences de Raviart, ce vol.

[3] Toutes les fonctions sont à valeurs réelles

Pour u, v, w $\in$ V, on pose

$$(1.6) \qquad b(u,v,w) = \sum_{i,j=1}^{n} \int_{\Omega} u_i \frac{\partial v_j}{\partial x_i} w_j \, dx \quad {}^{(1)}$$

Le problème (1.1) ...(1.4) peut alors se formuler de la façon suivante [2] : trouver $u \in L^2(0, T; V)$ [3] tel que

$$(1.7) \qquad \frac{d}{dt}(u(t), v) + \nu((u(t), v)) + b(uu(t), v) = (f(t), v), \quad v \in V,$$

$$(1.8) \qquad u(0) = u_o \quad {}^{(4)}$$

<u>Remarque</u> 1.1.

La condition "div u = 0" est <u>contenue</u> dans l'appartenance à V et la pression p n'a "disparu" que parce que l'on a pris le produit scalaire avec v de divergence nulle (et nul au bord).

On montre ceci (Leray [1], Ladyzenskaya [1], Lions [1], Lions-Prodi [1]) :

a) il existe une solution u de (1.7) (1.8), vérifiant en outre

$$(1.9) \qquad u \in L^{\infty}(o, T; H)$$

b) <u>si</u> n = 2, il y a unicité de u dans $L^2(0, ;V)$ vérifiant (1.7) (1.8) [5]

<u>On va maintenant approcher</u> u (dans un sens convenable) par <u>des fonctions</u> u <u>ne vérifiant plus</u> (1.2)

Après ce qu'on a dit au début de cette section, il est naturel

---

1) On définit ainsi une forme tri-linéaire continue sur V.

2) Suivant le travail classique de Leray [1]

3) Espace défini dans Raviart, ce vol.

4) On montre, Lions [1], que si $u \in L^2(0, T; v)$ <u>et</u> satisfait à (1.7) <u>alors</u> (1.8) a un sens.

5) Dans le cas "n = 3" on ignore s'il y a unicité ou non.

d'introduire

(1.10) $W = (H^1_o(\Omega))^n$

(1.11) $K = (L^2(\Omega))^n$

les produits scalaires dans W et K étant "les mêmes" que dans V et H, puis de chercher u vérifiant

$$(1.12)\quad \begin{cases} u_\varepsilon \in L^2(0,T; W) \\ \dfrac{d}{dt}(u_\varepsilon(t), v) + \nu((u_\varepsilon(t), v)) + b(u_\varepsilon(t), u_\varepsilon(t), v) + \dfrac{1}{\varepsilon}(\operatorname{div} u_\varepsilon(t), \operatorname{div} v) = \\ \qquad = (f(t), v) \;\forall v \in W \\ u_\varepsilon(0) = u_o \quad , \varepsilon > 0 . \end{cases}$$

Mais il y a là une difficulté : on vérifie sans peine que sur V , on a

(1.13) $b(u,u,u) = 0 \quad \forall u \in V$

ce qui conduit aux inégalités de l'énergie pour (1.7) et à l'existence d'une solution globale (i.e pour T quelconque) de (1.7) (1.8) [1)]; par contre (1.13) n'est pas vrai pour $u \in W$ de sorte que très probablement [2)] le problème (1.12) est mal posé . On doit donc le modifier de le facon suivant (cf. Temam [1] [2]) : on introduit

(1.14) $$\tilde{b}(u,v,w) = +\frac{1}{2}\int_\Omega \operatorname{div} u \left(\sum_{j=1}^{n} v_j w_j\right) dx + b(u,v,w),$$

forme trilinéaire qui évidemment coincide sur V avec b(u, v, w).

Mais $\forall u \in (H^1_o(\Omega))^n = W$ on a :

(1.15) $$\tilde{b}(u,u,u) = \frac{1}{2}\sum_{i,j}\int_\Omega u_i \frac{\partial}{\partial x_i}(u_j^2)dx + \frac{1}{2}\int_\Omega \operatorname{div} u \left(\sum_j u_j^2\right)dx = 0$$

---

1) Naturellement (1.13) a lieu "à cause de la physique"

2) Mais cela n'est pas démontré .

et on peut maintenant introduire le problème pénalisé suivant :

$$(1.16)\begin{cases} \text{trouver } u_\varepsilon \in L^2(0,T;W) \text{ vérifiant} \\ \frac{d}{dt}(u_\varepsilon(t), v) + ((u(t), v)) + b(u_\varepsilon(t), u_\varepsilon(t), v) + \frac{1}{\varepsilon}(\operatorname{div} u_\varepsilon(t), \operatorname{div} v) = \\ \qquad = (f(t), v) \quad \forall v \in W \\ u_\varepsilon(0) = u_0 \end{cases}$$

1.3. Résultats

On montre alors (Temam [1] [2]) :

Théorème 1.1. Le problème (1.16) admet une solution ; il y a unicité si $n = 2$.

En outre

Théorème 1.2. . Lorsque $\varepsilon \to 0$ on a

(i) si $n = 3$, on peut extraire u_ε, de façon que

(1.17) $u_{\varepsilon'} \to u$ dans $L^2(0,T;W)$ faible et $L^2(0,T;K)$ fort [1]

(ii) si $n = 2$, on a :

(1.18) $u_\varepsilon \to u$ dans $L^2(0,T;W)$ faible et $L^2(0,T;K)$ fort.

En outre

$$\frac{1}{\varepsilon} \operatorname{div} u_\varepsilon \to p \quad \text{dans} \quad \mathcal{D}'(Q)/R \quad ^{2)}$$

Remarque 1.2

Faisant $v = u_\varepsilon$ dans (1.16) on en déduit aussitot

(1.19) $\| \operatorname{div} u_\varepsilon \|_{L^2(Q)} \quad c$

---

1) Extraction inutile si le problème de Navier Stokes admet une solution unique.

2) Quotient de l'espace des distributions sur Q par R (avec extraction de suite si n=3) . On peut préciser ce résultat.

1.4 Applications.

Du point de vue numérique l'approximation se fait maintenant en deux temps :

1.4.1 . On choisit ε au moyen de (1.19) ;

1.4.2 On intégre numériquement (1.16) par les différences finies (ou Galerkin) ; il s'agit maintenant d'un problème non linéaire mais ou les méthodes standard s'appliquent. Cf. Temam [2]

Remarque 1.3

R. Temam a également adopté dans Temam [2] à (1.16) la méthode des pas fractionnaires (variante des directions alternées): cfr. Douglas (ce volume), Marchouk [1] , Yanenko [1], Temam [3], Lieutaud [1] .

2 Equations de Navier-Stokes ; méthode des dérivées artificielles.

2.1 Idée générale de la méthode .

L'idée est grosso modo de "remplacer" (1,2) par

$$\varepsilon \frac{\partial p}{\partial t} + \operatorname{div} u = 0$$

mais cela conduit à des difficultés avu les intégrales d'energie . Pour "récupérer" les intégrales d'énergie on modifie l'équation (1.1) comme suit : on cherche.[1)]

$$\text{(2.1)} \qquad \begin{cases} u^{\varepsilon} \in L^2 (0, T ; W) \\ q^{\varepsilon} \in L^{\infty}(0, T ; K) \end{cases}$$

avec

$$\text{(2.2)} \qquad \frac{\partial u^{\varepsilon}}{\partial t} - \nu \Delta u^{\varepsilon} + \sum_{i=1}^{n} u_i^{\varepsilon} \frac{\partial u^{\varepsilon}}{\partial x_i} + \operatorname{grad} \left(q^{\varepsilon} - \frac{|u^{\varepsilon}|^2}{2}\right) = f$$

1) Les notations sont celles du n.1 .

(2.3) $\varepsilon \frac{\partial q^{\varepsilon}}{\partial t} + \text{div } u^{\varepsilon} = 0 \quad (\varepsilon > 0)$

(2.4) $u^{\varepsilon}(x,t) = 0$ si $x \in \Gamma, \quad t \in]0, T[$,

(2.5) $\begin{cases} u^{\varepsilon}(x,0) = u_o(x) , \\ q^{\varepsilon}(x,0) = q_o(x) , \quad q_o \text{ choisi quelconque dans } L^2(\Omega) . \end{cases}$

2.2. Résultats.

On peut montrer (par le même genre de méthodes que dans Lions [1], Lions-Prodi [1])les résultats suivants :

Théorème 2.1. . Le problème (2.1) ... (2.5) admet une solution
Il y a unicité si $n = 2$

Théorème 2.2.. Lorsque $\varepsilon \to 0$, on peut extraire une sous suite encore notée u^{ε} , q^{ε}, telle que

(2.6) $u^{\varepsilon} \to u$ dans $L^2(0,T,W)$ faible et $L^2(0,T,K)$ fort

En outre

$\text{grad } (q^{\varepsilon} - \frac{1}{2} |u^{\varepsilon}|^2) \to \text{grad } p$ dans $\mathcal{D}'(Q)$.

2.3. Remarques et applications.

Si l'on compare les méthodes de pénalisation (N.1) et des dérivées artificielles on peut faire les observations suivantes :

(i) la méthode de pénalisation conduit à une mise en oeuvre numérique plus simple (notamment puisqu'il y a une équation de moins) ;

(ii) le choix de ε est plus simple dans la méthode de pénalisation ;

(iii) par contre la méthode des dérivées artificielles est de portée peut être plus générale comme le montre l'exemple donné au N. ci après , où une méthode de pénalisation ne semble pas commode.

3 . Equations avec conditions d'évolution au bord : méthode des dérivées artificielles.

3.1. Position du problème

Soit $\Omega \subset R^n$ ouvert borné de frontière Γ régulière .

On cherche une fonction $u = u\ (x,t)$ solution de

(3.1) $$A u = 0 \quad \text{pour} \quad x \in \Omega , \quad t > 0$$

(3.2) où $$\left\{ \begin{array}{l} A\varphi = -\sum \frac{\partial}{\partial x_i} (a_{ij}(x) \frac{\partial}{\partial x_j}) \quad {}^{1)} \\ a_{ij} = C^2(\Omega) , \quad \sum_{i,j=1}^{n} a_{ij}(x) \xi_i \xi_j > \alpha |\xi|^2 , \alpha > 0, \text{p.p. dans } \Omega \ {}^{2)} , \end{array} \right.$$

avec

(3.3) $$\frac{\partial u}{\partial \nu} + \frac{\partial u}{\partial t} = f \qquad \text{sur} \sum = \Gamma \times \left] 0 , \infty \right[$$

(où $\frac{\partial u}{\partial \nu} = + \sum a_{ij} \frac{\partial u}{\partial x_j} \cos (n, x_i)$, n = normale à Γ extérieure à Ω)

(3.4) et $$u(x, 0) = u_o (x) \quad , \quad x \in \Gamma .$$

Donc les derivées en t n'apparaissent que sur $\sum$ et à cause de (3.1) le problème n'est pas du type de Cauchy - Kowaleski.

---

1) Et donc A ne contient pas de dérivation en t

2) Pour un peu simplifier on prend des fonctions à valeurs réelles .

3.2 Solution "fontionnelle" du problème (3.1)(3.3) (3.4)

On utilise ici, sans les rappeler, quelques notations et résultats de Lions-Magenes [1] [2].

Soit $g \in H^{1/2}(\Gamma)$; soit w la solution de

(3.5) $$\begin{cases} A w = 0 & \text{dans } \Omega \\ w|_{\Gamma} = g \end{cases}$$

Alors

(3.6) $$\frac{\partial w}{\partial \nu} \text{ est } \underline{\text{défini}} \text{ et } \in H^{-1/2}(\Gamma)$$

d'où un opérateur

(3.7) $$g \longrightarrow \frac{\partial w}{\partial \nu} = B g$$

avec

(3.8) $$B \in \mathcal{L}(H^{1/2}(\Gamma) ; H^{-1/2}(\Gamma)) .$$

Cet opérateur B est coercitif au sens

(3.9) $$(Bg, g) \geq \beta \| g \|^2_{H^{1/2}} , \beta > 0 ,$$

où $(\varphi, \psi) = \int_\Gamma \varphi \psi \, d\Gamma$.

Le problème (3.1) (3.3) (3.4) équivaut au suivant : si w désigne la trace de u sur Σ, alors il faut trouver w solution de

(3.10) $$Bw + \frac{dw}{dt} = f ,$$

(3.11) $$w(0) = u_0 .$$

On peut alors appliquer les méthodes standard pour la résolution de (3.10) (3.11). (cf. par ex. Lions-Magenes [2], chap 3).

Du point de vue numérique, hormis des cas particuliers où l'on peut expliciter l'opérateur intégro différentiel singulier B (et où l'on peut alors appliquer les méthodes de Cherruault [1], chap. 5), l'intégration de (3.10) ne semble pas très facile.

D'où l'intérêt (éventuel) de la méthode des dérivées artificielles.

3.3. Méthode des dérivées artificielles.

On cherche u_ε solution de

(3.12) $$\varepsilon \frac{\partial u_\varepsilon}{\partial t} + A u_\varepsilon = 0 \quad , \varepsilon > 0 \, ,$$

(3.13) $$\frac{\partial u_\varepsilon}{\partial \nu} + \frac{\partial u_\varepsilon}{\partial t} = f \quad \text{sur } \Sigma \, ,$$

(3.14) $$u_\varepsilon(x,0) = U_o(x) \quad , x \in \Omega \, ,$$

où

U_o = prolongement de u_o à Ω(ou suppose que $u_o \in H^\alpha(\Gamma), \alpha > 0)$.

On peut alors montrer les résultats suivants :

Théorème 3.1. Le problème (3.12) (3.13) (3.14) admet une solution unique

$$u_\varepsilon \in L^2(0, T, H^1(\Omega)) \, .$$

Théorème 3.2. Lorsque $\varepsilon \to 0$ on a

(3.15) $$u_\varepsilon \to u \quad \text{dans } L^2(0,T; H^1(\Omega)) \, .$$

3.4 Applications.

On peut maintenant discrétiser (3.12) (3.13) (3.14). Il s'agit désormais d'un problème standard. Cette méthode a éte donnée dans Lions [2].

3.5 Remarques

Si A est self adjoint, ou peut considérer le problème

(3.16) $\qquad Au = 0 \quad \text{dans} \quad \Omega \times \{t > 0\}$,

(3.17) $\qquad \dfrac{\partial u}{\partial \nu} + \dfrac{\partial^2 u}{\partial t^2} = f \quad \text{sur} \quad \Sigma$,

(3.18) $\qquad \begin{cases} u(x,0) = u_o(x) \\ \dfrac{\partial u}{\partial t}(x,0) = u_1(x) \end{cases} \qquad u_i$ donné sur Γ.

Alors le méthode des dérivées artificielles conduit au problème suivant : on cherche u_ε solution de

(3.19) $\qquad \varepsilon \dfrac{\partial^2 u_\varepsilon}{\partial t^2} + A u_\varepsilon = 0$,

(3.20) $\qquad \dfrac{\partial u_\varepsilon}{\partial \nu} + \dfrac{\partial^2 u_\varepsilon}{\partial t^2} = f \quad \text{sur} \ \Sigma$,

(3.21) $\qquad \begin{cases} u_\varepsilon(x,0) = U_o(x) , \quad \dfrac{\partial u_\varepsilon}{\partial t}(x,0) = U_1(x) , \quad x \in \Omega \\ \text{où } U_i = \text{prolongement de } u_i . \end{cases}$

On a des résultats analogues à ceux de 3.2 .

Voir problème de ce type dans Garipov [1] .

BIBLIOGRAPHIE

Bramble-Payne [1] cf. Payne, dans Numerical Solution... Bramble ed. Acad. Press, 1967.

Y. Cherruault [1] Approximation d'opérateurs linéaires et applications.

R. Courant [1] Variational methods for the solution of problems of equilibrium and vibrations- Bull. A.M.S. 49 (1943), p. 1-23.

J. Douglas et Peaceman - Rachford [1] SIAM, 1955, p. 42-65 et 28-42.

R.M. Garipov [1] On the linear theory of gravity waves. Archive Rat. Mech. Anal. 24, 1967, p. 352-362.

O.A. Ladyzenskaya [1] Equations de l'hydrodynamique... Moscow 1962

R. Lattes et J.L. Lions [1] Méthode de quasi reversibilité et Applications. Dunod 1967.

J. Leray [1] J. Math. Pures et Appl. 12(1933), p. -82 ; 13(1934), p. 33-418; Acta Math. 63(1934), p. 193-248.

J. Lieutaud [1] Thése. Paris 1968.

J.L. Lions [1] Equations différentielles

[2] Cours CEA-EDF - Paris. 1965

J.L. Lions et E. Magenes [1] Problèmes aux limites non homogènes (II) Annales Inst. Fourier, 11(1961), p. 137-178.

[2] Problèmes aux limites non homogènes et applications-vol. 1 et 2 Dunod. 1968.

J.L. Lions et G. Prodi [1] C.R. Acad. Sc. Paris, 248(1959), p. 3519-3521

G.I. Marchuk [1] Méthodes numériques... Novosibirsk 1965

[2] Problèmes de Méthéorologie . Leningrad 1967. (en russe) Traduction Francaise, Armand Colin, Paris, 1968.

R. Teman [1] C.R. Acad. Sc. Paris, t. 263, p. 241-244.

[2] Article sur Navier S okes à Paraitre

[3] These , Paris, 1967.

N.N. Yanenko [1] Methode des pas fractionnaires .... (en russe) Novosibirsk. 1966 - Traduction Francaise, Armand Colin, Paris, 1968.

CENTRO INTERNAZIONALE MATEMATICO ESTIVO

(C.I.M.E.)

J.L. LIONS

"PROBLEMES AUX LIMITES NON HOMOGENES A DONNEES IRREGULIERES ; UNE METHODE D'APPROXIMATION "

Corso tenuto ad Ispra dal 3-11 Luglio 1967

PROBLEMES AUX LIMITES NON HOMOGENES A DONNEES IRREGULIERES ; UNE METHODE D'APPROXIMATION

par

J.L. Lions

Université de Paris

Introduction.

Pour les opérateurs elliptiques du 2eme ordre, une méthode d'approximation pou les problèmes aux limites homogènes à donnés irrégulières (le 2eme membre étant par ex. une masse de Dirac, la solution correspondante étant alors le noyau de Green) est donnée dans Bramble [1] et pour les données frontières irrégulières (la solution correspondante étant alors par exemple le noyau de Poisson) une autre méthode est donnée dans Jamet 1

Nous introduisons ici une méthode différente valable pour les opérateurs elleptiques d'ordre quelconque et des données au bord arbitrairement irrégulières et valable aussi pour les équations d'evolution. Mais supposant (à la différence des travaux cités de Bramble et Jamet) que le frontière et les coefficients dp, y "très réguliers". Les détails techiniques sont dans le cas général fort longs et utilisent les résultats de Lions-Magenes [2] ; nous expliquons ici la méthode sur des exemples simples.

§ 1. Cas elliptique

1. Example 1

1.1- Position du problème

Soit Ω un ouvert borné de $\mathbb{R}^n$ de frontière Γ assez régulière . Dans Ω ou se donne un opérateur A de la forme

$$(1.1) \qquad A = -\sum \frac{\partial}{\partial x_i}\left(a_{ij}\frac{\partial}{\partial x_j}\right) + a_o \; ^{1)}$$

a_o, a_{ij} $C^{\infty}(\Omega)$, avec

$$(1.2) \qquad \mathrm{Re}\left(\sum_{i,j=1}^{n} a_{ij}(x)\, \xi_j \, \overline{\xi_i}\right) \geq \alpha \sum_{j=1}^{n} |\xi_j|^2, \; \alpha > 0, \forall \xi_j \in \mathbb{C}, \; \mathrm{Re}\, a_o \geq \alpha > 0.$$

On considère le problème de Dirichlet

$$(1.3) \qquad A u = f$$

$$(1.4) \qquad u/_{\Gamma} = g$$

où l'on suppose que

$$(1.5) \qquad f \in L^2(\Omega)$$

$$(1.6) \qquad g \in H^{-1/2}(\Gamma).$$

On montre (cf. Lions-Magenes [1], [2] chap. 2) que le problème (1.3) (1.4) admet, sous les conditions (1.5) (1.6), une solution unique qui appartient à $L^2(\Omega)$.

Notre objet est de trouver une approximation de ce problème (dans $L^2(\Omega)$).

1.2. Idée générale de la méthode.

On va changer la nature du problème (1.3) (1.4) (du type "Dirichlet") en l'approchant par des problèmes de Neumann.

Soit $\frac{\partial}{\partial \nu}$ l'opérateur de dérivée conormale associé à A.

On remplace alors (1.4) par $\varepsilon \frac{\partial u}{\partial \nu} + u = g$ sur Γ, $\varepsilon > 0$.

On est ainsi conduit au problème (du type Neumann) suivant : trouver (pour $\varepsilon > 0$ donné) u_ε solution de

---

1) Tout ce qui suit s'étend aux opérateurs elliptiques d'ordre quelconque (on n'utilise jamais de principe du maximum).

(1.7) $$Au_\varepsilon = f \quad ,$$

(1.8) $$\varepsilon \frac{\partial u_\varepsilon}{\partial \nu} + u_\varepsilon = g \quad \text{sur } \Gamma .$$

On vérifie facilement que ce problème admet une solution unique dans $H^1(\Omega)$ [1].

Reste ensuite à faire $\varepsilon \to 0$.

1.3. Résultats de convergence lorsque $\varepsilon \to 0$.

Théorème 1.1. Lorsque $\varepsilon \to 0$, $u_\varepsilon \to u$ dans $L^2(\Omega)$.

Principe de la démonstration.

1) Le problème (1.7) (1.8) équivant à :

(1.9) $$a(u_\varepsilon , v) + \frac{1}{\varepsilon}(u_\varepsilon , v)_\Gamma = \frac{1}{\varepsilon}(g, v)_\Gamma + (f, v) \; \forall \, v \in H^1(\Omega)$$

où
$$a(u, v) = \sum_{i,j=1}^{n} \int_\Omega a_{ij} \frac{\partial u}{\partial x_j} \frac{\partial \bar{v}}{\partial x_i} \, dx + \int_\Omega a_0 \, u \, \bar{v} \, dx \quad ,$$

$$(f, v) = \int_\Omega f \bar{v} \, dx, \quad (g, v)_\Gamma = \int_\Gamma g \bar{v} \, d\Gamma .$$

On déduit de là l'existence et l'unicité de u_ε dans $H^1(\Omega)$.

2) Introduisons v_ε solution dans $H^1(\Omega)$ de

(1.10) $$\begin{cases} A^* v_\varepsilon = u_\varepsilon & \text{dans } \Omega , \\ \varepsilon \frac{\partial v_\varepsilon}{\partial \nu^*} + v_\varepsilon = 0 & \text{sur } \Gamma . \end{cases}$$

Puisque $u_\varepsilon \in H^1(\Omega)$, on a; $v_\varepsilon \in H^3(\Omega)$ (cf. par ex. Lions-Magenes [2] , chap. 2); si l'on prend dans (1.9) $v = v_\varepsilon$ et que l'on intègre par parties , on obtient [2]

---

[1] Formellement, (1.7) (1.8) est un problème voisin et régularisé de (1.3) (1.4)

[2] $|f| = (f, f)^{1/2}$

(1.11) $$|u_\varepsilon|^2 = (g, \frac{1}{\varepsilon} v_\varepsilon)_\Gamma + (f, v_\varepsilon) .$$

3) Mais en reprenant la <u>démonstration</u> de la régularité au bord, on montre que

(1.12) $$\|v_\varepsilon\|_{H^2(\Omega)} \leq c_1 |u_\varepsilon| , c_1 \text{ \underline{indépendant de} } \varepsilon .$$

Alors $\left\| \frac{\partial v_\varepsilon}{\partial \nu^*} \right\|_{H^{1/2}(\Gamma)} \leq c_2 |u_\varepsilon|$ et donc

$$\left\|\frac{1}{\varepsilon} v_\varepsilon\right\|_{H^{1/2}(\Gamma)} = \left\|\frac{\partial v_\varepsilon}{\partial \nu^*}\right\|_{H^{1/2}(\Gamma)} \leq c_2 |u_\varepsilon| .$$

Mais alors (1.11) donne

$$|u_\varepsilon|^2 \leq c_2 \|g\|_{H^{-1/2}(\Gamma)} |u_\varepsilon| + c_1 |f| |u_\varepsilon|$$

d'où

(1.13) $$|u_\varepsilon| \leq c_3 .$$

4) D'après (1.13) on peut extraire une suite encore notée u_ε telle que

(1.14) $$u_\varepsilon \longrightarrow w \quad \text{dans } L^2(\Omega) \text{ faible} .$$

Mais alors (Lions-Magenes [2], chap. 2), puisque $Au_\varepsilon = f$, $u_\varepsilon|_\Gamma \longrightarrow w|_\Gamma$ dans $H^{-1/2}(\Gamma)$, $\frac{\partial u_\varepsilon}{\partial \nu} \to \frac{\partial w}{\partial \nu}$ dans $H^{-3/2}(\Gamma)$ et (1.8) donne $w/_\Gamma = g$. Comme $Aw = f$, on a $w = u$ et donc

(1.15) $$u_\varepsilon \to u \quad \text{dans } L^2(\Omega) \text{ faible.}$$

5) Alors $v_\varepsilon \to v$ dans $H^2(\Omega)$ faible, où $A^* v = u$, $v/_\Gamma = 0$ et $|u_\varepsilon - u|^2 = |u_\varepsilon|^2 + |u|^2 - 2 \operatorname{Re} (u_\varepsilon, u)$

$$= (g, \frac{1}{\varepsilon} v_\varepsilon)_\Gamma + (f, v_\varepsilon) + |u|^2 - 2\operatorname{Re} (u_\varepsilon, u) ;$$

$\frac{1}{\varepsilon} v_\varepsilon = -\frac{\partial v}{\partial \nu^*} \to -\frac{\partial v}{\partial \nu^*}$ dans $H^{-1/2}(\Gamma)$ faible et donc

$$|u_\varepsilon - u|^2 \longrightarrow -(g, \frac{\partial v}{\partial \nu^*})_\Gamma + (f, v) - |u|^2 = 0 .$$

1.4. Applications.

On peut effectuer une approximation "en deux temps" : on choisit d'abord ε puis on approche (1.7) (1.8) par les méthodes générales de Aubin [1], Céa [1].

2. Exemple 2

2.1. Position du problème. Méthode.

On considère maintenant encore (1.3) (1.4) avec $f \in L^2(\Omega)$ mais avec cette fois

(2.1) $$g \in H^{-3/2}(\Gamma)$$ [1].

La solution u correspondante est cette fois dans $H^{-1}(\Omega)$ (Lions - Magenes [1] 2).

Cette fois la "régularisation" (1.7) (1.8) n'est plus suffisante et il faut une double régularisation.

On considère sur $H^2(\Omega)$ une forme $b(u,v)$ sesquilinéaire coercive choisie "la plus simple possible", par ex. $b(u,v)$ = produit scalaire sur $H^2(\Omega)$.

On définit alors u_ε (où $\varepsilon = \{\varepsilon_1, \varepsilon_2\}$). comme la solution dans $H^2(\Omega)$ de [2]

(2.2) $$\varepsilon_1 b(u_\varepsilon, v) + a(u_\varepsilon, v) + \frac{1}{\varepsilon_2}(u_\varepsilon, v)_\Gamma = \frac{1}{\varepsilon_2}(g, v)_\Gamma + (f, v) \quad \forall\, v \in H^2(\Omega).$$

On peut alors montrer le

Théorème 2.1. Lorsque $\varepsilon \to 0$, on a

(2.3) $$u_\varepsilon \to u \text{ dans } H^{-1}(\Omega).$$

---

1) Cela suffira pour comprendre comment résoudre le cas général : $g \in H^{-s}(\Gamma)$, $s > 0$ quelconque.

2) Comparer à J.P. Aubin. J.L. Lions [1] (ce volume).

La démonstration consiste à choisir dans (2.2) :

$$v = v_\varepsilon \quad \text{solution de}$$

$(\varepsilon_1 B^* + A^*)v_\varepsilon = (-\Delta+1)^{-1} u_\varepsilon$ où $(-\Delta+1)^{-1}$ est l'inverse de l'isomorphisme $(-\Delta+1)$ de $H_0^1(\Omega)$ sur $H^{-1}(\Omega)$, et avec des conditions aux limites convenables.

2.2. Application.

On a (théorème de Sobolev fractionnaire; cf J. Peetre [1])

$$H^{3/2}(\Gamma) \subset C^0(\Gamma) \quad \text{si} \quad \frac{1}{2} - \frac{3/2}{n-1} < 0$$

i.e. si $n \leq 3$.

Donc :

Si $n \leq 3$ on peut prendre dans (2.1) $g = \delta(x - x_0)$, $x_0 \in \Gamma$; le solution u correspondante est le noyau de Poisson .

Cf. pour cela une autre méthode (utilisant le principe du maximum) dans P. Jamet [1] .

2.3 Remarque

Par le genre de mthode donnée ici, on peut construire l'approximation ("en deux temps") des problèmes généraux non variationnels

(2.4) $$Au = f, \quad B_j u = g_j \, , \quad 1 \leq j \leq m \, ,$$

A elliptique d'ordre 2m, les B_j recouvrant A .

Il devait etre possible de trouver une approximation directe de ces problèmes , mais cela n'est pas fait à notre connaissance.

§ 2. Cas parabolique.

1. Nous nous bornons à un cas particulier très simple.

Dans le cylindre $Q = \Omega \times]0,T[$, on considère le problème

(1.1) $$\frac{\partial u}{\partial t} + A u = f$$

où A est donné comme au § 1 [1], $f \in L^2(Q)$, avec

(1.2) $\quad u\big|_{\Sigma} = g \quad , \quad \Sigma = \Gamma \times \left]0, T\right[$,

où

(1.3) $\quad g \in L^2(\Sigma)$

et avec

(1.4) $\quad u(x,o) = u_o(x) \ , \quad u_o \in L^2(\Omega)$.

Alors la solution u est <u>en particulier</u> dans $L^2(Q)$ (cf. Lions-Magenes [2], chap. 4) .

<u>On "approche" ce problème de la facon suivante</u> : on désigne par u_ε la solution de

(1.5) $\quad \dfrac{\partial u_\varepsilon}{\partial t} + Au_\varepsilon = f$,

(1.6) $\quad \varepsilon \dfrac{\partial u_\varepsilon}{\partial \nu} + u_\varepsilon = g \quad$ sur Σ,

(1.7) $\quad u_\varepsilon(x,o) = u_o(x)$.

On peut alors montrer le

<u>Théorème 1.1.</u> <u>Lorsque</u> $\varepsilon \to 0$, <u>on a</u>

(1.8) $\quad u'_\varepsilon \to u \quad$ dans $L^2(Q)$.

<u>Application.</u>

On peut , après avoir "régularisé" le problème initial en (1.5) (1.6) (1.7) . utiliser les méthodes de Raviart [1] .

2 <u>Remarque.</u>

Par double régularisation (analogue au N. 2 , § 1) on pourra attendre le cas où $g \in H^{-s}(\Sigma)$, $s > 0$ quelconque .

---

1) Les coefficients de A peuvent dépendre de t et tout ce qui suit s'etend aux opérateurs <u>d'ordre quelconque.</u>

BIBLIOGRAPHIE

J.P. Aubin [1] Approximation ... Bull. S.M.F. Mémoire N.12, 1967.

J.P. Aubin - J.L. Lions [1] Ce volume.

J.H. Bramble [1] Ce Volume.

J. Céa [1] Approximation variationnelle des problèmes aux limites. Annales Inst. Fourier 14 (1964), p. 345-444.

P. Jamet [1] A paraitre

J.L. Lions-E. Magenes [1] Problèmes aux limites non homogènes (II). Annales Inst. Fourier, 11(1961), p. 137-178.

[2] - Problèmes aux limites non homogènes et applications. Vol. 1 et 2, Paris, Dunod 1968.

J. Peetre [1] Espaces d'interpolation et théorème de Sobolev. Annales Inst. Fourier, 16(1966), p. 279-317.

P.A. Raviart [1] Ce Volume.

CENTRO INTERNAZIONALE MATEMATICO ESTIVO

(C.I.M.E.)

J.P. AUBIN et J.L. LIONS

REMARQUES SUR L'APPROXIMATION REGULARISEE DE PROBLEMES AUX LIMITES

Corso tenuto ad Ispra dal 3-11 Luglio 1967

REMARQUES SUR L'APPROXIMATION REGULARISCE DE PROBLEMES AUX LIMITES

par

J.P. Aubin et J.L. Lions

Introduction

On donne ici un procédé assez général permettant d'approcher dans la topologie la plus fine possible [1] la solution de certains problèmes aux limites variationnels linéaires de nature elliptique, par des solutions de problèmes aux différences finies (ou d'autres problèmes approchés) .

La méthode proposée utilise essentiellement :

(i) une "régularisation" des opérateurs différentiels ;

(ii) des opérateurs de prolongement et restriction (p_h, r_h) .

Une technique du type (i) a déjà été utilisée (mais non assortie de discrétisation) dans J.L.LIONS-G.STAMPACCHIA [1] (cf.(3.14) de ce dernier travail) .

La technique (ii) est utilisée dans J.P.AUBIN [1] [2] .

L'usage simultané de ces deux techniques conduit aux résultats de cette note; comme conséquence on obtient une estimation de l'erreur dans les espaces "usuels" de fonctions différentiables.

Le même genre de méthodes peut s'appliquer aux problémes non linéaires ou unilatéraux dans les cas (hélas rares) où l'on conduit un théorème de régularité.

1. Position du problème

1.1. Soit $\dot{V}$ un espace de Hilbert sur $\mathbb{C}$, de norme $\| \;\|$, et soit V' son anti-dual, (f, v) désignant la forme sésquilinéaire mettant V' et V en anti-dualité .

---

1) I.e. la topologie raisonnable de l'espace où est la solution du problème que l'on approche, sans perte de régularieté. Pour d'autres resultats dans ce sens, cf. les conférences de BRAMBLE (ce Volume) .

Soit $A \in \mathcal{L}(V;V')$ [1], tel que

(1.1) $$\operatorname{Re}(Av, v) \geq \alpha \|v\|^2, \ \alpha > 0, \ \forall v \in V.$$

L'adjoint A^* de A vérifie alors $\operatorname{Re}(A^* v, v) \geq \alpha \|v\|^2 \ \forall v \in V$ et on sait que A et A^* sont des isomorphismes de V sur V'.

On suppose maintenant connu un "théorème de régularité" (cf. section 1.2 ci après pour des exemples) : on suppose connus deux espaces de Hilbert W et F avec

$$W \subset V \qquad , F \subset V'$$

l'injection de chaque espace dans la suivant étant continue (mais W -resp. F - n'étant pas nécessairement dense dans V - resp. V'), de sorte que

(1.2)
(i) $A \in \mathcal{L}(W;\ F)$

(ii) si $f \in F$, la solution u dans V de "$Au = f$" appartient à W.

(mais A n'est pas nécessairement un isomorphisme de W sur F ! cf. aussi N° 5).

1.2. Example .

Si Ω est un ouvert borné de $\mathbb{R}^n$, de frontière Γ régulière, on prend

$$V = H^m(\Omega) = \left\{ v \mid D^p v \in L^2(\Omega),\ |p| \leq m \right\},$$

espace de Sobolev d'ordre m muni de la structure hilbertienne

$$\|v\| = \Big(\sum_{|p| \leq m} \int_\Omega |D^p v|^2 \, dx \Big)^{1/2}.$$

On considère l'operateur A défini par

(1.3) $$(Au, v) = a(u,v) = \sum_{|p|, |q| \leq m} \int_\Omega a_{pq}(x)\, D^q_u \, \overline{D^p v} \, dx ,$$

---

[1] De façon générale, $\mathcal{L}(X, Y)$ désigne l'espace des applications linéaires continues de X dans Y .

où les a_{pq} sont des fonctions données dans Ω et l'on suppose que (1.1) a lieu .

L'espace V' n'est pas un espace de distributions sur Ω. Si l'on prend $F = H^k(\Omega)$, alors la solution u dans V de " $Au = f$ " vérifie

$$(1.4) \quad \begin{cases} Au = f \text{ au sens des distributions dans } \Omega,\ A = \sum (-1)^{|p|} D^p (a_{pq} D^q), \\ \text{et } m \text{ conditions aux limites "du type Neumann" sur } \Gamma^{(1)}, \end{cases}$$

et sous des hypothèses de régularité convenables [1] on a alors

$$u \in H^{k+2m}(\Omega)$$

et si donc l'on prend $W = H^{k+2m}(\Omega)$ alors (1.2) a lieu .

1.3. Le problème est maintenant le suivant : si f est donné dans V' , on connait une méthode systématique d'approximation dans V de la solution de " $Au = f$" (cf. J.P. Aubin [1]) ; si maintenant f est donné dans F , peut-on approcher u dans W ? Nous allons donner un procédé assez général ayant cette propriété .

2 Schémas variationnels régularisés.

2.1 Opérateurs p_h, r_h

Soit h un paramètre de $\mathbb{R}^n$ destiné à tendre vers 0. De façon analogue à Aubin [1] , pour chaque h on suppose connus

(2.1)
- (i) un espace V_h de dimension finie
- (ii) un opérateur p_h injectif de V_h dans W
- (iii) un opérateur r_h (linéaire ou non) de V dans V_h

tels que , $|||\ \ |||$ désignant la norme dans W :

---

(1) Pour détails. cf. par ex. Lions-Magenes [1] , Chap. 2 .

(2.2)
(i) $\| u - p_h r_h u \| \leq \gamma(h) ||| u ||| \quad \forall u \in W$,
(ii) $\gamma(h) \to 0$ si $h \to 0$,
(iii) $p_h r_h v \to v$ dans V (resp W) $\forall v \in V$ (resp. W) lorsque $h \to 0$.

2.2. Example

Plaçons nous dans la situation de 1.2.

On peut alors construire cf. J. P. Aubin [2] p_h et r_h satisfaisant à (2.1) (2.2) et avec

(2.3) $\gamma(h) = O(|h|^{m+k})$ (si $W = H^{k+2m}(\Omega)$) [1]

2.3. Schéma variationnel régularisé.

Remarque préliminaire.

Une fois en possession des opérateurs p_h, r_h, l'approximation "naturelle" de l'équation

$$A u = f, \quad u \in V, \quad f \in V'$$

(ou encore, en posant $(Au, v) = a(u, v)$:

$$a(u, v) = (f, v) \quad \forall v \in V)$$

est :

(2.4) $$\begin{cases} a(p_h u_h, p_h v_h) = (f, p_h v_h) \quad \forall v_h \in V_h . \\ u_h \in V_h \end{cases}$$

Mais il n'y a pas de raison pour que, sans hypothèse supplémentaire, on ait alors $p_h u_h \to u$ dans W lorsque f est dans F.

La méthode proposée consiste à régulariser le schéma (2.4).

La forme b(n, v)

On choisit [2] $b(u, v)$ forme sesquilinéaire continue sur W, telle que

1) $h = \{h_1, \ldots h_n\}$ désigne alors la maille du réseau sur lequel on discrétise le problème aux limites

2) On aura intéret à prendre $b(u, v)$ le plus simple possible, par exemple produit scalaire le plus simple possible dans W.

(2.5) $\operatorname{Re} b(v,v) \geq \beta |||v|||^2 \quad \forall v \in W, \quad \beta > 0.$

On définit alors $u_h \in V_h$ comme la [1] solution de

(2.6) $\varepsilon(h)\, b(p_h u_h, p_h v_h) + a(p_h u_h, p_h r_h) = (f, p_h v_h) \quad \forall v_h \in V_h,$

où $\varepsilon(h) > 0$ et $\varepsilon(h) \to 0$ lorsque $h \to 0$ [2].

C'est le schéma variationnel régularisé qui a la propriété recherchée comme le montre le théorème suivant, qui sera établi au N. 3 :

Théorème 2.1. On suppose que A a les proprietés (1.1) (1.2), que p_h satisfait à (2.1) (2.2), et que b satisfait à (2.5). Si u_h est la solution du schéma (2.6) on a :

(i) si $f \in V'$ alors $p_h u_h \to u$ dans V lorque $h \to 0$ [3];

(ii) si $f \in F'$ et si $\varepsilon(h)$ est choisi de sorte que

(2.7) $$\frac{\gamma(h)}{\sqrt{\varepsilon(h)}} \to 0 \quad \text{lorsque } h \to 0,$$

alors $p_h u_h \to u$ dans W lorsque $h \to 0$.

Estimation de l'erreur

En utilisant la théorie de l'interpolation [4] on peut obtenir l'estimation de l'erreur suivante ; de façon générale, soit $[W, V]_\theta$ l'espace de Hilbert intermédiaire (J. L. LIONS [1]) de paramètre θ, $0 < \theta < 1$ [5].

Alors

---

(1) Il est immédiat de vérifier que (2.6) admet une solution unique.

(2) Il faudra préciser ce point - cf. (2.7) ci dessous.

(3) Cette propriété est vérifiée (comme on le montre sans peine)pour le schéma (2.4)

(4) On utilise ici seulement la théorie hilbertienne de l'interpolation, comme introduite dans Lions [1]. Un exposé détaillé en est donné dans Lions-Magenes [1], Chap. 1.

(5) Par exemple

Théorème 2.2

Sous les hypothèses du Théorème 2.1., on a :

(2.8) $$\|u-p_h u_h\|_{[W,V]_\theta} \leq c\, \mathcal{E}(h)^{\theta/2}$$

Le théorème est également démontré au N. suivant.

Remarque. 2.1.

La théorie de l'interpolation est un outil commode pour l'obtention de l'estimation de l'erreur. Pour un autre exemple d'application (utilisant cette fois l'interpolation d'espaces de Banach, comme dans Lions-Peetre [1]) est donné dans Peetre-Thomée [1] .

3. Démonstration des Théorèmes 2.1. et 2.2

3.1. Le cas (i) du Théorème 2.1.

On fait $v_h = u_h$ dans (2.6) et on prend les parties réelles des deux membres. Il vient

$$\mathcal{E}(h)\, \beta |||p_h u_h|||^2 + \alpha\|p_h u_h\|^2 \leq \|f\|_{V'}\|p_h v_h\|$$

(3.1) $$\sqrt{\mathcal{E}(h)}\,|||p_h u_h||| + \|p_h u_h\| \leq c^{(1)}$$

On peut alors supposer, par extraction d'une sous suite encore notée h, que

$$p_h u_h \longrightarrow u_* \quad \text{dans } V \text{ faible.}$$

Mais comme $(f, p_k v_h) = a(u, p_h v_h)$, on déduit de (2.6)

(3.2) $$\mathcal{E}(h)\, b(p_h u_h, p_h v_h) + a(p_h u_h - u, p_h v_h) = 0$$

Si l'on prend dans (3.2) $v_h = r_h v$, et puisque $p_h r_h v \longrightarrow v$ dans V fort on peut passer à la limite :

1) Les c désignent des constantes diverses.

$$a(u_*-u, v)=0 \quad \forall v \in V$$

donc $u_* = u$. Enfin, prenant $v_h = u_h$ dans (3.2) on obtient

$$\varepsilon(h)\, b(p_h u_h, p_h u_h) + a(p_h u_h - u, p_h u_h - u) = c(p_h u_h - u, u) \to 0$$

d'où (i) (et en outre $\sqrt{\varepsilon(h)}\, p_h u_h \to 0$ dans W) .

3.2 Le cas (ii) du Théorème 2.1.

Evidemment les estimations du cas 3.1 sont valables.

Prenons dans (3.2) $v_h = u_h - r_h u$ et posons

$$\delta_h = p_h u_h - p_h r_h u.$$

Il vient

(3.3) $$\varepsilon(h) b(p_h u_h, \delta_h) + a(\delta_h + p_h r_h u - u, \delta_h) = 0$$

d'où

$$\varepsilon(h) b(p_h u_h, p_h u_h) + a(\delta_h, \delta_h) = \varepsilon(h) b(p_h u_h, p_h r_h u) + a(u - p_h r_h u, \delta_h)$$

et comme u W on peut utiliser (2.2) (i) . Donc

$$\varepsilon(h) |||p_h u_h|||^2 + ||\delta_h||^2 \leq c\, \varepsilon(h) |||p_h u_h|||\, |||p_h r_h u||| + c\gamma(h) |||u|||\, ||\delta_h|| \leq$$

$$\leq \frac{1}{2} \varepsilon(h) |||p_h u_h|||^2 + \frac{1}{2} ||\delta_h||^2 + c(\varepsilon(h) + \gamma(h)^2)$$

d'où

(3.4) $$\varepsilon(h) |||p_h u_h|||^2 + ||\delta_h||^2 \leq c(\varepsilon(h) + \gamma(h)^2) \leq \text{(d'après (2.7))}\ c\,\varepsilon(h)$$

Donc

(3.5) $$|||p_h u_h||| \leq c$$

(3.6) $$||\delta_h|| \leq c\sqrt{\varepsilon(h)}$$

On peut donc d'après (3.5) supposer par extraction que $p_h u_h$ converge dans W faible et d'après (3.1) vers u, donc

(3.7) $p_h u_h \longrightarrow u$ dans W faible [1]

Mais on déduit de (3.3) que :

$$\varepsilon(h)\, b(\delta_h, \delta_h) + a(\delta_h, \delta_h) = -\varepsilon(h)\, b(p_h r_h u, \delta_h) - a(p_h r_h u - u, \delta_h)$$

d'où

$$b(\delta_h, \delta_h) + \frac{1}{\varepsilon(h)} a(\delta_h, \delta_h) \leq |b(p_h r_h u, \delta_h)| + c \frac{\gamma(h)}{\varepsilon(h)} \sqrt{\varepsilon(h)} \; .$$

et grâce à (2.7), $b(\delta_h, \delta_h) \longrightarrow 0$ d'où le théorème, car

$$p_h u_h - u = \delta_h + (p_h r_h u - u) \; .$$

3-3. Démonstration du théorème 2.2

On déduit de (3.6) et de (2.7) que

$$\| u - p_h u_h \| \leq c \sqrt{\varepsilon(h)} + \| u - p_h r_h u \| \leq c \sqrt{\varepsilon(h)}$$

(cas où $\theta = 1$) et par ailleurs, dans le cas où $\theta = 0$: $|||u - p_h u_h||| \leq c$.

On interpole entre ces deux majorations: on en déduit le résultat.

4. Exemple

4.1. On se place dans le cadre de 1.2 et 2.2. On considère donc un problème de Neumann

(4.1) $$Au = f,$$

avec A opérateur elliptique d'ordre 2m .

Si $F = H^k(\Omega)$, alors $W = H^{k+2m}(\Omega)$ de sorte qu'à $b(u, v)$ correspond un opérateur différentiel (soit B) d'ordre $2(k+2m)$.

Si on utilise les opérateurs p_h introduits dans J. P. Aubin [2] et [1] et chap II, § 2 et § 3 , conduisant à (2.3), on doit alors choisir dans (2.6)

(4.2) $$\varepsilon(h) = c|h|^{2\mu}$$

avec

(4.3) $$\mu < m+k.$$

---

(1) Cela, sous l'hypothèse $\gamma(h)/\sqrt{\varepsilon(h)} \leq c$. L'hypothèse plus restrictive (2.7) intervient pour la convergence forte

Le schéma variationnel régularisé (2.6) est alors un schéma aux différences finies, qui peut être explicité . (cf. J. P. Aubin, [1] , chap II , § 4).

Si $0 \leq \theta \leq 1$, l'espace $[W, V]_\theta$ est l'espace

(4.4) $$H^{k+2m(1-\theta)}(\Omega) .$$

Mais si $C^r(\overline{\Omega})$ désigne l'espace des fonctions r fois continûment différentiables dans $\overline{\Omega}$ on a, d'après le théorème de Sobolev fractionnaire, (S. L. Sobolev [1] , J. Peetre [1])

(4.5) $$\begin{cases} H^{k+2m(1-\theta)}(\Omega) \subset C^r(\overline{\Omega}) \text{ si} \\ k+2m(1-\theta) > \frac{n}{2} + r. \end{cases}$$

Soit k donné tel que $k+2m > \frac{n}{2} + r$. Alors si l'on choisit θ de façon que $\theta < (k - \frac{n}{2} - r)\frac{1}{2m} + 1$, on déduit du Théorème 2.2 et de (4.5) que

$$\| p_h u_h - u \|_{C^r(\Omega)} \leq c\, |h|^{\mu^\theta} , \quad \mu \text{ verifiant (4.3)} .$$

Par conséquent

(4.6) $$\| p_h u_h - u \|_{c^r(\Omega)} \leq c \, |h|^{m+k+(k-\frac{n}{2}-r)\frac{m+k}{2m} - \eta} > 0 \quad \forall \eta > 0.$$

5 - Transposition

5.1. Le problème

Supposons maintenant que

(5.1) $$\begin{cases} \text{i) } W \text{ est dense dans } V \\ \text{ii) } F \text{ est dense dans } V' \end{cases}$$

et que

(5.2) A et A^* est un isomorphisme de W sur F [1)]

Alors le transposé A^* est un isomorphisme de F' sur W' qui prolonge A donc :

(5.3) A est un isomorphisme de F' sur W'.

Soit f un élément de W'. Considérons la solution u dans F' de l'équation

(5.4) $$Au = f$$

Sans les hypothèses du théorème 2.1, le schéma variationnel régularisé (2.6) a encore un sens et l'on se pose le problème de savoir si $p_h u_h$ converge vers u dans l'espace F'.

5-2. Exemple.

On se place dans le cadre de 1.2. D'après J. L. Lions-E. Magenes [1], chap. 2, on peut choisir des espaces W et F tels que l'équation (5.4) conduise à des problèmes aux limites non homogènes pour l'opérateur $A = \sum_1 (-1)^{|p|} D^p (a_{pq} D^q)$.

5-3. Le théorème

Théorème 5-1.

Sous les hypothèses (5.1), (5-2) et celles du théorèmes 2-1, si $f \in W'$, alors $p_h u_h$ converge fortement vers u dans F'.

Démonstration

Soit $p_h' \; (W', V_h)$ le transposé de p_h et A_h l'opérateur de V_h sur lui même défini par la forme sesquilinéaire :

(5.5) $$(A_h u_h, v_h)_h = \mathcal{E}(h)\, b(p_h u_h, p_h v_h) + a(p_h u_h, p_h v_h)$$

1) L'espace W contient donc maintenant les conditions aux limites !

où (f_h, v_h) est la forme sesquilinéaire sur V_h mettant en antidualité V_h et V'_h .

La solution u_h de (2.6) est alors égale à

(5-6) $$u_h = A_h^{-1} \, p'_h \, f$$

et le théorème 2.1 affirme que si f appartient à F, alors.

(5.7) $$u - p_h u_h = T_h f = (A^{-1} - p_h A_h^{-1} p'_h) f$$

converge vers 0 dans W .

On montre de même qu'en remplaçant A par A^*,

(5.8) $$u - p_h u_h = T_h f = (A^{-1} - p_h A_h^{-1} p_h) f$$

converge vers 0 dans W pour tout f de F.

Les opérateurs T_h^* forment donc un ensemble équicontinue de $\mathcal{L}(F, W)$.

Par transposition, les opérateurs $T_h = T_h' = (A^{-1} - p_h A_h^{-1} p'_h)$ forment également un ensemble equicontinu de $\mathcal{L}(W'; F')$.

Pour que les opérateurs T_h convergent simplement vers o dans $\mathcal{L}(W'; F')$, il suffit alors que $\| T_h f \|_{F'}$ converge vers o pour tout élément f d'un ensemble dense dans W'. Prenons alors V' : V' est dense dans W' et, d'après la première partie du théorème 2-1 , on sait que $\| T_h f \|_{F'} \leq \| T_h f \|_V = \| u - p_h u_h \|_V$ tend vers 0 .

On en déduit donc que si $f \in V'$

(5.9) $$\| T_h f \|_{F'} = \| u - p_h u_h \|_{F'}$$

tend vers o avec h.

BIBLIOGRAPHIE

J.P. Aubin [1] Approximation des espaces de distributions et des opérateurs différentiels . Bull. S.M.F. Mémoire N. 12, 1967

[2] Evaluation des erreurs de troncature des approximations des espaces de Sobolev. A paraitre dans Journal of Math. Analysis and Applications.

Bramble [1] Conférences de ce Volume et Bibliographie de ce travail.

J.L. Lions [1] Espaces intermédiaires entre espaces hilbertiens ;applications. Bull. Math. R.P.R. t.2 , (1958) , p. 419-432

J.L. Lions-E. Magenes [1] Problèmes aux limites non homogènes et applications Vol. 1, 1968 .

J.L. Lions-J. Peetre [1] Sur une classe d'espaces d'interpolation. Institut des Hautes Etudes Scientifiques . N. 19 - 1964

J.L. Lions-G. Stampacchia [1] Variational Inequalities - Comm. Pure Applied Math . XX (1967), p. 493-519.

J. Peetre [1] Espaces d'interpolation et théorème de Soboleff. Annales Institut Fourier, 16, 1 (1966) , p. 279-317 .

J. Peetre-V. Thomée [1]. On the rate of convergence for discrete initial value problems. A paraitre.

S. L. Sobolev [1] Applications de l'analyse Fonctionnelle en Physique Mathématique - Leningrad. 1950 .

CENTRO INTERNAZIONALE MATEMATICO ESTIVO

(C.I.M.E.)

W. V. PETRYSHYN

ON THE APPROXIMATION-SOLVABILITY OF NONLINEAR FUNCTIONAL EQUATIONS IN NORMED LINEAR SPACES

Corso tenuto ad Ispra dal 3-11 Luglio 1967

ITERATIVE CONSTRUCTION OF FIXED POINTS OF CONTRACTIVE TYPE MAPPINGS IN BANACH SPACES

by

W. V. PETRYSHYN

(University-Chicago)

Introductory remarks. It is known that the problem of finding solutions of nonlinear functional equations (e.g. , nonlinear integral equations, boundary value problems for nonlinear ordinary and partial differential equations, etc.) can be formulated in terms of finding fixed points of a given nonlinear mapping defined on some subset of an infinite dimensional functional space. For compact mappings a general existence theory of fixed points based upon topological arguments has been constructed over a number of decades (associated with the names of Brouwer, Poincaré, Lefschetz, Schauder, Leray, Tichonoff, Rothe, Krasnoselsky, Altman, and others). More recently, there has begum a systematic study of fixed points (their existence and actual construction) of various classes of noncompact mappings (Brodski-Milman [2], De Marr [28], Browder [3a, 3b, 5 , 6, 4], Kirk 23 , Kachurovsky [21] , Edelstein [11, 12, 13] , Belluce-Kirk [1] , Göhde [19, 20], De Prima [10] , Lions and Stampacchia [27] , Browder-Petryshyn [8, 9] , Kaniel [22] , Shinbrot [40] , Petryshyn [31, 32, 33, 35, 36] , Opial [30] , Lees-Schultz [26], de Figueiredo [15, 16], Browder-de Figueiredo [7] , Petryshyn-Tucker [37] , and others) .

The purpose of this part of my talk is to survey, unify and extend a number of recent results concerning the iterative construction of fixed points of noncompact contractive and strictly pseudocontractive mappings acting in a Banach or a Hilbert space. We first devote a section to the definitions of various concepts used in this paper and to the historical development of the iterative method in the construction of fixed points of contractive mappings; then we devote a section to the complete proofs of the most recent results which, as will be shown below, include all those mentioned in the survey section. In the last section we discuss the convergence of fixed points of strictly contractive mappings kA studied in [6], where k is a real number such that $0 < k < 1$. It is worth noting that our proofs, which use the arguments analogous to those in [6, 8, 9, 30] are surprisingly simple.

1.1. Historical remarks and statements of results and definitions.

Let X be a Banach space, C a closed convex subset of X and A a (possibly) nonlinear mapping of C into X which is contractive (or nonexpansive), i.e.,

$$(1.1) \qquad \|Ax-Ay\| \leq \|x-y\| \text{ for all } x \text{ and } y \text{ in } C.$$

The study of the iterative construction of fixed points of contractive mappings is an extension of the classical theory of "the method of successive approximations" for strictly contractive mappings developed by Picard, Banach, Cacciopoli and others, where A is strictly contractive on C if

$$(1.2) \qquad \|Ax-Ay\| \leq k\|x-y\| \text{for all } x, y \; C \text{ and some } k<1$$

It is known 24 that if a strictly contractive operator A maps C into C, then the Picard sequence $\{x_{n+1}\}$ given by

$$(1.3) \qquad x_{n+1} = Ax_n = A^{n+1}x_0 \;(x_0 \text{ given in } C;\; n = 0, 1, 2, \dots)$$

converges to the unique fixed point of A in C.

In the case of contractive mappings A of C into C, the Picard sequence need not converge, there need not be any fixed points, nor need the fixed point be unique if it does exist. To obtain some positive results on the Picard sequence for contractive mappings, we need to impose further restrictions on the Banach spaces X, the subsets C and or the operator A. For the sake of completeness and clarity we recall the following definitions: X is uniformly convex if for any $\epsilon > 0$ there exists a $\delta(\epsilon)>0$ such that $\|x-y\| \geq \epsilon$ for $\|x\|\leq 1$ and $\|y\|\leq 1$ implies that $\|x+y\|\leq 2(1-\delta(\epsilon))$;

X is strictly convex if $\| tx+(1-t)y \| < 1$ for all $t \in (0,1)$ and all $x, y \in H$ such that $\|x\| = \|y\| = 1$; a mapping P of X into X is completely continuous if P is continuous and compact; P is weakly continuous if $x_n \rightharpoonup x$ implies $Px_n \rightharpoonup Px$, where "$\rightharpoonup$" and "$\rightarrow$" denote the weak and strong convergence in X, respectively; P is strongly continuous if $x_n \rightharpoonup x$ implies $Px_n \rightarrow Px$. Using Schauder's Existence Theorem 1 [38], Krasnoselsky [25] proved the following positive result for completely continuous contractive mappings.

Proposition 1.1. If X is uniformly convex, C a closed bounded convex subset of X and A a completely continuous contractive mapping of C into C, then the sequence $\{x_{n+1}\}$ determined by

(1.4) $$x_{n+1} = 1/2 Ax_n + 1/2 x_n \qquad (x_0 \text{ given in } C;\ n = 0, 1, 2, \ldots)$$

converges to a fixed point of A in C.

In 39 Schaefer extended Proposition 1.1 in two directions by proving the following.

Proposition 1.2. If X, C and A satisfy the conditions of Proposition 1.1 and λ is a fixed number such that $0 < \lambda < 1$, then the sequence $\{x_{n+1}\}$ determined by

(1.5) $$x_{n+1} = \lambda Ax_n + (1-\lambda)x_n \qquad (x_0 \in C;\ n = 0, 1, \ldots)$$

converges to a fixed point of A in C.

Proposition 1.3. If X is a real Hilbert space H, C a closed bounded convex subset of H and A a weakly continuous contractive map of C into C then x_{n+1} determined by (1.5) converges weakly to a fixed point of A in C.

Let us add that the proofs in [25, 39] were essentially based on the assertion that when X is uniformly convex and A is a contractive map of C into C, then the mapping $A_\lambda = \lambda A + (1-\lambda)I$ is asymptotically regular, i.e., $(A_\lambda^{n+1} x_0 - A_\lambda^n x_0) \rightarrow 0$ as $n \rightarrow \infty$ for any given x_0 in C (see Browder-Petryshyn [8]). In 14 Edelstein extended the validity of Proposition 1.2 to strictly convex Banach spaces X. In his proof Edelstein used Mazur's Theorem (see [29]) since in case X is strictly convex we cannot conclude that A_λ is asymptotically regular without first proving the strong convergence of the Picard sequence $\{A_\lambda^n x\}$.

As was already noted in Introductory remarks, the study of the existence and of the iterative construction of fixed points has been recently carried further for non-compact and contractive type mappings. Namely, suppose $C = B_r(0)$ denotes the closed ball of radius $r > 0$ in a real or complex Hilbert space H, S_r the boundary of B_r and A a contractive mapping of B_r into H which satisfies on S_r the Leray-Schauder condition

$$(1.6) \qquad Ax - \beta x \neq 0 \quad \text{for all } x \text{ in } S_r \text{ and any } \beta > 1.$$

Applying the recent theory of monotone operators Browder [3a] (see also De Prima [10]) proved that A has a fixed point in B_r. Using the above existence theorem the writer [35] has derived the following result for demicompact mappings A (a mapping P is said to be <u>demicompact</u> if whenever $\{u_n\}$ is a bounded sequence and $\{u_n - Pu_n\}$ is strongly convergent then there exists a strongly convergent subsequence $\{u_{n_i}\}$).

Proposition 1.4. Let A be a demicompact contractive mapping of B_r into H which satisfies (1.6). Then for any $x_0 \in B_r$ and any $\lambda \in (0,1)$ the sequence $\{x_{n+1}\}$ determined by the retraction-iteration method

$$(1.7) \qquad x_{n+1} = \lambda r_n Ax_n + (1-\lambda)x_n, \; r_n = \begin{cases} 1 & \text{if } \|Ax_n\| \leq r \\ \dfrac{r}{\|Ax_n\|} & \text{if } \|Ax_n\| \geq r \end{cases} \qquad (n=0,1,2,\ldots)$$

converges to a fixed point of A in B_r.

Proposition 1.4 extends in two directions the iterative construction of fixed points of contractive mappings. First, we replace the assumption of compactness of A by demicompactness; second, for such mappings, instead of assuming that A maps B_r into B_r we assume only that A satisfies the Leray-Schauder condition on S_r. In our view the latter result for demicompact operators is the main advancement of the theory. Note that if A is compact, then it is clearly demicompact but the converse is not true. For example, it is easy to see that $A = T + S$, where T is strictly contractive and S is compact, is demicompact but it is certainly not compact. Clearly, if A maps B_r into B_r then $r_n = 1$ for each n and (1.7) reduces to the method (1.5). In this case, Proposition 1.4. is valid for an arbitrary closed bounded convex subset C without any change in its proof not only in Hilbert spaces but also in uniformly convex Banach spaces. Thus, in this case, Propositions 1.1. and 1.2 follow from our Proposition 1.4. Let us add that (1.6) is implied by the following conditions:

$$(Ax, x) \leq \|x\|^2 \quad \text{or} \quad \|Ax\| \leq \|x\| \quad \text{for all } x \text{ in } S_r$$

Remark 1.1. It was shown by de Figueiredo-Karlovitz [18] that if the retraction R_C of X onto $C = B_1(0)$ is contractive for a Banach space X of dimension > 2, then X is a Hilbert space. This fact indicates that the retraction-iteration method (1.7) is essentially restricted to Hilbert spaces.

Further progress in the theory of the iterative construction of fixed points was made in Browder-Petryshyn [8] where a number of results concerning the Picard sequence $\{A^n x\}$ was established. We mention here only two results from [8] which are directly relevant to our discussion. We first reeall that a mapping P of X into X is said to be strongly closed(•) if for any sequence $\{u_n\}$ in X with $u_n \rightharpoonup u$ and $Pu_n \rightharpoonup v$ we have $Pu = v$.

(•)Some authors refer to the above property as "demiclosedness." We prefer the term "strongly closed"(employed also in [36]) since it is more in harmony with the concept of "strong continuity"used, especially, by Soviet mathematicians On the other hand, the therm "demiclosedness" of P was used in [41, 34, 36] to describe the situation" $u_n \to u$" and $Pu_n \rightharpoonup v$ implies that $Pu = v$" which seems to be more in harmony with the concept of "democontinuity"(i. e., $u_n \to u$ implies $Pu_n \rightharpoonup Pu$)introduced by Browder and accepted by the mathematical community.

Proposition 1.5. Let A be an asympotically regular (i.e. $A^{n+1}x-A^{n}x \to 0$ as $n \to \infty$ for every x in X) contractive mapping of X into X such that I-A is strongly closed. Suppose that the set F(A) of fixed points of A is not empty. Then for any $x_0 \in X$ the weak limit of any weakly convergent subsequence $\{A^{n(j)}x_0\}$ of the Picard sequence $\{A^n x_0\}$ lies in F(A). In particular, if X is reflexive and F consists of a single point y, then $A^n x_0 \rightharpoonup y$.

We remark that Propostion 1.5 is valid, in particular, when X is a Hilbert space or a Banach space having a weakly continuous duality mapping (•) J. In this case it has been shown in [36] (see also [5, 9]) using the theory of monotone and J-monotone operators that if A is contractive then (I-A) is strongly closed. Hence in this case we drop in Proposition 1.5 the assumption that (I-A) is strongly closed (see Theorem 4 in [8]).

Remark 1.2. Proposition 1.5 was stated for mappings A of X into X instead of C into C in order to employ the theory of monotone and J-monotone operators in the proof of strong closedness of (I-A). This is not a restriction in case of a Hilbert space since a contractive mapping defined on C can be extended to a contractive mapping defined on all of H. But it is a restriction in case of a Banach space since, in general, such an extension theorem is not available.

Proposition 1.6. Let A be an asymptotically regular contractive mapping of X into X with a nonempty set F(A) of fixed points. Suppose further that A satisfies the following condition.

(α) (I-A) maps bounded closed sets into closed sets.

Then for each x_0 in X the Picard sequence $\{A^n x_0\}$ convergence to a fixed point

(•) Let X be the dual space of X and (f, x) the value of the linear functional f in X^* as the element x of X. Let μ be a continuous strictly increasing real values function defined on $R^+ = \{t : t \geq 0\}$ with $\mu(0) = 0$. A mapping J of X into X^* (see [5]) is called a duality mapping with a gauge function μ if $(Jx, x) = \|Jx\| \|x\|$ and $\|Jx\| = \mu(\|x\|)$ for every x in X.

in F(A).

It was shown in [8, 39] that for any fixed $\lambda \in (0,1)$ the mapping $A'_\lambda = \lambda I + (1-\lambda) A$ (or $A_\lambda = \lambda A + (1-\lambda) I$) is a contractive asymptotically regular mapping with $F(A'_\lambda) = F(A)$ (or $F(A_\lambda) = F(A)$) provided that A is a contractive mapping of a uniformly convex Banach space X into X with a nonempty set F(A) of fixed points. Since, furthermore, for $\lambda \in (0,1)$ we have $I - A'_\lambda = (1-\lambda)(I-A)$ (or $I - A_\lambda = \lambda(I-A)$) we see that A satisfies condition (α) if and only if A'_λ (or A_λ) does also. Using the above remarks we have the following corrollary of Proposition 1.6.

Proposition 1.7. Let A be a contractive mapping of a uniformly convex Banach space X into X with a nonempty set F(A) of fixed points. Suppose that A satisfies condition (α). Then the sequence $\{x_{n+1}\}$ determined by (1.5) or by

$$(1.5') \qquad \lambda x_{n+1} = x_n + (1-\lambda) A x_n \quad (0<\lambda<1;\ x_0 \in X;\ n = 0,1,2,\ldots)$$

converges to a fixed point of A.

It will be shown in the next section that Proposition 1.7 remains valid when A is assumed only to be a contractive mapping of a closed bounded convex subset C of a uniformly convex Banach space X into C. In this case we drop the assumption that A has a nonempty set F(A) of fixed points since in this case the existence of fixed points of A in C follows from the recent existence theory proved independently by Browder [4], Kirk [23] and Gohde [19]. Since, as is not hard to see, the demicompactness (and, in particular, compactness) of A implies the validity of condition (α)(•), it follows the modified Proposition 1.7 is an extension of Proposition 1.4 for the case when A map C into C.

(•) It was stated in [8] that condition (α) is equivalent to the demicompactness. This is clearly not the case since the mapping A=I satisfies condition (α) but certainly is not demicompact. The authors of [8] caught this inaccuracy too late for the correction to be inserted in the paper 8

When in Proposition 1.5 the space X is a Banach space with a weakly continuous duality mapping (and, in particular, a Hilbert space) Opial [30] strengthened the result of Proposition 1.5 (or rather of Theorem 4 in 8) by showing that the entire sequence $\{A^n x_0\}$ is necessarily weakly convergent (see also Theorems 7 and 8 in [9]). More specifically, the following was proved in [30].

Proposition 1.8. *Let* C *be a closed set in a uniformly convex Banach space* X *having a weakly continuous duality mapping and let* A *be a contractive mapping of* C *into* C *with at least one fixed point in* C . *Then for any* x_0 *in* C *and any* $\lambda \in (0,1)$ *the Picard sequence* $\{A_\lambda^n x_0\}$ *is weakly convergent to a fixed point of* A *in* C.

We recall that if C is bounded, then the existence of at least one fixed point of A in C follows from the results in [4, 19, 23]. Let us add that Opial did not use the theory of monotone or J-monotone operators to prove the strong closedness of (I-A). His arguments were similar to Schaefer's [39]. We recall that for weakly continuous contractions in real Hilbert spaces the weak convergence of $\{A_\lambda^n x_0\}$ was first proved by Schaefer [39]. The extension of this result to general contractions was thus carried out in two stages, the proof of Proposition 1.5 by Browder-Petryshyn [8], and the proof of Proposition 1.8 by Opial [30].

Remark 1.3. It was already noted in [9] that Gohde's assertion in [19] that the weak convergence result in [39] follows already from the existence theorem is inaccurate since the proof by Schaefer uses two facts, the existence of at least one fixed point and the fact that if a subsequence of $\{A_\lambda^n x_0\}$ converges weakly to y, then y is a fixed point of A. This latter fact follows as in the proof of Theorem 7 in [9] and Opial has utilized this argument together with a simplified form of Schaefer's proof.

Let us add that under the stronger assumption that the strong limit set of the iterates is nonempty, convergence results for contractions have been given by Edelstein [11, 12, 13] and Gohde [20]. However, there seems to be no explicit way of extracting convergent subsequence and therefore the methods are not really constructive.

In [9] Browder and the writer studied various classes of nonlinear operators (contractive, strictly pseudocontractive, psedocontractive) mapping a closed bounded convex subset C of a real Hilbert space H into C or into H. For the purposes of our present discussion we shall mention only those results from [9] which bear direct relation to the iterative construction of fixed points by means of the processes (1.5) or (1.7).

As a first example we state the following proposition which for a Hilbert space is an extension of Proposition 1.8.

Proposition 1.9. Let A be a contractive mapping of C into H. Suppose that if $u \in C$ and if $u = R_C Au$, then u is a fixed point of A. Then A has fixed points in C and the sequence $\{x_{n+1}\}$ determined by the retraction-iteration method

$$x_{n+1} = \lambda R_C(Ax_n) + (1-\lambda)x_n \qquad (0<\lambda< 1; x_0 \in C; n=0,1,2\dots) \tag{1.7A}$$

converges weakly to a fixed point of A in C.

Note that if $C = B_r(0)$ then our second assumption reduces to the Lerady-Schauder condition (1.6) and in this case the retraction-iteration method is practically constructible.

Following [9] we say that a bounded closed convex set C is uniformly smooth with smoothing constant $R > 0$ if and only if for each boundary point x_1 in C

1) C has only one supporting hyperplane at x_1 (i.e., $(x_1, v) = c_0$ and $(x_1, u) \geq c_0$ for all u in C)

2) There exists u_0 in H such that $\|u_0 - x_1\| = R$ and $\|u_0 - u\| \leq R$ for all $u \in C$ (i.e., $C \subset B_R(u_0)$ and $x_1 \in \partial B(u_0)$).

Our next result is the following extension of Proposition 1.4.

Proposition 1.10. Suppose that in addition to the conditions of Proposition 1.9 we assume that A is a demicompact map of C into H, where C is uniformly smooth with smoothing constant R. Then the sequence $\{x_{n+1}\}$ determined by the retraction-iteration method (1.7A) converges to a fixed point of A in C.

Let us note that the main difficulty in proving Proposition 1.10 lies in the proof of the demicompactness of $R_C A$. The rather tricky proof of this in [9] utilizes the idea (which, essentially, was used for the first time by the writer in [35]) that outside some narrow band about the surface of C the retraction is strictly contractive. Needless to say the chief difficulty in the application of Propositions 1.9 and 1.10 to actual problems lies in the fact that unless C is a ball there is no efficient way of construction the retraction mappings R_C.

Following [9] we say that the mapping A of C into H is strictly pseudo-contractive if there exists a constant $k < 1$ such that

$$\|Ax-Ay\|^2 \leq \|x-y\|^2 + k\|(I-A)x - (I-A)y\|^2 \quad \text{for all } x, y \in C. \tag{1.8}$$

It has been shown in [9] that for every fixed t such that $0 < t \leq 1-k$ the mapping $A_t = tA + (1-t)I$ is contractive, A_t has the same fixed points as A in C and $(A_t)_\lambda = \lambda I + (1-\lambda)A_t = A_\gamma$ where $\gamma = 1-(1- \quad)t$ with $t < 1-k$ for any fixed $\lambda \in (0,1)$ if and only if $\gamma > k$. Consequently we have the following extensions of Propositions (1.9) and (1.10) for A mapping C into C.

Proposition 1.11. Let A be a strictly pseudocontractive mapping of C into C. Then, for any $x_0 \in C$ and any γ such that $k < \gamma < 1$, $\{A^n x_0\}$ converges weakly to a fixed point of A in C. If additionally we assume that A is demicompact, then $\{A^n x_0\}$ converges strongly.

Similar result holds for the retraction-iteration method (1.7) when it is applied to strictly pseudocontractive mappings A of $B_r(o)$ into H which on S_r satisfy the Leray-Schauder condition (1.6).

Remark 1.4. In [9] the authors also discuss the construction of fixed points of a pseudocontractive mapping A of $B_r(o)$ into H with A satisfying the Leray-Schauder condition on S_r, where A is said to be <u>pseudocontractive if</u>

$$\| Ax - Ay\|^2 \leq \|x - y\|^2 + \|(I-A)x - (I - A)y\|^2 \quad \text{for all} \quad x, y \in C.$$

The mothod used in the construction of fixed points of such mappings is essentially based on the strict contraction mapping principle combined with the radial retraction R. The idea is that for each fixed s with $0 < s < 1$ we construct the fixed points u_s in $B_r(o)$ of the mapping $A_s = sA + (1-s)u_o$ for any u_o in $B_r(o)$ by the method

$$(1.10) \qquad u_s = \lim_n (u_s^n) = \lim_n \left\{ R(1 - \quad (I-A_s)) \right\}^n w_0 , \qquad w_o \in B_r(o)$$

where the real number μ lies in a certain interval (see [9] and then show that $u_s \to u$, where u is a fixed point of A closest to u_o. We will not investigate the method (1.10) since this is outside the scope of our discussion.

Let X be a real separable Banach space. A pair of sequence $(\{X_n\}, \{P_n\})$, where each X_n is a finite dimensional subspace of X and P_n is a linear projection of X onto X_n, is said to be a <u>projectionally complete system</u> if $P_n x \to x$ for each x in X (see [34]).

<u>Definition.</u> A nonlinear mapping A of $D(A)$ ($\subset X$) into X will <u>be called P_γ-compact</u> if $P_n A$ is continuous in X_n for all sufficiently large n and if there exists a constant $\gamma \geq 0$ such that for any p dominating γ (i.e., $p \geq \gamma$ if $\gamma > 0$ and $p > \gamma$ if $\gamma = 0$) and any bounded sequence $\{x_n\}$ with $x_n \in X_n \cap D(A)$ the strong convergence of the sequence $\{P_n Ax_n - px_n\}$ implies the existence of a strongly convergent subsequence $\{x_{n_i}\}$ and an element x in $D(A)$ such that $x_{n_i} \to x$ and $P_{n_i} Ax_{n_i} \to Ax$ as $n_i \to \infty$.

Note that our present definition of P_γ -compactness is a generalization of the concept of P-compactness introduced and studied by the writer in [31, 32, 33]. Indeed, when $\gamma = 0$, $D(A) = X$ and the sequences $\{X_n\}$ and P_n are such that

(1.11) $X_n \subset X_{n+1}$ $(n = 1, 2, 3, \ldots)$, $\overline{\bigcup_n X_n} = X$ and $\|P_n\| \leq K$, $K \geq 1$, then the P_0-compactness is just the P-compactness. In what follows we shall retain the the term P-compactness whenever $\gamma = 0$. Let us remark that upon a closer examination of the proofs of fixed point theorems in [31, 32, 33] it becomes obvious that in those proofs we have only used the properties of A and X which are embodied in the requirement that A be P_1-compact (according to our extended definition) as a mapping from $B_r(0)$ into X. However, in the applications of fixed point theorems in [32, 33] to monotone operators we still have to use the definition as given in 31,32. It turns out that our present extended definition of P_γ -compactness is more suitable for theoretical and applicational purposes and, particularly, for the purpose of unification of various results.

Returning to our discussion of iterative methods we would like to finish this section by stating the following result proved in [37] for P-compact mappings.

Proposition 1.12. Let C be a closed bounded convex subset of a real uniformly convex Banach space having a projectionally complete system. If A is a contractive P_1-compact mapping of C into C, then the Picard sequence $\{x_{n+1}\}$ determined by (1.5) converges to a fixed point of A in C.

2.1 Convergence proofs. In this section we prove three theorems concerning the strong and weak convergence of the Picard sequences $\{A_\lambda^n x\}$ under certain conditions on A and X, where A is a contractive or a strictly pseudocontractive mapping of C into C. We have seen in the previous section that in each case the strong or the weak convergence of $\{A^n x\}$ depended upon the existence of fixed points of A in C. It turns out that in case A is a contractive mapping (which is the case that we are interested in this discussion) the properties and the additional conditions on A which insure either the strong or the weak convergence of the iterative method (1.5) admit at the same time a simple proof of an existence theorem which, as will be seen below, is general enough to cover all cases surveyed in Section 1.1. Thus, there is no necessity to refer the reader to the more general existence theorems proved in [4, 19, 23] by rather complicated arguments .

Unless explicitly stated otherwise in this section the set C will always be assumed to be a closed bounded convex subset of a given space X .

Theorem 1. Let X be a general Banach space and let A be a contractive mapping of a subset C of X into C such that (I-A)(C) is a closed set in X. Then A has a fixed point in C.

Proof. Assume without loss of generality that the origin $0 \in C$. Let r_n be a sequence of numbers such that $0 < r_n < 1$ for each n and $r_n \to 1$ as $n \to \infty$. It is obvious that $A_n = r_n A$ is a strictly contractive mapping of C into C. Hence, by the strict contraction mapping principle, for each n there exists a unique point u_n in C such that $A_n u_n = u_n$. For the sequence $\{u_n\}$ thus determined we have

$$u_n - Au_n = r_n Au_n - Au_n = (r_n - 1) Au_n \to 0 \text{ as } n \to \infty$$

since $r_n \to 1$ and $\{Au_n\} \subset C$ is bounded. Hence 0 lies in the closure of (I - A)(C) . Since, by assumption, (I-A)(C) is closed, $0 \in (I-A(C)$. and therefore A has a fixed point in C.

Lemma 1. If in addition to conditions of Theorem 1 we assume that X is strictly convex, then the set F(A) of fixed points of A in C is a closed convex set.

Proof. By Theorem 1, $F(A) \neq \emptyset$. Suppose that x and y are in $F(A)$ and $z = tx + (1-t)y$ for $t \in (0,1)$. Then, by the contractivity of A and convexity of C, z C and

$$\|x-y\| = \|Ax-Ay\| \leq \|Ax-Az\| + \|Az-Ay\| \leq \|x-z\| + \|z-y\| = \|x-y\| .$$

Since X is strictly convex the above inequality implies that $Ax - Az = a(Az - Ay)$ for some real number a. Therefore, the vector Az is on the line through Ax and Ay and, hence, on the line through x and y. But we also have that $\|Az-Ax\| \leq \|z-x\|$ and $\|Az-Ay\| \leq \|z-y\|$. Hence we must have that $Az = z$. The closedness of $F(A)$ is obvious.

Remark 2.1. Lemma 1 is valid under the assumption that X is strictly convex and the contractive mapping A has a nonempty set $F(A)$ of fixed points.

Lemma 2. Let X *be a uniformly convex Banach space and* A *a contractive mapping of* C $(\subset X)$ *into* C. *Then for any fixed* $\lambda \in (0,1)$ *the mapping* $A_\lambda = \lambda I + (1-\lambda)A$ *is asymptotically regular (i.e.,* $A_\lambda^{n+1} x_0 - A_\lambda^n x_0 \to 0$ *as* $n \to \infty$ *for* $x_0 \in C$) *and* A_λ *has the same fixed points as* A.

Proof. It is obvious that $F(A) = F(A_\lambda)$ and that A_λ maps C into C and is contractive. Now let u be a fixed point of A in C and, for any given x_0 in C, let $\{A_\lambda^n x_0\}$ be the Picard sequence. Then

$$\|A_\lambda^{n+1} x_0 - u\| = \|A_\lambda^{n+1} x_0 - Au\| \leq \|A_\lambda^n x_0 - u\| \quad \text{for all } n .$$

Thus $\|A^n x_0 - u\|$ converges to some $d_0 \geq 0$. If $d_0 = 0$, then $A_\lambda^{n+1} x_0 - A_\lambda^n x_0 \to 0$ and the proof is finished. Suppose now that $d_0 > 0$. Since $Au = u$ and

$$A_\lambda^{n+1} x_0 - u = \lambda A_\lambda^n x_0 + (1-\lambda)AA^n x_0 - u = \lambda(A_\lambda^n x - u) + (1-\lambda)(AA_\lambda^n x_0 - u)$$

and since $\|A^{n+1} x_0 - u\| \to d_0, \|A^n_\lambda x_0 - u\| \to d_0$ and $\|AA^n x_0 - u\| \leq \|A^n x_0 - u\|$ it follows from the uniform convexity of X that

$$\|(A^n_\lambda x_0 - u) - (AA^n_\lambda x_0 - u)\| \to 0, \text{ i.e., } A^{n+1}_\lambda x_0 - A^n_\lambda x_0 \to 0$$

since $A_\lambda - I = (\lambda - 1)(I - A)$; hence the proof of Lemma 2 is complete.

We now prove the first main theorem (compare it with Proposition 1.6) of this section by the simple arguments analogous to those used in [8].

<u>Theorem 2. Let</u> X <u>be a uniformly convex Banach space and</u> A <u>a contractive mapping of the subset</u> C <u>of</u> X <u>into</u> C. <u>Suppose further that</u> (I-A) <u>satisfies condition</u> (α), <u>i.e.</u>, (I-A) <u>maps every (bounded) closed subset of</u> C <u>into a closed subset. Then the set</u> F(A) <u>of fixed points of</u> A <u>in</u> C <u>is a nonempty closed convex set and for each</u> x_0 <u>in</u> C <u>and</u> $\lambda \in (0, 1)$ <u>the sequence</u> $\{A^n x_0\}$ <u>converges (strongly) to a fixed point of</u> A <u>in</u> C.

<u>Proof.</u> Since, by condition (α), (I-A) (C) is a closed set it follows from Theorem 1 that $F(A) \neq \emptyset$. Since uniform convexity of X implies its strict convexity, Lemma 1 implies that F(A) is a closed convex set. If u is a fixed point of A, and therefore of A_λ the sequence $\{\|A^n_\lambda x_0 - u\|\}$ is decreasing .

It suffices therefore to show that there exists a subsequence of $\{A^n_\lambda x_0\}$ which converges strongly to a fixed point of A in C. Let G be the strong closure of the set $\{A^n_\lambda x_0\}$. By Lemma 2, $(I-A_\lambda)(A^n_\lambda x_0) \to 0$ as n ∞ . Hence 0 lies in the strong closure of $(I-A_\lambda)(C)$. But , since for any $\lambda \in (0, 1)$, $I-A_\lambda = (1-\lambda)(I-A)$ we see that $(1-A_\lambda)$ satisfies condition (α) if and only if (I-A) does also. Hence, by condition (α), the set $(I-A_\lambda)(C)$ is closed since G is closed (and necessarily bounded) and therefore 0 lies in $(1-A_\lambda)(C)$. Hence there exists a strongly convergent subsequence of $\{A^n_\lambda x_0\}$ which converges to an element y in G such that $(I-A_\lambda)_y = 0$, i.e., y is a fixed point of A in C .

We shall now prove that in addition to completely continuous operators the two classes of mappings which satisfy condition (α) are demicompact and P_1-compact mappings studied by the writer.

<u>Lemma 3. Let</u> X <u>be a general Banach space and</u> A <u>a continuous demicompact mapping of</u> $D(A)(\subset X)$ <u>into</u> X . <u>Then</u> A <u>satisfies condition</u> (α).

<u>Proof.</u> Let Q be a bounded closed set in D(A) and let $\{u_k\}$ be a sequence in Q such that $(I-A)u_n \to v$ as $n \to \infty$. Since u_n is a bounded sequence, and $\{(I-A)u_n\}$ is strongly convergent, by demicompactness of A, there exists a strongly convergent subsequence $\{u_{n_i}\}$. Hence the closedness of Q and the continuity of A imply that $u_{n_i} \to u$ for some u in Q with $(I-A)u_{n_i} \to (I-A)u = v$ and Lemma 3 is proved.

<u>Lemma 4. Let</u> X <u>be a real Banach space with a projectionally complete</u> $(\{X_n\}, \{P_n\})$. <u>Let</u> A <u>be a Lipschitzian</u> P_1<u>-compact mapping of</u> $D(A) \subset X$ <u>into</u> X . <u>Then</u> A <u>is demicompact and hence satisfies condition</u> (α) .

<u>Proof.</u> Let Q be a bounded closed set in D(A) and let $\{u_k\}$ be a sequence in Q such that $(I-A)u_k \to v$ as $k \to \infty$. Since the system $(\{X_n\}, \{P_n\})$ is projectionally complete in X , for each integer k and $_k = 1/k$, there exists an integer n(k) (which we can and shall assume that $n(k) > k$) such that $\| u_k - P_{n(k)} u_k \| < \varepsilon_k$. This and the relation $(I-A)u_k \to v$ imply that with $w_{n(k)} = P_{n(k)} u_k$ in $X_{n(k)}$ we have

$$\| P_{n(k)} A w_{n(k)} - w_{n(k)} + v \| \leq \| P_{n(k)} A w_{n(k)} - P_{n(k)} A u_k \|$$

$$+ \| P_{n(k)} A u_k - w_{n(k)} + P_{n(k)} v \| + \| v - P_{n(k)} v \| .$$

Since $\{P_n\}$ is uniformly bounded, say, by $K \geq 1$ and A is Lipschitzian, say with constant $L > 0$, it follows from the above inequality that

$$\| P_{n(k)}Aw_{n(k)}-w_{n(k)}-v \| \leq KL \|w_{n(k)}-u_k\| K\|Au_k-u_k+v\| + \|v-P_{n(k)}v\|$$

$$\longrightarrow 0 \quad \text{as} \quad k \longrightarrow \infty.$$

Thus, since A is P_1-compact, there exists a subsequence $\{w_{n(j)}\}$ of $\{w_{n(k)}\}$ and an element u in D(A) such that $w_{n(j)} \to u$ and $P_{n(j)}Aw_{n(j)} \longrightarrow Au$ as $n(j) \longrightarrow \infty$. Consequently, $\|u_j-u\| \leq \|u_j-w_{n(j)}\| + \|w_{n(j)}-u\| \to 0$ as $n(j) \longrightarrow \infty$. Since $\{u_j\}$, being a subsequence of $\{u_n\}$, lies in Q and Q is closed we must have that $u \in Q$ and $Au = v$, i.e., A is demicompact. The last assertion of Lemma 4 follows from the above proof and Lemma 3.

Remark 2.2. In virtue of Lemmas 3 and 4 and the observations following Proposition 1.4, Theorem 2 implies the validity of all propositions of the previous section asserting the (strong) convergence of the sequence $\{A_\lambda^n x_0\}$ provided that in these propositions A maps a closed bounded convex subset C of a uniformly convex Banach space X into C.

To derive the strong convergence of the retraction-iteration method, Remark 1.1 implies that we need consider only Hilbert spaces H. In virtue of Lemma 3, the strong convergence of the retraction-iteration method (1.7) or (1.7A) will follow from the following lemma.

Lemma 5. *Let A be a contractive demicompact mapping of a closed bounded convex subset C of real space H into H, where C is uniformly smooth with smothing constant R. If R_C denotes the retraction of H onto C, then the mapping $R_C A$ is demicompact.*

Note. In case $C = B_r(0)$, Lemma 5 was first proved by the writer in [35]. For C, which is uniformly smooth with smoothing constant R, Lemma 5 was proved in [9] by the arguments which we use here. It is essentially based on the following lemma whose rather tricky proof is given in [9].

Lemma 6. *If C is uniformly smooth with smoothing constant R and* the distances $d(x, C) \geq r_0$ and $d(y, C) \geq r_1$, then

$$(2.1) \qquad \|R_C x - R_C y\| \leq \left(1 + \frac{r_0 + r_1}{2R}\right)^{-1} \|x-y\|$$

Proof of Lemma 5. Assuming the validity of Lemma 6 we now prove Lemma 5. Suppose $\{u_n\}$ is a sequence in C such that $g_n = u_n - R_C Au_n \to g$ in H. If there exists an infinite subsequence $\{u_{n_i}\}$ such that $d(Au_{n_i}, C) \to 0$ as $i \to \infty$, then $R_C Au_{n_i} - Au_{n_i} \to 0$, and, hence, $Au_{n_i} - u_{n_i} \to g$. Since A is demicompact, there exists a subsequence $\{u_{n_j}\}$ of $\{u_{n_i}\}$ which is strongly convergent and, hence, in this case Lemma 5 is proved. Otherwise, there exists a constant $\delta > 0$ such that $d(Au_n, C) \geq \delta$ for all n. In this case, by Lemma 6,

$$\|R_C Au_n - R_C Au_m\| \leq (1 + \delta/R)^{-1} \|Au_n - Au_m\| \leq k\|u_n - u_m\|,$$

with $k = (1 + \delta/R)^{-1} < 1$. Hence

$$\|u_n - u_m\| \leq \|(u_n - R_C Au_n) - (u_m - R_C Au_m)\| + \|R_C Au_n - R_C Au_m\|$$

whence we obtain the relation $(1-k)\|u_n - u_m\| \leq \|g_n - g_m\| \to 0$ as $n, m \to \infty$, i.e., $\{u_n\}$ is strongly convergent and the proof of Lemma 5 is complete.

Remark 2.3. In virtue of Lemma 4 all results proved for contractive and strictly pseudocontractive demicompact mappings remain valid for similar P_1-compact mappings.

Finally, to prove the second assertion of Proposition 1.11 concerning the (strong) convergence of the method (1.5) for strictly pseudocontractive mappings, as was already noted, it is not hard to show that for every fixed t such that $0 < t \leq 1 - k$ the mapping $A_t = tA + (1-t)I$ is contractive maps C into C and has the same fixed points as A. Furthermore, $(A_t)_\lambda = \lambda I + (1-\lambda)A_t = (1-t(1-\lambda))I + t(1-\lambda)A = \gamma I + (1-\gamma)A$, where we have put $\gamma = 1-t(1-\lambda)$. It is not hard to see that, for any fixed

$\lambda \in (0,1)$, $t < 1-k$ if and only if $\gamma > k$. Indeed, if $t < 1-k$ and $1 > \gamma > 0$ is to hold for any $\lambda \in (0,1)$ it follows that γ must satisfy the inequality $\gamma = 1-t(1-\lambda) > 1-(1-k)(1-\lambda) > k$. On the other hand, if $k < \gamma = 1-t(1-\lambda)$ is to hold for any $\lambda \in (0,1)$ the parameter t must satisfy the inequality $k < 1-t$, i.e., $t < 1-k$. Now, since $A_\gamma - I = (1-\gamma)(A-I)$, it follows that if A is demicompact then so is A_γ and consequently, by Theorem 2, the Picard sequence $\left\{ (A_t)_\lambda^n x_0 \right\} = \left\{ A_\gamma^n x_0 \right\}$ converges to a fixed point of A in C.

To justify the observation following Proposition 1.11, concerning the convergence of the retraction-iteration method for a strictly pseudocontractive mapping of $B_r(0)$ into H, by the second part of Proposition 1.11 and Lemma 5, it suffices to show that A_γ satisfies the Leray-Schauder condition on S_r, i.e., $A_\gamma x - \beta x \neq 0$ for all x in S_r and any $\beta > 1$. Suppose, to the côntrary, that $A_\gamma x_0 - \beta_0 x_0 = 0$ for some x_0 in S_r and some $\beta_0 > 1$. Then $A_\gamma x_0 = \gamma x_0 + (1-\gamma) A x_0 = \beta_0 x_0$ or $A x_0 = \frac{\beta_0 - \gamma}{1-\gamma} > 1$, since $\beta_0 > 1$, in contradiction to (1.6).

Suppose now that we drop the additional assumption that A satisfies condition (α). Without this condition, thus far it was only possible (see [39, 8, 30, 9]) to prove the weak convergence of $\left\{ A^n x_0 \right\}$. In discussing the weak convergence of $\left\{ A_\lambda^n x_0 \right\}$ we will use the arguments similar to those applied in [8, 30, 9]. The two succeeding lemmas were proved in a somewhat different form in [30].

Lemma 7. *Let X be a uniformly convex Banach space having a weakly continuous duality mapping J of X into X^*. Let $\{x_n\}$ be a sequence in X such that $x_n \rightharpoonup x_0$. Then there exists a subsequence $\{x_m\}$ of $\{x_n\}$ such that for any x in X*

(2.2)
$$\lim_m \|x_m - x\| \geq \lim \|x_m - x_0\| .$$

with equality holding in (2.2) *if and only if* $x = x_0$.

Proof. It is obvious that there exists a subsequence $\{x_m\}$ of $\{x_n\}$ such that $\|x_m - x_0\| \to d_0$ and $\|x_m - x\| \to d$ as $m \to \infty$, where d_0 and d are some real numbers. Hence, since $x_m - x_0$ 0 implies $J(x_m - x_0) \rightharpoonup 0$, the definition of J and the passage to the limit in the inequality

$$(J(x_m - x_0), x_m - x_0) \leq \|J(x_m - x_0)\| \, \|x_m - x\| + |(J(x_m - x_0), x - x_0)|$$

imply that $\mu(d_0)d_0 \leq \mu(d_0)d$, from which (2.2) follows. To complete the proof suppose $d_0 = d$ in (2.2). Since $\|x_m - (tx + (1-t)x_0)\| \leq t\|x_m - x\| + (1-t)\|x_m - x_0\|$ for any t in $[0,1]$ it follows from the first part of Lemma 7 and our assumption $d_0 = d$ that

$$\|x_m - x_0\| \to d_0, \; \|x_m - x\| \to d_0 \text{ and } \|x_m - (tx + (1-t)x_0)\| \to d_0.$$

Hence, by uniform convexity of X, it follows that $\|(x_m - x_0) - (x_m - x)\| \to 0$, i.e., $x = x_0$.

An immediate consequence of Lemma 7 is the following result which is essential in our discussion.

Lemma 8. Let X be a uniformly convex Banach space with a weakly continuous duality mapping J of X into X^*. If A is a contractive mapping of a closed convex set C of X into X, then the mapping (I-A) is strongly closed.

Proof. Let $\{x_n\}$ be a sequence in C which converges weakly to some x_0 in X and let $\{x_n - Ax_n\}$ converge strongly to some y_0 in X. Since C is closed and convex and hence weakly closed, x_0 lies in C. Hence it suffiees to show that $(I-A)x_0 = y_0$. Let $\{x_m\}$ be a subsequence of $\{x_n\}$ such that $\lim \|x_m - x_0\|$ exists. Since $x_m - Ax_m \to y_0$, there exists a sequence $z_m \to 0$ such that $Ax_m = x_m - y_0 + z_m$ and

$$\|x_m - x_0\| \geq \|Ax_m - Ax_0\| = \|x_m - y_0 - Ax_0 + z_m\| \geq \|x_m - y_0 - Ax_0\| - \|z_m\|$$

which implies that $\| \lim_m x_m - x_0 \| \geq \lim_m \| x_m - y_0 - Ax_0 \|$. It follows from this and Lemma 7 that $x_0 = y_0 - Ax_0$.

Remark 2.4. As was already noted, Lemma 8 was first proved in a more general setting in [5] using the theory of J-monotone operators under the assumption that A is defined on all of X. In case of a Hilbert space a somewhat different proof is given in [9] .

Remark 2.5. Using Lemma 8 it is easy to show that if X has properties assumed in Lemma 8 and A is a contractive mapping of a closed bounded convex subset C of X into X , then the set $(I-A)(C)$ is closed . Indeed, let $\{u_n\}$ be a sequence in C with $(I-A)u_n \to v_n$ as $n \to \infty$. We need to show that v_0 lies in $(I-A)(C)$. Since X, being uniformly convex, is reflexive, we may replace $\{u_n\}$ by a subsequence, which we again denote by $\{u_n\}$, such that $u_n \rightharpoonup u_0$ for some u_0 in X. Since C is closed and convex and hence weakly closed, u_0 lies in C. Hence, by Lemma 8, $(I-A)u_0 = v_0$. Consequently by Theorem 1, if A maps C into C, then A has fixed points in C.

We now prove the second main theorem (compare it with Proposition 1.8) in this section by the arguments similar to those used in [39, 30, 9] in the case of Hilbert space.

Theorem 3. Let C be a closed bounded convex subset of a uniformly convex Banach space having a weakly continuous duality mapping J of X into X^* and let A be a contractive mapping of C into C. Then for any x_0 in C and any $\lambda \in (0,1)$ the sequence $\{A^n x_0\}$ converges weakly to a fixed point of A in C.

Proof. By Theorem 1, Remark 2.4 and Lemma 1, A has a nonempty closed convex set F(A) of fixed points in C and $F(A) = F(A\ )$. Since, for any u in F(A), the sequence $\{\| A^n x_0 - u \|\}$ is decreasing, we can define the nonnegative real valued function g(u) from F(A) into R^+ by

$$g(u) = \inf_n \| A^n x_0 - u \| = \lim_n \| A_\lambda^n x_0 - u \| .$$

It is easy to see that $g(u)$ is a continuous convex function and, thus, weakly lower semicontinuous on $F(A)$. Since $F(A)$ is weakly compact, there exists an element u_0 in $F(A)$ such that

$$g(u_0) = d_0 = \inf \{ g(u) : u \in F(A) \} .$$

Furthermore, u_0 is unique. Indeed, suppose there exists another point v_0 in $F(A)$ such that $d_0 = g(v_0)$. Then, by the convexity of $g(u)$, it follows that, for any $t \in [0, 1]$, $g(tu_0 + (1-t)v_0) \leq tg(u_0)+(1-t)g(v_0) = d_0$. Hence, as $n \longrightarrow \infty$, we have

$$\|A^n x_0 - u_0\| \longrightarrow d_0,\ \|A_\lambda^n x_0 - v_0\| \rightarrow d_0 \text{ and } \|A_\lambda^n x_0 - (tu_0 + (1-t)v_0\| \rightarrow d_0$$

from which, on account of uniform convexity of X, it follows that

$$\|(A^n x_0 - u_0) - (A^n x_0 - v_0)\| \longrightarrow 0, \text{ i.e., } u_0 = v_0 .$$

We now prove that $A_\lambda^n x_0 \rightharpoonup u_0$. If not, then since $\{A_\lambda^n x_0\}$ is bounded and X is reflexive, there exists a subsequence $\{A^{n_i} x_0\}$ and an element u in C such that $A_\lambda^{n_i} x_0 \rightharpoonup u$ and $u \neq u_0$. Since, by Lemma 2, A_λ is asymptotically regular it follows that $(I-A)(A^{n_i} x_0) \rightarrow 0$ and $A^{n_i} x_0 \rightharpoonup u \in C$. Hence, by Lemma 8, $(I-A)u = 0$, i.e., $u \in F(A)$. Finally, by Lemma 7,

$$d_0 = g(u_0) = \lim_{n_i} \| A^{n_i} x_0 - u_0 \| > \lim_{n_i} \| A_\lambda^{n_i} x_0 - u \| = g(u) .$$

This yields the contradiction of the definition of d_0. Hence we must have that $u_0 = u$, i.e., every weakly convergent subsequence of $\{A_\lambda^n x_0\}$ converges weakly to u_0. Hence $\{A_\lambda^n x_0\}$ converges weakly to u_0.

Remark 2.6. For a somewhat different proof of Theorem 3 in case X is a Hilbert space see [9]. We remark that the class of Banach spaces X having weakly continuous duality mappings J includes all Hilbert spaces as well as the sequence spaces ℓ^p for $1 < p < \infty$ but, as was shown in [7], it does not include any of the L^p spaces except

L^2. As was already noted, when F(A) has just one point, Theorem 3 was first proved in [8] by using the theory of J-monotone operators. In its present form it was proved in [30].

We remark that the weak convergence of the retraction-iteration method (1.7A) (i.e., Proposition 1.9) and of the sequence $\{A_\gamma^n x_0\}$ (i.e., Proposition 1.11) follows from Theorem 3 and the observations following Remark 2.3 since, as is not hard to prove the mappings $R_C A$ and A_γ are contractive from C to C and satisfy the needed conditions on the boundary of C.

For the sake of completeness we end this section by proving Edelstein's extension of Proposition 1.2 to strictly convex Banach spaces (see [14, 17]).

Theorem 4. *If A is a completely continuous contractive mapping of a subset C of a strictly convex Banach spaces X into C, then for any x_0 in C and any $\lambda \in (0,1)$ the sequence $\{A_\lambda^n x_0\}$ converges to a fixed point of A in C.*

Proof. It suffices to show that there exists a subsequence of $\{A_\lambda^n x_0\}$ which converges to a fixed point of A in C. First note that, by Schauder's fixed point principle or by Theorem 1 above, A has nonempty set F(A) of fixed points in C which is closed, convex and $F(A) = F(A_\lambda)$. Next, let M be the convex closure of the set $A(C) \cup \{x_0\}$. By Masur's Theorem [29], M is compact. Since $\{A_\lambda^n x_0\} \subset M$, there exists a subsequence $\{A_\lambda^{n_i} x_0\}$ and u in C such that $A^{n_i} x_0 \to u$. If u F(A), then Theorem 4 follows. Suppose, to the contrary, that $u \notin F(A)$. Then, for each x in F(A), the strict convexity of X implies that for some $a > 0$, depending on x, $a \equiv \|u-x\| - \|A_\lambda u - x\| > 0$. Since $A_\lambda^{n_i} x_0 \to u$, it follows that, for all sufficiently large n_i,

$$\left\| A_\lambda^{n_i + 1} x_0 - x \right\| \leq \left\| A_\lambda u - x \right\| + a/2 .$$

Since $x \in F(A\ )$ and A_λ is contractive, the latter two inequalities and the strict convexity of X imply that $\|A_\lambda^p x_0 - x\| \leq \|u - x\|_{n_i} - a/2$ for all sufficiently large p which contradicts the convergence $A_\lambda x_0 \to u$.

3.1. Convergence of fixed points of strictly contractive mappings kA.

Let A be a contractive mapping of a subset C (closed bounded and convex) of a Banach space X into C and let $\{k_n\}$ be a sequence of real numbers such that $k_n \to 1$ $(n \to \infty)$ and $0 < k_n < 1$. Let v_0 be an arbitrary point of C. Then for each n the mappings $A_n(x) = k_n Ax + (1-k_n)v_0$ is strictly contractive mapping of C into C. Hence, for each n, there exists a unique point u_n in C such that $A_n u_n = u_n$. Our problem is to discuss the conditions under which, as $n \to \infty$, the sequence $\{u_n\}$ converges to a fixed point of A in C. In case X is a uniformly convex Banach space (in particular, a Hilbert space) with X^* strictly convex and with a weakly continuous duality mapping J of X into X^* and A is a contractive mapping of X into X such that A maps C into C, this problem was completely settled in an interesting paper by Browder [6]. Browder's discussion rests heavily upon the theory of J-monotone operators introduced in [5] and further developed in [7] since, as shown by Browder, there exists a nice nonnection between such operators and contractive mappings defined on the whole space. X.

The purpose of this section is to rederive Browder's results [6] for a contractive mapping A which is defined only onthe subset C, i.e., A maps C into C. It is worth noting that one cannot use the theory of J-monotone operators when A is defined only on C. Our proofs utilize the assertions of Lemmas 7 and 8. Let us add that our proofs thus become quite simple. At the end of this section we consider a practically useful choice of the sequence $\{k_n\}$ suggested by de Figueiredo [17] for which a single iteration process converges.

We now reformulate and prove somewhat generalized versions of Browder's results in [6].

Theorem 5. Let X be a uniformly convex Banach space having a weakly continuous duality mapping J of X into X^* and let A be a contractive mapping of a subset C of X into C. Let v_0 be an arbitrary point of C. For each k_n in (0, 1) let A_n be the strictly contractive mapping of C into C defined by $A_n x = k_n Ax+(1-k_n)v_0$ for all x in C. Let u_n be the unique fixed point of A_n in C.

Them from each sequence $\{u_{n(j)}\}$ with $k_{n(j)} \to 1$ as $n(j) \to \infty$ we can extract a strongly convergent subsequence converging to a fixed point of A in C.

Proof. It suffices to assume for a given sequence $k_j = k_{n(j)} \to 1$ as $j \to \infty$, that $v_j = u_{n(j)} \rightharpoonup v$ and to prove that $v_j \to v$. Since C is weakly closed, v lies in C and since v_j is the fixed point of $A_{n(j)} = k_{n(j)} A + (1-k_{n(j)})v_0 = k_j A + (1-k_j)v_0$ in C we have

$$v_j - av_j = v_j - k_j Av_j + k_j Av_j = (1-k_j)v_0 - (1-k_j) Av_j \to 0$$

as $j \to \infty$ since $k_j \to 1$ and $\{Av_j\}$ is bounded. Hence, by Lemma 8, v is a fixed point of A in C, i.e., $v \in F(A)$. Now for each j we have

$$(3.1) \qquad (i-k_j)v_j + k_j \{v_j - Av_j\} = (1-k_j)v_0$$

and, since v is a fixed point of A in C,

$$(3.2) \qquad (1-k_j)v + k_j \{v - Av\} = (1-k_j)v.$$

Subtracting (3.2) from (3.1) and taking its inner product with $J(v_j - v)$ we get

$$(3.3) \quad (1-k_j)(v_j-v, J(v_j-v)) + k_j((I-A)v_j - (I-A)v, J(v_j-v) = (1-k_j)(v_0-v, J(v_j-v)).$$

Since $v_j, v \in C$ and A is contractive on C, it follows from this and the definition of J that

$$((I-A)v_j - (I)A)v, J(v_j-v)) = (v_j-v, J(v_j-v)) - (Av_j - Av, J(v_j-v)) = \|v_j - v\| \, \|J(v_j-v)\| - $$
$$- (Av_j - Av, J(v_j-v)) \geq \|v_j - v\| \, \|J(v_j-v)\| - \|Av_j - Av\| \, \|J(u_j - v)\| \geq 0$$

Thus, it follows from (3.3) and the definition of J that

$$(1-k_j)\,\|v_j-v\|\,\|J(v_j-v)\| \leq (1-k_j)(v_0-v, J(v_j-v))$$

Cancelling the positive factor $(1-k_1)$, we obtain

$$\|v_j-v\| \quad (\,\|v_j-v\|\,) \leq (v_0-v, J(v_j-v)) \,. \tag{3.4}$$

Since $v_j \rightharpoonup v$ and J is weakly continuous, $J(v_j-v) \rightharpoonup 0$. Hence it follows from (3.4) that $\|v_j-v\| \to 0$, i.e., $v_j \to v$.

Theorem 6'. *If in addition to conditions of Theorem 5 we assume that there exists a point u_0 in F(A) such that*

$$(v_0-u_0, J(u_0-v)) \geq 0 \text{ for all } v \text{ in } F(A)\,, \tag{3.5}$$

then $u_n \to u_o$ as $k_n \to 1$ $(n \quad \infty)$.

Proof. Let u_0 be a point of F(A) such that (3.5) holds for all v in F(A) . Then, since u_0 is a fixed point of A in C, by repeating a part of the proof of Theorem 5 we get

$$\|v_j-u_0\| \,(\,\|y_j-u_0\|\,) \leq (v_0-u_0, J(u_j-u_0)) \,. \tag{3.6}$$

Since $v_j-u_0 \rightharpoonup v - u_0$, and J is weakly continous , $J(v_j-u_0) \rightharpoonup J(v-u_0)$. Hence, by (3.5), $(v_0-u_0, J(v_j-u_0)) \to (v_0-u_0, J(v-u_0)) \leq 0$. Consequently, (3.6) implies that $v_j \to u_0$ as $j \to \infty$. Since $\{v_j\}$ was any weakly convergent sequence gotten from $\{u_n\}$, it follows that $u_n \to u_0$.

Note. Observe that we did not require for X to be strictly convex.

Corollary 1. *Let A be a contractive mapping of a subset C of a Hilbert space H into C. For any fixed point v_0 in C and any $k_n \in (0,1)$ let A_n be the mapping of C into C defined by $A_n(x)=k_nAx+(1-k_n)v_0$ for all x in C. Let u_n be the unique fixed point of A_n in C . Then $u_n \to u_0$ as $n \to \infty$, where u_0 is the fixed point of A in C nearest to v_0 .*

Proof. To obtain Corollary 1 from Theorem 6, first note that in case X is a Hilbert space H we can identify its dual with H by the inner product and the simplest weakly continuous duality mapping is the identity mapping I. Thus, all that remains to prove is that to any point v_0 in C there exists a (unique) point u_0 in F(A) nearest to v_0 such that $(v_0-u_0, u_0-v) \geq 0$ for all v in F(A). This fact was proved by Browder [6] (see also [9, 27]). But, for the sake of completeness we reproduce this proof here.

It is well known that, since F(A) is a closed bounded convex set in H, to each point v_0 in C or in H there exists a unique point u_0 in F(A) nearest to v_0. If, for any other point v in F(A), we put $u_t = (1-t)u_0+tv$ for $0 \leq t \leq 1$, then by definition of u_0 we have the relation

$$\|v_0-u_0\|^2 \leq \|v_0-u_t\|^2 = \|v_0-u_0\|^2 + 2t(v_0-u_0, u_0-v) + t^2\|u_0-v\|^2 .$$

The above inequality implies that for any v in F(A)

(3.7) $$(v_0-u_0, u_0-v) \geq 0.$$

On the other hand, suppose there exists another point u_1 in F(A) such that for all v in F(A)

(3.8) $$(v_0-u_1, u_1-v) \geq 0.$$

Setting $v = u_1$ in (3.7) and $v = u_0$ in (3.8) and adding we get

$$0 \leq (u_1-u_0, u_0-u_1) = -\|u_0-u_1\|^2,$$

i.e., $u_0 = u_1$ and, therefore, u_0 is uniquely characterized by (3.7).

Special choice of the sequence $\{k_n\}$

Let k_n be a sequence of real numbers such thah $k_n \to 1$ as $n \to \infty$ and $0 < k_n < 1$. The partial disadvantage of Theorem 5 and Corollary 1, from the computational point of view, is that to obtain a fixed point

u_0 of A in C we need to go through two limiting processes, one to obtain the fixed point u_n of $A_n = k_n A$ in C (we assume heer, as in [17], that $0 \in C$ and take $v_0=0$) for each given k_n, the second to construct u_0 as the limit of u_n as $n \to \infty$ (i.e., $k_n \to 1$).

In [17] de Figueiredo, using Browder's Theorem 1 in [6] (asserting the strong convergence of u_n to u_0 in Hilbert space H) shows that by choosing $k_n = n/n+1$ and, for each y_0 in C , determining the sequence $\{y_n\}$ by the process

$$(3.9) \qquad y_n = A_n^{n^2} y_{n-1} \qquad (n=1, 2, \ldots, A_n = n/n+1 \ A)$$

we get the strong convergence of y_n to u_0. Using our Theorem 5 we now prove the convergence of the method (3.9) for mappings A acting in certain Banach spaces.

Theorem 7. Let X be a Banach space having a weakly continuous duality mapping J of X into X^* and let A be a contractive mapping of $C(\subset X)$ into C. Assume that 0 C and that there exists $u_0 \in F(A)$ such that $(u_0, J(u_0-v)) \leq 0$ for all v in F(A). If $A_n = k_n A (k_n = n/n+1;\ n=1, 2, 3, \ldots)$, then for each y_0 in C the sequence $\{y_n\}$ determined by the process (3.9) converges to a fixed point of A in C.

Proof. Let x_n be the fixed point of A_n in C. Then since A_n is strictly contractive for any initial approximation x_0 in C we have the error estimate

$$(3.10) \qquad \left\|A^k x_0 - x_n\right\| \leq \frac{\left(\frac{n}{n+1}\right)^k}{1-\frac{n}{n+1}} \left\|A_n x_0 - x_0\right\| \leq m \frac{n^k}{(1+n)^{k-1}},$$

where m is the diameter of C. Since, by Theorem 5, $x_n \to u_0$ and $u_0 \in F(A)$ it is easy to see that $y_n \to u_0$. Indeed, it follows from (3.9) and (3.10) that

$$\left\|y_n - u_0\right\| \leq \left\|A_n^{n^2} y_{n-1} - x_n\right\| + \left\|x_n - u_0\right\| \leq m\left(n^{n^2} / (1+n)^{n^2-1}\right) + \left\|x_n - u_0\right\| .$$

Since, as is not hard to verify, $n^{n^{\alpha}}/(1+n)^{n^{\alpha}-1} \rightarrow 0$ as $n \rightarrow \infty$ for any $\alpha > 1$ and $x_n \rightarrow u_0$ it follows from the above inequality that $y_n \rightarrow u_0$.

Note. In case X is a Hilbert space, in virtue of Corollary 1, Theorem 7 reduces to the result contained in [17].

W. V. Petryshyn

References

1. L.P. Belluce and W.A. Kirk, Fixed point theorems for families of contraction mappings, Pac. J. Math. 18(1966), 213-217.

2. M.S. Brodsky and D.P. Milma, On the center of a convex set, Dokl. Akad. Nauk SSSR 59 (1948), 837-840.

3a. F.E. Browder, Existence of periodic solutions for nonlinear equations of evolution, Proc. Nat. Acad. Sci, USA, 53 (1965), 1100-1103.

3b. __________, Fixed point theorems for noncompact mappings in Hilbert space, Proc. Nat. Acad. Sci., USA, 53 (1965), 337-342.

4. __________, Nonexpansive nonlinear operators in Banach space, Proc. Nat. Acad. Sci., USA, 54 (1966), 1041-1044.

5. __________, Fixed point theorems for nonlinear semicontractive mappings in Banach spaces, Arch. Rat. Mech. And Anal., 21 (1966), 259-269.

6. __________, Convergence of approximants to fixed points of nonexpansive nonlinear mappings in Banach spaces, Arch. Rat. Mech. and Anal. (1967), 82-90.

7. F.E. Browder and D.G. de Figueiredo, J-monotone nonlinear operators in Banach spaces, Konkl. Nederl. Akad. Wetensch, 69 (1906), 412-420.

8. F.E. Browder wna W.V. Petryshyn, The solution by iterattion of nonlinear functional equations in Banach spaces, Bull. Amer. Math. Soc., 72 (1966), 571-575.

9. ____________________, Construction of fixed points of nonlinear mappings in Hilbert space, (to appear in J. Math. Anal. and Appl.).

10. C.R. De Prima, Nonexpansive mappings in convex linear topological spaces (to appear).

11. M. Edelstein, An extension of Banach's contraction principle, Proc. Amer. Math., 12 (1961), 7-10.

12. __________, On predominantly contractive mappings, J. London Math. Soc., 37(1962), 74-79.

13. __________, On nonexpansive mappings in Banach spaces, Proc. Cambridge Phil. Soc., 60 (1964), 439-447.

14. __________, A remark on a theorem of M.A. Krasnoselsky, Amer. Math. Monthly, 13 (1966), 509-510.

15. D.G. De Figueiredo, Fixed point theorems for weakly continuous mappings, Math.Res. Center, Univ. of Wisconsin, Technical Rep. No. 638 (1966).

16. ________________, Fixed point theorems for nonlinear operators and Galerkin approximations, J. of Differenzial Equations, 3(1967), 271-281.

17. ________________, Topics in nonlinear functional analysis, Inst. for fluid Dynamics and Appl. Math. , Univ. of. Maryland, Lecture Notes N_0 48 (1967).

18. ________________, and L.A. Karlovitz, On the radial projection in normed spaces, Bull. Amer. Math. Soc. 73 (1967), 364-368.

19. D. Gohde Zum Prinsip der kontraktiven Abbildung, Math. Nachr. 30 (1965), 251-258.

20. __________, Uber Fixpunkte bei steligen Selbstabbildungen mit kompakten Itericrten, Math. Nachr. 28 (1964), 45-55.

21. R.I. Kachurovsky, On some fixed point principles, Usen Zap. Moskov Reg. Ped. Inst. 96(1960), 215-219.

22. S. Kaniel, Quasi-compact nonlinear operators in Banach space and applications, Arch. Rat. Mech. and Anal., 21(1966), 259-278.

23. W.A. Kirk, A fixed point theorem for mappings which do not increase distance, Amer, Math. Monthly, 72(1965), 1004-1006.

24. A.N. Lolmogoroff and S.V. Fømin, Elements of the theory of functions and functional analysis, Graylock Press, 1957.

25. M.A. Krasnoselsky, Two observations about the method of successive approximations, Usp. Math. Nauk 10(1955), 123-127.

26. M. Lees and M.H. Schultz, A Leray-Schauder principle for A-Compact mappings and the numerical solution of nonlinear two-point boundary value problems, John Wiley and Sons, Inc. , New York 1966, 167-179.

27. J.L. Lions and G. Stampacchia, Variational inequalities, (to appear).

28. R.De Marr, Common fixed points for commuting contraction mappings, Pac. J. Math. 13(1963), 1139-1141.

29. S. Mazur, Uber die kleinste konvex Menge, die eine gegebene kompakte, Menge enthalt, Studia Math. 2(1930), 7-9.

30. Z. Opial, Weak convergence of the sequence of successive approximations for nonexpansive mappings (to appear).

31. W.V. Petryshyn, On a fixed point theorem for nonlinear P-compact operators in Banach space, Bull. Amer. Math. Soc. 72(1966), 329-334.

32. ________________, On nonlinear P-Compact operators in Banach space with applications to constructive fixed point theorems, J. Math. Anal. Appl. 15(1966), 228-242.

33. ________________, Further remarks on nonlinear P-compact operators

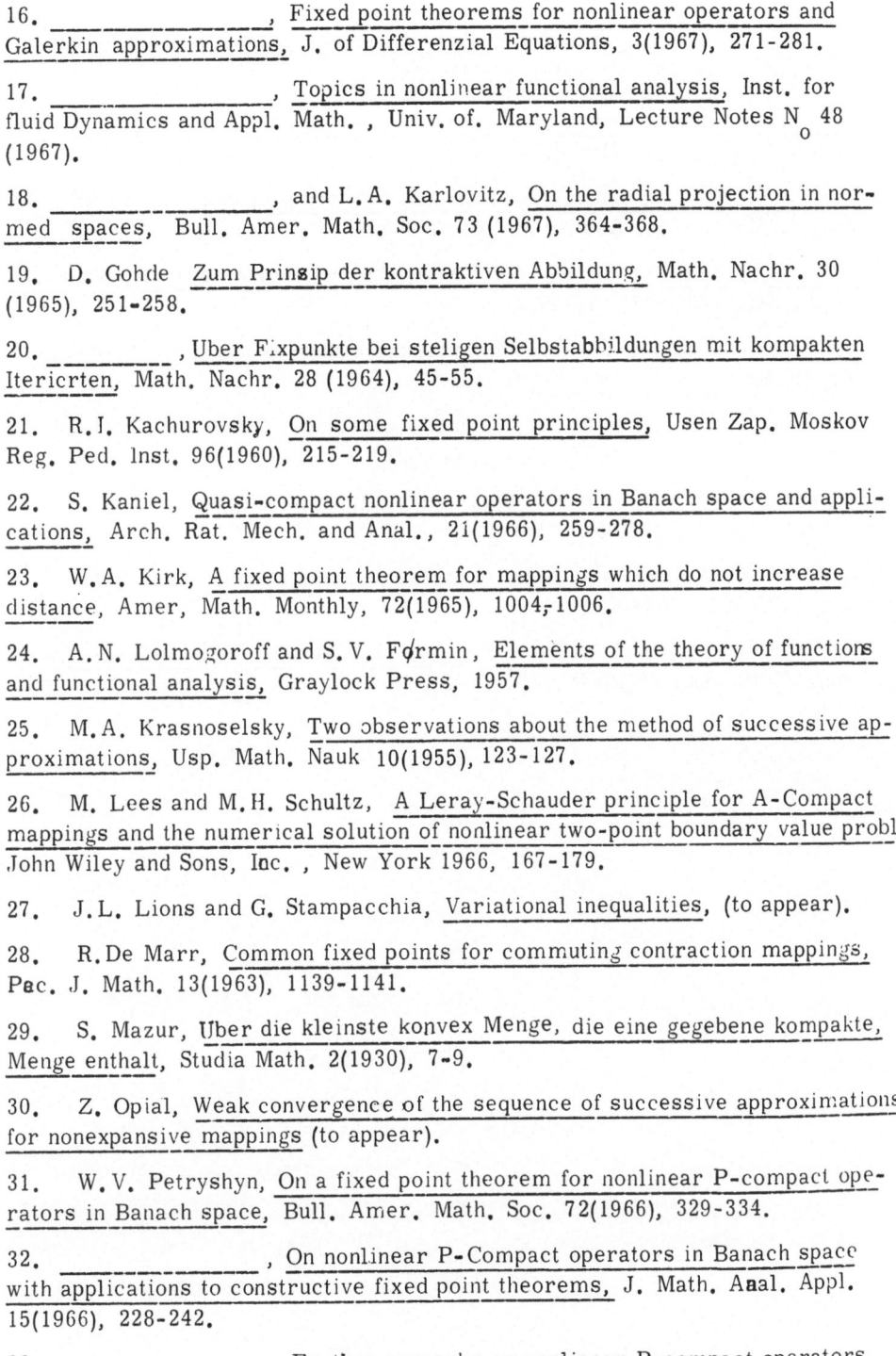

in Banach space, Proc. Nat. Acad. Sci. USA, 55(1966), 684-687 (extended version in J. Math. Anal. Appl. 16(1966), 243-253).

34. __________, On the extension and the solution of nonlinear operator equations, Illinois J. Math., 10(1966), 255-274.

35. __________, Construction of fixed points of demicompact mappings in Hilbert space, J. Math. Anal. Appl. 14(1966), 276-284.

36. __________, Remarks on fixed point theorems and their extensions, Trans. Amer. Math. Soc. 126(1967), 43-43.

37. __________, and T.S. Tucker, On functional equations involving P -compact operators (to appear).

38. J. Schauder, Der Fixpunktsatz in Funktionalraumen, Studia Math. 2(1930), 171-180.

39. H. Schaefer, Uber die Methode sukzessiver Approximationen, Jahresber. Deutsch. Math. Verein. 59(19579, 131-140.

40. M. Shinbrot, A fixed point theorem and some applications, Anch. Rat. Mech. Anal. 17(1964), 255-271.

41. E. Zarantonello, The closure of the numerical range contains the spectrum, Bull. Amer. Math. Soc. 70(1964), 781-787.

CENTRO INTERNAZIONALE MATEMATICO ESTIVO

(C.I.M.E.)

W. V. PETRYSHYN

ITERATIVE CONSTRUCTION OF FIXED POINTS OF CONTRACTIVE TYPE MAPPINGS IN BANACH SPACES

Corso tenuto ad Ispra dal 3-11 Luglio 1967

ON THE APPROXIMATION-SOLVABILITY OF NONLINEAR FUNCTIONAL EQUATIONS IN NORMED LINEAR SPACES

by

W. V. Petryshyn

1. 1. Introduction. The purpose of this part of my talk is to outline some of the recent results concerning the projection-and the approximation-solvability of nonlinear functional equations in normed linear spaces. We first state a general theorem on projectional-solvability proved in [15] which at the same time unites the earlier results on linear equations obtained in [11, 18, 10, 6, 13] (*) with the recent results on nonlinear monotone operatirs obtained in [7, 12, 2, 3, 19, 14].

Let X be a separable Banach space, X_n a finite dimensional subspace of X and P_n a linear projection of X onto X_n. Let "$\rightarrow$" and "$\rightharpoonup$" denote respectively the strong and the weak convergence in X. The sequence $\{X_n\}$ is called projectionally complete in X if and only if $P_n x \rightarrow x$ for each x in X. Let A be a (possibly) nonlinear mapping of X into X and $\{A_n\}$ a sequence of mappings defined by $\{A_n\} = \{P_n A P_n\}$. In [15] the author considered the problem of giving a constructive proof of the existence and the uniqueness of solutions of the equation

(1) $$Ax = f \quad , \qquad f \in X ,$$

as strong (+) limits of solutions x_n in X_n of approximate equations

(2) $$A_n x_n = P_n(f)$$

(*) For a complete and self-contained theory of the projection method see [15]; for a more complete list of references and contributions to the theory of projection and Galerkin methods see [11, 18, 15, 5].

(+) In [15] the author also considered the problem of constructing solutions of Eq. (1) as weak limits of solutions x_n of Eq. (2) in case the operator A is demicontinuous or weakly continuous and monotone or K-monotone.

Definition 1. Eq. (1) is said to be strongly projectionally-solvable if and only if there exists an integer $N > 0$ such that for each $n \geq N$ and each f in X Eq. (2) has a unique solution x_n in X_n such that $x_n \to x$ and x is the unique solution of Eq. (1).

One of the results obtained in 15 was the following theorem.

Theorem I. Suppose there exists $N > 0$ and a continuously strictly increasing function $\alpha(r)$ of $R^+ = \{r \geq 0\}$ into R^+ with $\alpha(0) = 0$ and $\lim_{r \to \infty} \alpha(r) = \infty$ such that A_n is continuous in X_n for $n \geq N$, $A_n x \to Ax$ for each x in X and

$$\|A_n x - A_n y\| \geq \alpha(\|x-y\|) \text{ for each } n \geq N \text{ and all } x \text{ and } y \text{ in } X_n. \tag{3}$$

Then Eq. (1) is strongly projectionally-solvable provided A satisfies the following condition (c): If $\{X_m\}$ is any subsequence of $\{X_n\}$ and $\{x_m\}$ is an arbitrary sequence in X with $x_m \in X_m$ such that $x_m \to x$ and $A_m x_m \to g$, then $Ax = g$.

We omit here the proof of Theorem 1 since below we shall prove a much more general theorem. Theorem I was then used in [15] to establish the strong projectional-solvability of various classes of linear and nonlinear equations. To illustrate the generality of Theorem 1 we now deduce from it the following two important corollaries.

Corollary 1. If X is a Hilbert space H and $A = T + S$, where T and S are linear bounded operators such that S is completely continuous and T is definite, i.e., T is such that

$$(Tx, x) \geq c\|x\|^2, \quad c > 0, \quad x \in H, \tag{4}$$

then Eq. (1) is strongly projectionally-solvable provided that $N(A) = 0$, where $N(A)$ is the null space of A.

Proof. Since A satisfies condition (c), A_n is continuous in X_n and $A_n x \to Ax$, it suffices to show that A satisfies (3), i.e., that there exists $N>0$ and (in fact, $\alpha(\rho) \leq$ or) such that

(3°) $\|A_n x\| \geq \alpha(\|x\|)$ $(=c\|x\|)$ for all x in X_n and $n \geq N$.

Assume, to the contrary, that (3°) does not hold. Then we can find a sequence $\{x_m\} \subset H$ with $\|x_m\| = 1$ such that $x_m \rightharpoonup x$ and $A_m x_m \to 0$ as $m \to \infty$. First, the complete continuity of S implies that $S_m x_m \to Sx$ and

$$(5) \qquad T_m x_m = A_m x_m - S_m x_m \to -Sx \quad , \qquad m \to \infty .$$

But, $T_m x_m \rightharpoonup Tx$. Hence, $Tx = -Sx$ or $Ax = 0$. Now, since T is definite, we have

$$c\|x_m - x\|^2 \leq |(Tm_x - Tx, x_m - x)| = |(Tx_m, x_m) - (Tx, x_m) - (Tx_m, x) + (Tx, x)|$$

$$\to - (Sx, x) - (Tx, x) - (Tx, x) + (Tx, x) = 0$$

si ce $Ax = 0$. Thus, $x = 1$ and $Ax = 0$ in contradiction to $N(A) = \{0\}$.

Remark. Corollary 1 was proved in [6]. An independent proof for T satisfying slightly weaker conditions was given in [10]. See also [18]. For other applications of Theorem I to linear bounded and unbounded operators see [15].

Corollary 2. *If* $X = H$ *and* A *is continuous nonlinear complex monotone operator, i.e.,*

$$(5') \qquad |(Ax - Ay, x - y)| \geq c\|x - y\|^2 , \qquad c > 0 , \quad x, y \in H,$$

then Eq. (1) is strongly projectionally-solvable.

Proof. The continuity of A implies that A_n is continuous in X_n and that $A_n x \to Ax$. Since $P_n x = x$ for all x in X_n, the Buniakovski-Schwartz inequality, when applied to (5) shows that (3) is satisfied with $\alpha(r) = cr$. Hence by Theorem I, it suffices to show that A satisfies condition (c). Suppose $\{x_m\} \subset H$ with $x_m \in X_m$ and $\|x_n\| = 1$ such that $x_m \rightharpoonup x$ and $A_m x_m \to g$. Then, by (5) ,

$$c\|x_m - P_m x\|^2 \le |(Ax_m - AP_m x, x_m - P_m x)| = |(Ax_m, x_m) - (AP_m x, x_m)$$
$$- (Ax_m, P_m x) + (AP_m x, P_m x)| \to |(g, x) - (Ax, x)$$
$$- (g, x) + (Ax, x) = 0.$$

Hence, $x_m \longrightarrow x$ and $A_m x_m \longrightarrow Ax$ = i.e., A satisfies condition (c) .

Remark. Corollary 2 is certainly true when A is strongly monotone, i.e. , $\mathrm{Re}(Ax-Ay, x-y) \geq c\|x-y\|^2$. When $A=I-F$, where F is monotone, the result was first proved by Minty [12] (see also Kachurowski [7]) . For general continuous A, it was proved by Browder [2]. For complex monotone, operator A with the additional assumption that A is bounded, Corollary 2 was proved by Zarantonello [19] and by Browder [3] with the boundedness assumption dropped. All the above proofs are nonconstructive and rather complicated.

In his forthcoming paper [4] Browder notes that for certain classes of nonlinear opertaros (e.g. J-monotone operators) a direct verification of condition (c) seems sometimes to require additional hypothesis, which thus might in general restrict the class of equations to which Theorem I is applicable. He then shows that if instead of condition (c) we assume that Eq. (1) possesses a solution, then somewhat weaker assumptions ensure that Eq. (1) is strongly projectionally-solvable. In his subsequent paper [5] Browder exploits further the solvability conditions (s) by deriving general results on the relation between solvability condition and the approximation-solvability (to be defined below) of Eq. (1). In [16] the speaker showed that at least when X is a reflexive Banach space then, under the assumption of the other hypotheses of Theorem I, the condition (c) and the solvability condition (S) are equivalent. In his subsequent paper [17] the speaker showed the equivalence between the solvability condition (S) and the modifed condition (c) (the so-called, condition (H)) without the assumption that the underlying spaces are Banach spaces. Some of the results mentioned above are given in the next section.

2. Approximation-solvability. Let X and Y be two normed linear spaces and let $\{X_n\} \subset X$ and $\{Y_n\} \subset Y$ be two sequences of finite dimensional subspaces. Let P_n be a mapping of X into X_n and Q_n a mapping of Y into Y_n. Let A_n be a sequence of nonlinear mappings of X_n into Y_n defined by $A_n = Q_n A \big|_{X_n}$, where A is a nonlinear map of X into Y.

As before we give a constructive proof of the existence and uniqueness of solutions of the equation

$$Ax=f, \qquad f \in Y \tag{6}$$

as strong limits of solutions $x_n \in X_n$ of equations

$$A_n x_n = Q_n(f). \tag{7}$$

To make the nature of our problem more precise we need the following definitions (see also [16, 17] and Browder [4, 5]).

Definition 2. A quadruple of sequences $(\{X_n\}, \{Y_n\}, \{P_n\}, \{Q_n\})$ is called an approximation scheme for Eq. (6) if and only if $\dim X_n = \dim Y_n$ for each n, P_n and Q_n are continuous, and $P_n x \to x$ for each x in X and $Q_n y \to y$ for y in Y.

Definition 3. Eq. (6) is said to be uniquely approximation-solvable with respect to the given scheme $\Gamma_n = (\{X_n\}, \{Y_n\}, \{P_n\}, \{Q_n\})$ if there exists an integer $N > 0$ such that for each $n \geq N$ and each f in Y, Eq. (7) has a unique solution x_n in X_n such that $x_n \to x$ in X and x is the unique solution of Eq. (6).

Note. We emphasize that the mappings P_n and Q_n are not assumed to be projections or even linear though for most of the direct methods (7) for the approximate solution of Eq. (6) they are linear mappings induced by any one of the the following groups of methods: the projection or Galerkin type methods, and the finite difference methods(*) (assuming, as in Lax-Richtmyer [8], that exact and approximate solutions belong to the same space).

The main result of this section is the following theorem.

Theorem 1. Let X and Y be two normed linear spaces and let Γ_n be an approximation scheme for Eq. (6). Suppose there exists an integer $N>0$ and a function $\alpha(r)$ (as in Theorem I) such that A_n is continuous in X_n for each $n \geq N$, $Q_n A P_n x \to Ax$ for each x in X and

(8) $$\|A_n x - A_n y\| \geq \alpha(\|x-y\|) \text{ for } n>N \text{ and all } x \text{ and } y \text{ in } X_n.$$

Then Eq. (6) is uniquely approximation-solvable if and only if A satisfies the following condition (H): If Γ_m is any subscheme of the approximation scheme Γ_n for Eq. (6) and $\{x_m\}$ is any bounded sequence in X with $x_m \in X_m$ such that $A_m x_m \to g$ for some g in Y, then there exists a $\{x_{m_i}\}$ of $\{x_m\}$ and ana element x in X such that $x_{m_i} \to x$ and $A_{m_i} x_{m_i} \to Ax(=g)$ as $m_i \to \infty$.

Proof. (Uniqueness.) Suppose that $u \neq v$ and $Au=Av$. Because, for each u and v in X, $P_n u$ and $P_n v$ lie in X_n, (8) implies that for each $n>N$

$$\|Q_n A P_n u - Q_n A P_n v\| \geq \alpha(\|P_n u - P_n v\|).$$

Since $P_n u \to u$, $P_n v \to v$ and $Q_n A P_n x \to Ax$ for each x in X, the passage to the limit in the above inequality implies that $0 \geq \alpha(\|u-v\|)$. The latter inequality contradicts the properties of $\alpha(r)$ and shows that A is one-to-one.

( )If we do not assume that X_n and Y_n are subspaces of X and Y respectively (as is usually the case when Eq. (7) is a finite difference analog of Eq. (6)), then all results outlined in this section remain vaild without any substantial chahge in the proof provided that, as in Aubin [1] (see also the papers discussed during this meeting by Lions, Reviart) we introduce the mappings $r_n : X \to X_n$, $s_n : Y \to Y_n$ and $p_n : X_n \to X$ subject to certain conditions (for details see 1 and the paper of Raviart) guaranteeing that the operators A_n defined by $A_n = s_n A p_n$ have the needed properties.

(Existence.) By (8) and our assumptions, A_n is a one-to-one continuous mapping of X_n into Y_n for each $n \geq N$, and hence, by the Brouwer Theorem on Invariance of Domain, the range $R(A_n)$ is an open set in Y_n; furthermore, it follows from (8) and the continuity that $R(A_n)$ is also a closed set in Y_n for each $n \geq N$. Indeed, if $\{y_m\} \subset R(A_n)$ and $y_m \to y$ in Y_n, then there exists a sequence $\{x_m\} \subset X_n$ such that $y_m = A_n x_m$ and, by (8) $\|y_\ell - y_k\| \geq$ $\geq \alpha(\|x_\ell - x_k\|)$. If $\beta(r)$ denotes the continuous strictly increasing function of R^+ into R^+ with $\beta(0) = 0$ which is the inverse of $\alpha(r)$, then since y_m is a convergent sequence

$$\|x_\ell - x_k\| \leq \beta(\|y_\ell - y_k\|) \to 0, \qquad (\ell, k \to \infty).$$

Thus there exists an element x in X_n such that $x_m \to x$ and $A_n x_m \to A_n x = y \in R(A_n)$ by the continuity of A_n for $n \geq N$, i.e., $R(A_n)$ is closed. Since $R(A_n)$ is a nonempty set in Y_n, which is both open and closed in Y_n, it follows that $R(A_n) = Y_n$.

Hence for each $n \geq N$ and for each given f in Y there exists a unique element x_n in X_n such that $A_n x_n = Q_n(f)$. For the sequence $\{x_n\}$ thus determined, (8) and the fact that $\|Q_n A(0)\| < K_0$ for some constant K_0 imply that

$$\|A_n x_n\| \geq \|A_n x_n - A_n(0)\| - \|A_n(0)\| \geq \alpha(\|x_n\|) - K_0$$

whence, in view of (7) and the fact that $\|Q_n(f)\| \leq K$ for some constant K

$$\alpha(\|x_n\|) \leq K + K_0 \text{ which implies that } \|x_n\| \leq \beta(K_0 + K),$$

i.e., $\{x_n\}$ is a bounded sequence with $x_n \in X_n$ and such that $A_n x_n = Q_n(f) \to f$, as $n \to \infty$. Hence, condition (H) implies that there exists a subsequence $\{x_{n_i}\}$ of $\{x_n\}$ and an element x in X such that $x_{n_i} \to x$ and $A_{n_i} x_{n_i} \to Ax$, as $n_i \to \infty$. Because $A_{n_i} x_{n_i} = Q_{n_i}(f) \to f$ as $n_i \to \infty$, we obtain that $Ax = f$, i.e., x is a solution of Eq. (6). Since, as was shown above, x is the unique solution of Eq. (6) for a given f in Y we conclude <u>a posteriori</u> that the selection of the subsequence was not necessary. Consequently, the entire sequence $\{x_n\}$ converges strongly to x, i.e., Eq. (6) is uniquely approximation-solvable.

<u>Converse</u>. Let $\Gamma_m = (\{X_m\}, \{Y_m\}, \{P_m\}, \{Q_m\})$ be an arbitrary subscheme for Eq. (6) and let x_m be any bounded sequence in X with x_m in X_m such that

$$A_m x_m = g_m \to g, \qquad (m \to \infty) \tag{9}$$

for some g in Y. To prove that A satisfies condition (H) note first that, by our conditions on Q_m,

$$\| g_m - Q_m(g) \| \leq \| g_m - g \| + \| g - Q_m(g) \| \to 0, \quad (m \to \infty) . \tag{10}$$

Since, by hypothesis, Eq. (6) is uniquely approximation solvable, for every $m \geq N$ and every g in Y, there exists a unique element y_m in X_m such that

$$A_m y_m = Q_m(g) , \tag{11}$$

$y_m \to y$ in X, as $m \to \infty$, and y is the unique solution of the equation $Ay = g$. Substracting (9) from (11) we obtain the equation

$$Q_m y_m - A_m x_m = Q_m(g) - g_m$$

whence, in virtue of (8), for all $m \geq N$ we derive the inequality

$$\alpha(\|y_m - x_m\|) \leq \|A_m y_m - A_m x_m\| = \|Q_m(g) - g_m\| .$$

This, (10) and the properties of β (r) imply that

$$\|y_m - x_m\| \leq (\|Q_m(g) - g_m\|) \to 0, \quad (m \to \infty) \tag{12}$$

Since $y_m \to y$, (12) implies that $x_m \to y$, as $m \to \infty$, and $Ay = g$. Consequently, (9) implies that $A_m x_m \to Ay$ and thus shows that A satisfies condition (H).

The following theorem gives the relation between the solvability condition (S) and condition (H).

Theorem 2. Under the hypothesis of Theorem 1 the following equivalent assertions are valid:

(A_1) A satisfies condition (H)

(A_2) Eq. (6) is uniquely approximation-solvable.

(A_3) For any given f in Y, Eq. (6) has a solution x in X.

Remark. Looking over the proofs of Theorem 1 and 2 we note that nowhere did we use the continuity property of the mappings P_n and Q_n assumed in Definition 2. Consequently, Theorems 1 and 2 remain valid without the assumption that P_n and Q_n are continuous. This fact may on occasion prove to be important and useful particularly in the application of these theorems to the collocation and finite difference methods.

Let us remark in passing that combining the results of Theorem 2 with those of Theorem 2 in [16] we have the following useful theorem.

Theorem 3. If additionally we assume that X and Y are separable reflexive Banach spaces, then under the hypotheses of Theorem 1 the following equivalent assertions are valid;

(A_1) A satisfies condition (H).

(A_2) Eq. (6) is uniquely approximation-solvable.

(A_3) For any given f in Y, Eq. (6) has a solution x in X.

(A_4) A satisfies condition (C): If Γ_m is any subscheme of Γ_n and $\{x_m\}$ is an arbitrary sequence in X with x_m in X_m such that $x_m \to x$ and $A_m x_m \to g$, then $Ax = g$.

Remark. Theorem I follows as a corollary (sufficiency part) of Theorem 3 (A_4).

An immediate and practively useful corrollary of Theorem 2 is the following theorem.

Theorem 4. Let X and Y be two normed linear spaces and let Γ_n be an approximation scheme for Eq. (6): Suppose that A is a continuous mapping of X into Y and that $Q_n(y)$ converges to y strongly in Y and uniformly for y in a compact subset of Y. Suppose further that there exists an integer $N>0$ and a function (r) as in Theorem 1 such that the inequality (8) holds. Then , under the above hypothesis, assertions (A_1), (A_2) and (A_3) of Theorem 2 are equivalent.

Remark. Under the hypotheses of Theorem 4 with $\alpha(r)$ satisfying slightly weaker conditions, the equivalence of assertions (A_2) and (A_3) was first obtained by Browder in [5] .

3. Perturbed problems. Our next theorems deal with the problem of constructive solutions of perturbed equations

$$(13) \qquad \tilde{A}x = Ax + Bf = f \qquad f \in Y,$$

as strong limits of solutions $x_n \in X_n$ of approximate equations

$$(14) \qquad A_n x_n + B_n x_n = Q_n(f) \qquad (A_n = Q_n A|_{X_n}, \; B_n = Q_n B|_{X_n}),$$

where B is a completely continuous mapping of X into Y(i.e., B is continuous and maps bounded sets in X into relatively compact sets in Y) and A satisfies certain conditions to be specified below.

Theorem 5. Suppose that X and Y are normed linear spaces , $Q_n(y)$ converges to y strongly in Y and uniformly for y in a compact subset of Y , B is completely continuous and A satisfies condition (H). Suppose further that for a given f in Y and each n N the solutions x_n of Eq. (14) exist and are uniformly bounded by a constant independent of n.

If for any given f in Y Eq. (13) has at most one solution x in X, then Eq. (8) is approximation-solvable (i.e., Eq. (14) has a solution x_n in

X_n for each $n \geq N$ such that $x_n \to x$, as $n \to \infty$, and x is the unique solution of Eq. (14)).

Let us remark that though the second hypothesis of Theorem 5 is rather restrictive, nevertheless it is not artificial and often can be verified in applications. Condition (a) which was introduced by Browder [5] is essentially of the same strength. Furthermore, if X is a reflexive Banach space, $Y = X^*$ and P_n and $Q_n = P_n^*$ are linear projections in X and X^*, respectively, then the second hypothesis of Theorem 5 is implied by the coerciveness condition

$$\frac{|(\tilde{A}u, u)|}{\|u\|} \to \infty , \qquad \text{as } \|u\| \to \infty ,$$

which has been successfully used by a number of authors (see, for example, Browder [5, 3] and Lions [9]).

Remark. For the proofs of Theorem 2 to 5 as well as for other results and the relation of condition (H) to the concept of P-compactness see [17].

References

1. J.P. Aubin, Nonlinear stability-approximation of nonlinear operational equations (to appear).

2. F.E. Browder, On the solvability of nonlinear functional equations, Duke Math. J. 39 (1963), 557-566.

3. ______, Remarks on nonlinear functional equations II, Illinois J. Math. 7 (1965), 608-618.

4. ______, Nonlinear accretive operators in Banach space (to appear).

5. ______, Approximation-solvability of nonlinear functional equations in normed linear spaces (to appear in Archive Rat. Mech. Anal.).

6. S. Hildebrandt and E. Wienholtz, Constructive proofs of representation theorems in Hilbert space, Comm. Pure Appl. Math. 17 (1964), 369-373.

7. R.I. Kachurovski, Monotone nonlinear operators in Banach spaces, Dokl. Akad. Nauk. SSR, (1965), 679-698.

8. P.D. Lax and R.D. Richtmyer, Survey of stability of linear finite difference equations, Comm. Pure Appl. Math. 9(1956), 267-293.

9. J.-L. Lions, Sur certaines equations paraboliques non lineaires, Bull. Soc. Math. France, 93(1965), 155-175.

10. V.E. Medvedev. On the convergence of the Bubnov-Galerkin method, Prikl. Mat. Mekh. 17(1963), 1148-1151.

11. S.G. Mikhlin, Variationsmethoden der Mathematischen Physik, Akademie-Verlag, Berlin, 1962.

12. G.J. Minty, Monotone (nonlinear) operators in Hilbert space, Duke Math. J., 29 (1962), 344-346.

13. W.V. Petryshyn, On a class of K-p.d. and non-K-p. d. operators and operator equations, J. Math. Anal. Appl. 10(1965), 1-24.

14. ______, On the extension and the solution of nonlinear operator equations, Illinois J. Math. 10 (1966), 255-274.

15. ______, Projection methods in nonlinear numerical functional analysis, (to appear in J. Math. Mech.)

16. ______, Remarks on the approximation-solvability of nonlinear functional equations (to appear in Archive Rat. Mech. Anal.)

17. ______, On the approximation-solvability of nonlinear equations (to appear in Math. Ann.)

18. N.I. Polsky, Projection methods in applied mathematics, Dokl. Akad. Nauk, SSSR 143 (1962), 787-790.

19. E.H. Zarantonello, The closure of the numerical range contains the spectrum, Bull. Amer. Math. Soc. 70 (1964), 781-787.

CENTRO INTERNAZIONALE MATEMATICO ESTIVO

(C.I.M.E.)

P.A. RAVIART

"APPROXIMATION DES EQUATIONS D'EVOLUTION PAR DES METHODES VARIATIONNELLES"

Corso tenuto ad Ispra dal 3-11 Luglio 1967

APPROXIMATION DES EQUATIONS D'EVOLUTION PAR DES METHODES VARIATIONNELLES

par

P. A. Raviart

(Université de Rennes)

Introduction .

Le présent cours est une introduction à l'approximation par les différences finies des solutions des équations aux dérivées partielles d'évolution. On y traite des équations linéaires du 1er ordre et du 2ème ordre en t à l'aide des méthodes variationnelles : cette étude reprend avec quelques améliorations techniques un certain nombre de résultats de [10]. Dans chaque cas, après un bref rappel théorique, on examine deux schémas classiques, l'un implicite et l'autre explicite, et on donne des théorèmes de stabilité et de convergence. En appliquant ces résultats à un certain nombre d'exemples, on obtient en particulier des résultats de stabilité pour des équations paraboliques et hyperboliques à coefficients mesurables et bornés dans des domaines cylindriques.

Les méthodes utilisées peuvent se généraliser à l'étude de l'approximation d'équations d'évolution couplées [7] , d'équations à coefficients non bornés [8] , ou d'équations prises dans des domaines non cylindriques [9] . Elles permettent également d'étudier dans des cas généraux les méthodes de directions alternées et à pas fractionnaires 16 , 3 . Enfin , elles peuvent s'appliquer avec succès à certaines équations non linéaires [10] , [12] .

1. Equations d'évolution du 1er ordre en t.

1.1. Formulation abstraite.

Soient V et H deux espaces de Hilbert sur C. On suppose que V est séparable, que $V \subset H$ avec injection continue et que V est dense dans

H. On désigne par $\| \ \|$ la norme dans V, par (,) et $| \ |$ respectivement le produit scalaire et la norme dans H . Soit V' l'antidual fort de V, de norme $\| \ \|^*$; si on identifie H à son antidual H' , on a $V \subset H \subset V'$ avec injections continues. Si $f \in V'$ et $v \in V$, (f, v) est le produit scalaire qui met V' et V en antidualité. Si f H, on retrouve le produit scalaire dans H .

Soit $u, v \longrightarrow a(t;u,v)$ une famille de formes sesquilinéaires continues sur $V \times V$ dépendant du paramètre $t \in [0, T]$, $T < \infty$, avec les propriétés suivantes :

(1.1) $\begin{cases} \forall u, v \in V \text{ , la fonction } t \longrightarrow a(t;u,v) \text{ est mesurable et} \\ a(t;u,v) \leq K \|u\| \|v\|, \ K = \text{constante indépendante de } t ; \end{cases}$

(1.2) $\begin{cases} \text{il existe deux constantes réellles } \lambda \text{ et } \alpha > 0 \text{ telles que} \\ \operatorname{Re} a(t;v,v) + \lambda |v|^2 \geq \alpha \|v\|^2 \ , \forall v \in V \text{ , p.p. en } t. \end{cases}$

Soit $A(t) \in \mathcal{L}(V, V')$ [1] l'opérateur défini pour presque tout t par:

(1.3) $$(A(t)\ u, v) = a(t;u,v) \ , \qquad \forall u, v \in V .$$

De la séparabilité de l'espace V et de l'hypothèse (1.1) , on déduit que l'opérateur A(.) $(t \longrightarrow A(t))$ est linéaire continu de $L^2(0, T ; V)$ dans $L^2(0, T; V')$ [2] .

Théorème 1.1.

Sous les hypothèses précédentes et pour f donné dans $L^2(0, T; V')$, u_o donné dans H , il existe une fonction u et une seule vérifiant

---

[1] Si X et Y sont deux espaces vectoriels topologiques, $\mathcal{L}(X, Y)$ est l'espace des applications linéaires continues de X dans Y .

[2] Si X est un espace de Banach de norme $\| \ \|_X$, $L^p(0, T;X)$, $1 \leq p < \infty$ désigne l'espace des (classes de) fonctions $t \longrightarrow u(t)$ fortement mesurables sur (0, T) pour la mesure de Lebesgue dt à valeurs dans X et telles que $\|u\|_{L^p(0,T;X)} = \left(\int_o^T \|u(t)\|_X^p \, dt \right)^{1/p} < \infty$. Modification usuelle si $p = \infty$.

(1.4) $$u \in L^2(0,T;V) \quad , \quad \frac{du}{dt} \in L^2(0,T;V')$$ (3)

(1.5) $$\frac{d}{dt} u(t) + A(t)\, u(t) = f(t) \quad \text{p.p. en } t ,$$

(1.6) $$u(0) = u_o .$$

Pour la démonstration de ce théorème, voir [4], [5].

Remarque 1.1.

D'après [5], toute fonction u satisfaisant à (1.4) est p.p. égale à une fonction continue de [0, T] dans H et la condition (1.6) a un sens. Notons également que si u et v sont deux fonctions satisfaisant à (1.4), on a la formule de Green.

(1.7) $$\int_o^{\sigma} \left\{ (u(t), \frac{dv}{dt}(t)) + (\frac{du}{dt}(t), v(t)) \right\} dt = $$
$$= (u(\sigma), v(\sigma)) - (u(0), v(0)) .$$

Dans la suite, on décomposera la forme $a(t;u,v)$ en

(1.8) $$a(t;u,v) = a_o(t;u,v) + a_1(t;u,v)$$

où $a_o(t;u,v)$ est une forme sesquilinéaire continue sur $V \times V$ vérifiant

$(1.1)_o$ $$\begin{cases} \forall u, v \in V, \text{ la fonction } t \longrightarrow a_o(t;u,v) \text{ est mesurable et} \\ a_o(t;u,v) \leq K_o \|u\| \|v\| , \; K_o = \text{constante indépendante de } t , \end{cases}$$

$(1.2)_o$ $$\operatorname{Re} a_o(t;v,v) \geq \alpha \|v\|^2 \quad , \forall v \in V ,$$

---

(3) $\frac{du}{dt}$ désigne la dérivée de u au sens des distributions sur]0, T[à valeurs dans V' (cf. [15]). Autrement dit, l'hypothèse $\frac{du}{dt} \in L^2(0,T;V')$ signifie qu'il existe une fonction $\frac{du}{dt} = g \in L^2(0,T;V')$ telle que $\int_o^T (g(t),v)\varphi(t)\,dt = -\int_o^T (u(t),v)\varphi'(t)\,dt, \quad \forall v \in V, \forall \varphi \in C_o^{\infty}(]0,T[)$.

tandis que $a_1(t;u,v)$ est une forme sesquilinéaire continue sur $V \times H$ avec

$$(1.1)_1 \quad \begin{cases} \forall u \in V \ , \forall v \in H \ , \text{ la fonction } t \rightarrow a_1(t;u,v) \text{ est mesurable} \\ \text{et } a_1(t;u,v) \leq K_1 \|u\| |v| \ , \ K_2 = \text{constante indipendante} \\ \text{de } t \ . \end{cases}$$

Il est clair qu'une telle décomposition (1.8) est toujours possible et les propriétés (1.1), (1.2) et (1.8), $(1.1)_o$, $(1.2)_o$, $(1.1)_1$ sont équivalentes.

1.2. <u>Exemple</u> I.

Soit Ω un ouvert de l'espace euclidien R^n de frontière $\Gamma = \partial \Omega$. On désigne par $L^2(\Omega)$ l'espace des (classes de) fonctions définies (p.p) sur Ω à valeurs complexes et de carré sommable sur Ω pour la mesure de Lebesgue dx. On munit $L^2(\Omega)$ de la norme hilbertienne

$$(1.9) \qquad \|u\|_{L^2(\Omega)} = \left(\int_\Omega |u(x)|^2 \ dx \right)^{1/2} .$$

On désigne par $H^1(\Omega)$ l'espace de Sobolev des (classes de) fonctions $u \in L^2(\Omega)$ dont les dérivées premières $\frac{\partial u}{\partial x_i}$, $i = 1, \ldots, n$, sont dans $L^2(\Omega)$, ces dérivées étant prises au sens des distributions sur Ω (cf. [14]). On munit $H^1(\Omega)$ de la norme hilbertienne

$$(1.10) \qquad \|u\|_{H^1(\Omega)} = \left\{ \sum_{i=1}^n \int_\Omega \left| \frac{\partial u}{\partial x_i}(x) \right|^2 dx + \int_\Omega |u(x)|^2 \ dx \right\}^{1/2} .$$

Si la frontière Γ est "assez régulière", on peut définir la trace $\gamma_o u$ de u sur Γ lorsque $u \in H^1(\Omega)$: c'est un élément de $L^2(\Gamma)$, espace des (classes de) fonctions complexes de carré sommable sur Γ pour la mesure superficielle $d\sigma$. En outre l'application $u \rightarrow \gamma_o u$ est continue de $H^1(\Omega)$ dans $L^2(\Gamma)$.

On note $H_o^1(\Omega)$ l'adhérence dans $H^1(\Omega)$ de $C_o^\infty(\Omega)$, sous espace des fonctions indéfiniment différentiables à support compact dans Ω. Si la frontière Γ est "assez régulière", $H_o^1(\Omega)$ est exactement l'espace des fonctions $u \in H^1(\Omega)$ telles que $\gamma_o u = 0$. On définit $H^{-1}(\Omega)$ comme étant l'antidual fort de $H_o^1(\Omega)$; il est aisé de voir (à l'aide du théorème de Hahn Banach) que $H^{-1}(\Omega)$ se compose des distributions T sur Ω de la forme

$$T = f_o + \sum_{i=1}^{n} \frac{\partial f_i}{\partial x_i}, \quad f_i \in L^2(\Omega), \quad i = 0, 1, \dots, n. \tag{1.11}$$

Ceci étant posé, on prend

$$\begin{cases} H = L^2(\Omega) \\ V = \text{sous espace fermé de } H^1(\Omega) \text{ avec } H_o^1(\Omega) \subset V \subset H^1(\Omega). \end{cases} \tag{1.12}$$

La famille de formes sesquilinéaires $a(t;u, v)$ est donnée par

$$\begin{aligned} a(t;u,v) = & \sum_{i,j=1}^{n} \int_\Omega a_{ij}(x,t) \frac{\partial u}{\partial x_j}(x) \frac{\partial \bar{v}}{\partial x_i}(x)\, dx + \\ & + \sum_{i=1}^{n} \int_\Omega a_i(x,t) \frac{\partial u}{\partial x_i} \bar{v}(x)\, dx + \int_\Omega a_o(x,t) u(x) \bar{v}(x) dx, \end{aligned} \tag{1.13}$$

où les $a_{ij}, a_i, a_o \in L^\infty(Q_T)$, $Q_T = \Omega \times]0, T[$ avec

$$\text{Re} \sum_{i,j=1}^{n} a_{ij}(x,t) \xi_j \bar{\xi}_i \geq \alpha \sum_{i=1}^{n} |\xi_i|^2, \quad \alpha > 0, \forall \xi_i \in C, \tag{1.14}$$

p.p. dans Q_T.

On considère la décomposition suivante de la forme $a(t;u, v)$:

$$\begin{aligned} a_o(t;u,v) = & \sum_{i,j=1}^{n} \int_\Omega a_{ij}(x,t) \frac{\partial u}{\partial x_j}(x) \frac{\partial \bar{v}}{\partial x_i}(x)\, dx + \\ & + \alpha \int_\Omega u(x)\, \bar{v}(x)\, dx, \end{aligned} \tag{1.15}$$

(1.16) $$a_1(t;u,v) = \sum_{i=1}^{n} \int_\Omega a_i(x,t) \frac{\partial u}{\partial x_i}(x)\, v(x)\, dx + \int_\Omega (a_o(x,t) - \alpha)\, u(x)\, \bar{v}(x)\, dx .$$

Il est clair que $a_o(t;u,v)$ vérifie les hypothèses $(1.1)_o$ et $(1.2)_o$ avec

$$K_o = \| \mathcal{A} \|_{L^\infty(Q_T)}$$

où $\mathcal{A}(x,t)$ est la norme euclidienne de la matrice $(a_{ij}(x,t))_{i\leq i,\, j\leq n}$, tandis que $a_1(t;u,v)$ vérifie l'hypothèse $(1.1)_1$.

Nous sommes dans les conditions d'application du théorème 1.1. Il suffit maintenant d'interpréter le problème résolu par ce théorème . Examinons deux choix simples pour l'espace V.

a) $\underline{V = H^1_o(\Omega)}$.

On prend $f \in L^2(0,T;H^{-1}(\Omega))$, $u_o \in L^2(\Omega)$. Soit $\varphi \in C^\infty_o(Q_T)$ une fonction complexe indéfiniment différentiable à support compact dans Q_T . Si u est la solution de (1.4) , (1.5) et (1.6) , on a

$$\int_o^T \left(\frac{du}{dt}(t), \varphi(t)\right)dt + \int_o^T a(t;u(t), \varphi(t))\, dt = \int_o^T (f(t), \varphi(t))\, dt.$$

Ainsi au sens des distributions sur Q_T , la solution u vérifie

(1.17) $$\frac{\partial u}{\partial t} - \sum_{i,j=1}^{n} \frac{\partial}{\partial x_i}\left(a_{ij} \frac{\partial u}{\partial x_j}\right) + \sum_{i=1}^{n} a_i \frac{\partial u}{\partial x_i} + a_o u = f .$$

La condition u $L^2(0,T;H^1_o(\Omega))$ signifie que u, $\frac{\partial u}{\partial x_i}$ (i = 1,...,n) appartient à $L^2(Q_T)$ et que $\gamma_o u = 0$ p.p. en t si la frontière Γ de Ω est "assez régulière". En résumé la solution $u = u(x,t)$ de (1.4) , (1.5) , (1.6) vérifie

(i) l'équation (1.17) au sens des distributions sur Q_T ,

(ii) la condition initiale $u(x;0) = u_o(x)$,

(iii) la condition aux limites $u(x,t) = 0$, $x \in \Gamma$, p.p. en t .

On a ainsi résolu le problème de Cauchy avec conditions aux limites de Dirichlet pour l'opérateur

$$(1.18)\qquad \frac{\partial}{\partial t} - \sum_{i,j=1}^{n} \frac{\partial}{\partial x_i}\left(a_{ij}\frac{\partial}{\partial x_j}\right) + \sum_{i=1}^{n} a_i \frac{\partial}{\partial x_i} + a_o .$$

b) $\underline{V = H^1(\Omega)}$.

On prend ici $f \in L^2(Q_T)$, $u_o \in L^2(\Omega)$. Comme au cas précédent, la solution $u = u(x,t)$ de (1.4) , (1.5) , (1.6) vérifie l'équation (1.17) au sens des distributions sur Q_T . On en déduit que pour tout $v \in H^1(\Omega)$

$$\int_\Omega \left\{ \frac{\partial u}{\partial t}(x,t) - \sum_{i,j=1}^{n} \frac{\partial}{\partial x_i}\left(a_{ij}(x,t)\frac{\partial u}{\partial x_j}(x,t)\right) + \right.$$

$$\left. + \sum_{i=1}^{n} a_i(x,t)\frac{\partial u}{\partial x_i}(x,t) + a_o(x,t)u(x,t)\right\} \bar{v}(x)dx =$$

$$= \int_\Omega f(x,t)\,\bar{v}(x)\,dx \quad \text{p.p. en } t ;$$

et ceci doit être égal à

$$\left(\frac{du}{dt}(t), v\right) + a(t;u(t),v) \qquad \text{p.p. en } t.$$

<u>Formellement,</u> on en deduit que pour tout v $H^1(\Omega)$ et p.p. en t

$$a(t;u(t),v) = \int_\Omega \left\{ - \sum_{i,j=1}^{n} \frac{\partial}{\partial x_i}\left(a_{ij}(x,t)\frac{\partial u}{\partial x_j}(x,t)\right) + \right.$$

$$\left. + \sum_{i=1}^{n} a_i(x,t)\frac{\partial u}{\partial x_i}(x,t) + a_o(x,t)u(x,t)\right\} \bar{v}(x)\,dx$$

d'où en utilisant <u>toujours formellement</u> la formule de Green

$$(1.19)\qquad \frac{\partial u}{\partial A(t)}(x,t) = \sum_{i,j=1}^{n} a_{ij}(x,t)\cos(n,x_i)\frac{\partial u}{\partial x_j}(x,t) = 0 ,$$

$$x \in \Gamma, \text{p.p. en } t$$

où n désigne la normale extérieure en x à Γ et $\cos(n,x_i)$ désigne le $i^{ème}$ cosinus directeur de cette normale. En résumé, dans ce cas, la solution $u = u(x,t)$ de (1.4), (1.5) et (1.6) vérifie

(i) l'équation (1.17) au sens des distributions sur Q_T,

(ii) la condition initiale $u(x,0) = u_o(x)$,

(iii) la condition aux limites formelle (1.19).

Ceci résoud le problème de Cauchy avec conditions aux limites de Neumann pour l'opérateur (1.18).

Signalons que les raisonnements formels effectués peuvent être justifiés avec des hypothèses de régularités (cf. [6]).

1.3. Exemple II

On désigne par $H(\Delta;\Omega)$ l'espace des fonctions $u \in L^2(\Omega)$ telles que $\Delta u = \sum_{i=1}^{n} \frac{\partial^2 u}{\partial x_i^2} \in L^2(\Omega)$. On munit $H(\Delta;\Omega)$ de la norme hilbertienne

$$(1.20) \qquad \|u\|_{H(\Delta;\Omega)} = \left(\int_\Omega (|\Delta u(x)|^2 + |u(x)|^2)\, dx\right)^{1/2} .$$

On prend alors par exemple

$$(1.21) \qquad \begin{aligned} &H = L^2(\Omega) \quad , \quad V = H(\Delta;\Omega) . \\ &a(t;u,v) = \int_\Omega a(x,t) \,.\, \Delta u(x) . \Delta \bar{v}(x)\, dx \end{aligned}$$

où $a = a(x,t) \in L^\infty(Q_T)$ avec

$$(1.22) \qquad \operatorname{Re} a(x,t) \geq \alpha > 0 \qquad \text{p.p. dans } Q_T .$$

Les formes $a_o(t;u,v)$ et $a_1(t;u,v)$ définies par

$$(1.23) \qquad a_o(t;u,v) = a(t;u,v) + \alpha \int_\Omega u(x) \,.\, \bar{v}(x)\, dx,$$

$$(1.24) \qquad a_1(t;u,v) = -\alpha \int_\Omega u(x)\, \bar{v}(x)\, dx ,$$

vérifient les hypothèses $(1.1)_o$, $(1.2)_o$ et $(1.1)_1$ avec en particulier

$$K_o = \|a\|_{L^\infty(Q_T)} \quad .$$

On interprète le problème résolu par le théorème 1.1 de la même façon que précédemment. Soient $f \in L^2(Q_T)$ et $u_o \in L^2(\Omega)$; la solution $u = u(x,t)$ de (1.4) , (1.5) , (1.6) vérifie

(i) (1.25 $\frac{\partial u}{\partial t} + \Delta(a\Delta u) = f$ au sens des distributions sur Q_T ,

(ii) la condition initiale $u(x,0) = u_o(x)$

(iii) les conditions aux limites formelles

(1.26) $$\Delta u(x,t) = 0 \ , \ \frac{\partial}{\partial n}(a\Delta u)(x,t) = 0, \quad x \in \Gamma \ ,$$

où $\frac{\partial}{\partial n}$ désigne la dérivée normale à .

On pourrait bien entendu multiplier les exemples (cf. [4]) .

2. Formulation abstraite de l'approximation des équations du 1er ordre en t.

Soit $\mathcal{H}$ une partie bornée de $R^p - \{0\}$ dont l'adhérence contient l'origine ; on notera h le point générique de $\mathcal{H}$. Dans la suite, h sera un paramètre destiné à tendre vers 0.

Soit V_h un espace de Hilbert sur C dépendant du paramètre h (4) . On note $(\ , \ )_h$ le produit scalaire dans V_h et $| \ |_h$ la norme correspondante. On se donne une autre norme sur V_h , soit $\| \ \|_h$, équivalente à la norme $| \ |_h$; elle vérifie donc

(2.1) $$\|v_h\|_h \leq C(h) \, |v_h|_h \quad , \quad C(h) = \text{constante}^{(5)}$$

(4) En pratique V_h sera une espace de dimension finie tel que dim $V_h \to \infty$ lorsque $h \to 0$.

(5) Le cas intéressant est celui où $C(h) \to \infty$ lorsque $h \to 0$.

On désigne par $\| \; \|_h^*$ la norme duale de $\| \; \|_h$, i.e.

(2.2) $$\| f_h \|_h^* = \sup_{v_h \; V_h} \frac{(f_h, v_h)_h}{\| v_h \|_h} \quad , \quad f_h \in V_h \; .$$

On a trivialement

(2.3) $$| f_h |_h \leq C(h) \; \| f_h \|_h^* \quad , \; \forall f_h \in V_h \; .$$

Soit S un paramètre entier > 0 destiné à tendre vers $+\infty$ et soit $k = (k_o; k_1, \ldots, k_{S-1}) \in R^S$ satisfaisant à

(2.4) $$\begin{cases} k_s > 0 \; , \quad s = 0, 1, \ldots S-1 \; , \\ \sum_{s=o}^{S-1} k_s = T \; , \\ \lim_{S \to \infty} \; \max_{o \leq s \leq S-1} k_s = 0 \end{cases}$$

On va considérer deux schémas type d'approximation de

(1.5) $$\frac{d}{dt} u(t) + A(t) \, u(t) = f(t) \qquad \text{p.p. en } t \; ,$$

(1.6) $$u(0) = u_o \; ,$$

et nous dégagerons sur ces deux schémas les méthodes générales d'étude de l'approximation.

Pour chaque couple $\{h, k\}$ on se donne

(2.5) $$\begin{cases} \left\{ A_{h,k}^{(s)} \in \mathcal{L}(V_h; V_h) \; ; \quad s = 0, 1, \ldots, S-1 \right\} , \\ \left\{ f_{h,k}^{(s)} \in V_h \; ; \; s = 0, 1, \ldots, S-1 \right\} , \\ u_{h,k}^{(o)} \in V_h \; . \end{cases}$$

On considère le schéma implicite à un pas de temps variable. Trouver $u_{h,k} = \left\{ u_{h,k}^{(s)} \in V_h \ ; s = 0, 1, \ldots, S \right\}$ solution de

$$(2.6) \qquad \frac{1}{k_s} (u_{h,k}^{(s+1)} - u_{h,k}^{(s)}) + A_{h,k}^{(s)} u_{h,k}^{(s+1)} = f_{h,k}^{(s)} , \quad s = 0, 1, \ldots, S-1 .$$

Comme deuxième schéma, on considère le schéma explicite également à un pas de temps variable. Trouver $u_{h,k} = \left\{ u_{h,k}^{(s)} \in V_h \ ; s = 0, 1, \ldots, S \right\}$ solution de

$$(2.7) \qquad \frac{1}{k_s} (u_{h,k}^{(s+1)} - u_{h,k}^{(s)}) + A_{h,k}^{(s)} u_{h,k}^{(s)} = f_{h,k}^{(s)} , \quad s = 0, 1, \ldots, S-1 .$$

On va étudier successivement la stabilité de ces deux schémas et la convergence des solutions $u_{h,k}$ vers la solution u de (1.4), (1.5), (1.6) dans des sens qui seront précisés. Bien entendu, on peut considérer des schémas d'approximation plus élaborés. (cf. [11] par exemple).

3. Etude de la stabilité

3.1. Quelques lemmes.

Donnons d'abord deux lemmes faciles qui seront d'un usage constant dans la suite.

Lemme 3.1.

Soit X un espace de Hilbert sur C et soit $B \in \mathcal{L}(X;X)$ d'adjoint B^* tel que

$$(3.1) \qquad \operatorname{Re} (Bx, x)_X \geq \beta \|x\|_X^2 , \quad \beta > 0 , \quad x \in X .$$

Alors pour tout $x \in X$, on a

$$(3.2) \qquad \| Bx \|_X \leq \left\| B \left(\frac{B+B^*}{2}\right)^{-1/2} \right\|_{\mathcal{L}(X;X)} \left\{ \operatorname{Re} (Bx, x)_X \right\}^{1/2} .$$

Démonstration. D'après (3.1) l'opérateur $(\frac{B+B^*}{2})^{-1/2}$ existe et appartient à $\mathcal{L}(X;X)$. On a donc

$$\|Bx\|_X = \|B(\frac{B+B^*}{2})^{-1/2}(\frac{B+B^*}{2})^{1/2}x\|_X \leq \|B(\frac{B+B^*}{2})^{-1/2}\|_{\mathcal{L}(X;X)}\|(\frac{B+B^*}{2})^{1/2}x\|_X$$

d'où le résultat puisque

$$\|(\frac{B+B^*}{2})^{1/2}x\|_X = \{\mathrm{Re}\,(Bx,x)\}^{1/2}.$$

Lemme 3.2. (Lemme de Gronwall discret).

Si les φ_s, $s = 0,1,\ldots,S$, sont des nombres ≥ 0 qui vérifient

$$\varphi_s \leq C + \delta \sum_{r=o}^{s-1} k_r \varphi_r\,, \quad r = 1,\ldots,S\,, \quad \varphi_o \leq C \tag{3.3}$$

avec $C > 0$, $\delta > 0$, on a

$$\varphi_s \leq C \exp\,(\sum_{r=o}^{s-1} k_r) \leq C \exp\,(\delta T)\,, \quad s = 0,1,\ldots,S. \tag{3.4}$$

Démonstration. On montre par récurrence que

$$\varphi_s \leq C + \sum_{r=o}^{s-1} k_r \varphi_r \leq C \prod_{r=o}^{s-1} (1+\delta k_r).$$

On obtient le résultat en remarquant que $1+\delta k_r \leq \exp\,(\delta k_r)$.

3.2. Stabilité du schéma implicite.

Pour l'étude de la stabilité, on va faire sur les opérateurs $A^{(s)}_{h,k}$ les hypothèses suivantes :

$$A^{(s)}_{h,k} = A^{(s)}_{h,k,o} + A^{(s)}_{h,k,1}\,, \quad A^{(s)}_{h,k,i} \in \mathcal{L}(V_h;V_h)\,, \qquad i = 0,1\,, \tag{3.5}$$

(3.6) $\begin{cases} \text{il existe une constante } \beta > 0 \text{ indépendante de } h, k, s \text{ telle} \\ \text{que } \operatorname{Re}\,(A^{(s)}_{h,k,o}\, v_h, v_h)_h \geq \beta \|v_h\|_h^2 \;, \forall v_h \in V_h \;, \\ s = 0, 1, \ldots, S-1 \;, \end{cases}$

(3.7) $\begin{cases} \text{il existe une constante } P > 0 \text{ indépendante de } h, k, s \text{ telle que} \\ |A^{(s)}_{h,k,1}\, v_h|_h \leq P \|v_h\|_h \;, \forall v_h \in V_h \;, \forall s = 0, 1, \ldots, S-1. \end{cases}$

L'opérateur $A^{(s)}_{h,k,o}$ peut être considéré comme la "partie principale" de $A^{(s)}_{h,k}$. Considérons d'abord le schéma implicite (2.6).

Théorème 3.1.

Sous les hypothèses précédentes, le schéma (2.6) admet une solution unique $u_{h,k} \in V_h^{S+1}$ lorsque

(3.8) $$\max_s k_s < \frac{4\beta}{P^2} .$$

Démonstration. Connaissant $u^{(s)}_{h,k} \in V_h$, le schéma (2.6) donne pour $u^{(s+1)}_{h,k}$ l'équation

$$u^{(s+1)}_{h,k} + k_s A^{(s)}_{h,k}\, u^{(s+1)}_{h,k} = k_s f^{(s)}_{h,k} + u^{(s)}_{h,k} .$$

Pour que cette équation admette une solution unique, il suffit d'après le théorème de Lax-Milgram que

$$|v_h|_h^2 + k_s \operatorname{Re}\,(A^{(s)}_{h,k} v_h, v_h)_h > \rho |v_h|_h^2 \;, \rho > 0 \;, \forall v_h \in V_h .$$

Mais en vertu de (3.6) et (3.7)

$$\operatorname{Re}\,(A^{(s)}_{h,k}\, v_h, v_h)_h \geq \beta \|v_h\|_h^2 - P\|v_h\|_h |v_h|_h \geq -\frac{P^2}{4\beta} |v_h|_h^2 .$$

Il suffit donc que l'on ait pour $s = 0, 1, \ldots, S-1$

$$1 - \frac{P^2}{4\beta} k_s > 0 ,$$

d'où le résultat.

Théorème 3.2.

On fait les hypothèses précédentes. Pour tout choix des constantes ε et η telles que $0 < \varepsilon \leq 2\beta$, $0 < \eta < 1$, la solution $u_{h,k}$ du schéma implicite (2.6) vérifie une inégalité de l'énergie discrète lorsque k est assez petit $(1 - \frac{2P^2}{\varepsilon} \max_s k_s \geq \eta)$

$$|u_{h,k}^{(s)}|_h^2 + \frac{2\beta - \varepsilon}{\eta} \sum_{r=o}^{s-1} k_r \|u_{h,k}^{(r+1)}\|_h^2 \leq$$

$$\leq \frac{1}{\eta}\left[|u_{h,k}^{(o)}|_h^2 + \frac{2}{\varepsilon} \sum_{r=o}^{s-1} k_r \|f_{h,k}^{(r)}\|_h^{*2}\right] \exp\left(\frac{2P^2}{\varepsilon\eta} \sum_{r=o}^{s-2} k_r\right).$$

Le schéma implicite (2.6) est donc inconditionnellement stable au sens de (3.9) .

Démonstration. En multipliant scalairement (2.6) par $u_{h,k}^{(s+1)}$ (et en supprimant les indices h et k pour alléger l'écriture), on obtient

$$(u^{(s+1)} - u^{(s)}, u^{(s+1)}) + k_s (A^{(s)} u^{(s+1)}, u^{(s+1)}) = k_s (f^{(s)}, u^{(s+1)}).$$

On prend deux fois la partie réelle de l'équation obtenue et on remarque que

$$2\mathrm{Re}\,(u^{(s+1)} - u^{(s)}, u^{(s+1)}) = |u^{(s+1)}|^2 - |u^{(s)}|^2 + |u^{(s+1)} - u^{(s)}|^2 ,$$

$$2\mathrm{Re}\,(A^{(s)} u^{(s+1)}, u^{(s+1)}) \geq 2\beta\|u^{(s+1)}\|^2 - 2P\|u^{(s+1)}\| |u^{(s+1)}| ,$$

$$|2\mathrm{Re}\,(f^{(s)}, u^{(s+1)})| \leq 2\|f^{(s)}\|^* \|u^{(s+1)}\| .$$

On en déduit donc que pour $s = 0, 1, \dots, S-1$

$$|u^{(s+1)}|^2 - |u^{(s)}|^2 + 2\beta k_s \|u^{(s+1)}\|^2 \leq 2k_s \left[\|f^{(s)}\|^* + P|u^{(s+1)}|\right] \|u^{(s+1)}\|$$

d'où pour tout $\varepsilon > 0$

(3.10) $$|u^{(s+1)}|^2 - |u^{(s)}|^2 + (2\beta - \varepsilon)\, k_s \| u^{(s+1)} \|^2 \leq$$
$$\leq \frac{2}{\varepsilon}\, k_s \| f^{(s)} \|^{*2} + \frac{2P^2}{\varepsilon}\, k_s\, |u^{(s+1)}|^2 .$$

En sommant (3.10) de 0 à s-1 , on trouve

(3.11) $$|u^{(s)}|^2 + (2\beta - \varepsilon) \sum_{r=o}^{s-1} k_r \| u^{(r+1)} \|^2 \leq$$
$$\leq |u^{(o)}|^2 + \frac{2}{\varepsilon} \sum_{r=o}^{s-1} k_r \| f^{(r)} \|^{*2} + \frac{2P^2}{\varepsilon} \sum_{r=o}^{s-1} k_r\, |u^{(r+1)}|^2 .$$

Si k est choisi de facon que

$$\frac{2P^2}{\varepsilon} \max_s k_s \leq 1 - \eta \;,\quad 0 < \eta < 1 ,$$

on a

$$\frac{2P^2}{\varepsilon} \sum_{r=o}^{s-1} k_r |u^{(r+1)}|^2 \leq \frac{2P^2}{\varepsilon} \sum_{r=o}^{s-2} k_r\, |u^{(r+1)}|^2 + (1-\eta)\, |u^{(s)}|^2 .$$

L'inégalité (3.11) devient alors en divisant par η

(3.12) $$|u^{(s)}|^2 + \frac{2\beta - \varepsilon}{\eta} \sum_{r=o}^{s-1} k_r \| u^{(r+1)} \|^2 \leq$$
$$\leq \frac{1}{\eta} \Big[|u^{(o)}|^2 + \frac{2}{\varepsilon} \sum_{r=o}^{s-1} k_r \| f^{(r)} \|^{*2} \Big] + \frac{2P^2}{\varepsilon \eta} \sum_{r=o}^{s-2} k_r |u^{(r+1)}|^2 ,$$

pour s = 1 , ..., S .

Il est alors possible d'appliquer le lemme de Gronwall discret : le résultat s'en déduit trivialement.

3.3. Stabilité du schéma explicite .

Considérons maintenant le schéma explicite (2.7) . Nous allons étudier la stabilité sous les mêmes hypothèses que précédemment. Fixons d'abord une notation. Si $B_h \in \mathcal{L}(V_h; V_h)$, nous désignons par $|B_h|_h$ la norme suivante de l'opérateur B_h

(3.13) $$|B_h|_h = \sup_{v_h \in V_h} \frac{|B_h v_h|_h}{|v_h|_h}$$

Théorème 3.3.

On fait les hypothèses (3.5), (3.6) et (3.7). Sous les conditions de stabilité

(3.14) $$k_s \left| A^{(s)}_{h,k,o} \left(\frac{A^{(s)}_{h,k,o} + (A^{(s)}_{h,k,o})^*}{2} \right)^{-1/2} \right|_h^2 \leq 2(1-\eta), \quad s=0,1,\dots,S-1$$

où $\eta > 0$ est une constante arbitrairement petite et indépendante de h, k, s,

(3.15) $$k_s C(h)^2 \leq \rho \quad , \quad s = 0,1,\dots, S-1 \ ,$$

où $\rho > 0$ est une constante arbitrairement grande et indépendante de h, k, s, la solution $u_{h,k}$ du schéma explicite (2.7) vérifie l'inégalité de l'énergie discrète

(3.16) $$|u^{(s)}_{h,k}|_h^2 + \beta\eta^2 \sum_{r=o}^{s-1} k_r \| u^{(r)}_{h,k} \|_h^2 \leq$$
$$\leq \left[|u^{(o)}_{h,k}|_h^2 + L \sum_{r=o}^{s=1} k_r \| f^{(r)}_{h,k} \|_h^{*\,2} \right] \exp(LP^2 \sum_{r=o}^{s-1} k_r)$$

avec

(3.17) $$L = 2 \left(\frac{1}{\beta\eta^2} + \rho \left(1 + \frac{1}{\eta}\right) \right) .$$

Le schéma explicite est donc conditionnellement stable.

Démonstration. On multiplie scalairement (2.7) par $u^{(s)}_{h,k}$; il vient pour $s = 0,1,\dots, S-1$ (après suppression des indices h et k)

$$(u^{(s+1)} - u^{(s)}, u^{(s)}) + k_s (A^{(s)} u^{(s)}, u^{(s)}) = k_s (f^{(s)}, u^{(s)}) .$$

En prenant deux fois la partie réelle de cette équation et en remarquant que

$$2\mathrm{Re}\,(u^{(s+1)}-u^{(s)},\ u^{(s)}) = |u^{(s+1)}|^2 - |u^{(s)}|^2 - |u^{(s+1)} - u^{(s)}|^2 ,$$

on obtient

$$|u^{(s+1)}|^2 - |u^{(s)}|^2 - |u^{(s+1)} - u^{(s)}|^2 + 2\,k_s\,\mathrm{Re}\,(A_o^{(s)} u^{(s)}, u^{(s)}) =$$

$$= 2\,k_s\,\mathrm{Re}\,(f^{(s)}, u^{(s)}) - 2k_s\,\mathrm{Re}\,(A_1^{(s)} u^{(s)}, u^{(s)}) .$$

Mais d'après (3.7)

$$2\left| -(A_1^{(s)} u^{(s)}, u^{(s)}) + (f^{(s)}, u^{(s)}) \right| \leq 2(P|u^{(s)}| + \|f^{(s)}\|^*)\|u^{(s)}\| \leq$$

$$\leq \beta\varepsilon\|u^{(s)}\|^2 + \frac{1}{\beta\varepsilon}(P|u^{(s)}| + \|f^{(s)}\|^*)^2$$

pour tout $\varepsilon > 0$, ce qui donne en utilisant (3.6)

$$2\left| -(A_1^{(s)} u^{(s)}, u^{(s)}) + (f^{(s)}, u^{(s)}) \right| \leq$$

$$\leq \varepsilon\,\mathrm{Re}\,(A_o^{(s)} u^{(s)}, u^{(s)}) + \frac{1}{\beta\varepsilon}(P|u^{(s)}| + \|f^{(s)}\|^*)^2 .$$

On obtient ainsi pour $s = 0, 1, \dots, S-1$

$$(3.18)\qquad |u^{(s+1)}|^2 - |u^{(s)}|^2 - |u^{(s+1)} - u^{(s)}|^2 + (2-\varepsilon)k_s\,\mathrm{Re}(A_o^{(s)} u^{(s)}, u^{(s)}) \leq$$

$$\leq \frac{1}{\beta\varepsilon}\,k_s\,(P\,|u^{(s)}| + \|f^{(s)}\|^*)^2 .$$

Il faut maintenant majorer $|u^{(s+1)} - u^{(s)}|^2$. D'après (2.7) on a

$$|u^{(s+1)} - u^{(s)}| \leq k_s\,(|A_o^{(o)} u^{(s)}| + |A_1^{(s)} u^{(s)}| + |f^{(s)}|$$

d'où en utilisant le lemme 3.1 avec $B = A_o^{(s)}$

$$(3.19)\qquad |u^{(s+1)} - u^{(s)}| \leq k_s \Big[\Big| A_o^{(s)} \Big(\frac{A_o^{(s)} + (A_o^{(s)})^*}{2}\Big)^{-1/2} \Big|$$

$$\Big\{ \mathrm{Re}\,(A_o^{(s)} u^{(s)}, u^{(s)}) \Big\}^{1/2} + P\|u^{(s)}\| + |f^{(s)}| \Big] .$$

Remarquons maintenant que d'après (2.1) et (2.3)

$$P\|u^{(s)}\| + |f^{(s)}| \leq C(h)\left[P|u^{(s)}| + \|f^{(s)}\|\right]. \tag{3.20}$$

De (3.19) et (3.20) on deduit que pour s = 0, 1, ..., S-1

$$|u^{(s+1)}-u^{(s)}|^2 \leq k_s^2(1+\delta)\left|A_o^{(s)}\left(\frac{A_o^{(s)}+(A_o^{(s)})^*}{2}\right)^{-1/2}\right|^2 \text{Re}(A_o^{(s)}u^{(s)},u^{(s)}) + k_s^2(1+\frac{1}{\delta})C(h)^2\left[P|u^{(s)}|+\|f^{(s)}\|^*\right]^2, \tag{3.21}$$

où δ est une constante > 0 arbitraire. Reportant la majoration (3.21) de $|u^{(s+1)}-u^{(s)}|^2$ dans l'inégalité (3.18), on trouve que

$$|u^{(s+1)}|^2-|u^{(s)}|^2+(2-\varepsilon-(1+\delta)k_s\left|A_o^{(s)}\left(\frac{A_o^{(s)}+(A_o^{(s)})^*}{2}\right)^{-1/2}\right|^2)k_s \text{Re}(A_o^{(s)}u^{(s)},u^{(s)})$$
$$\leq \left(\frac{1}{\beta\varepsilon}+(1+\frac{1}{\delta})k_s C(h)^2\right)k_s\left[P|u^{(s)}|+\|f^{(s)}\|^*\right]^2.$$

Sous les conditions de stabilité (3.14) et (3.15) et en choisissant par exemple $\delta = \eta$, $\varepsilon = \eta^2$, on a pour s = 0, 1, ..., S-1

$$|u^{(s+1)}|^2 - |u^{(s)}|^2 + \beta\eta^2\|u^{(s)}\|^2 \leq \tag{3.22}$$
$$\leq 2\left(\frac{1}{\beta\eta^2}+\rho\left(1+\frac{1}{\eta}\right)\right)\left(P^2|u^{(s)}|^2+\|f^{(s)}\|^{*2}\right).$$

En sommant (3.22) de 0 à s-1, (s ≤ S), on obtient

$$|u^{(s)}|^2+\beta\eta^2\sum_{r=o}^{s-1}k_r\|u^{(r)}\|^2 \leq \tag{3.23}$$
$$\leq |u^{(o)}|^2 + 2\left(\frac{1}{\beta\eta^2}+\rho\left(1+\frac{1}{\eta}\right)\right)\left[P^2\sum_{r=o}^{s-1}k_r|u^{(r)}|^2+\sum_{r=o}^{s-1}k_r\|f^{(r)}\|^{*2}\right].$$

Il suffit d'appliquer le lemme de Gronwall discret pour obtenir le résultat.

Remarque 3.1.

Les conditions de stabilité ne dépendent que des parties principales $A^{(s)}_{h,k,o}$ des opérateurs $A^{(s)}_{h,k}$.

Corollaire 3.1.

Sous les hypothèses (3.5), (3.6) , (3.7) et lorsque pour tout s $A^{(s)}_{h,k,o} = (A^{(s)}_{h,k,o})^*$ (i.e. la partie principale de chaque opérateur $A^{(s)}_{h,k}$ est auto-adjointe), le schéma explicite (2.7) est stable sous les conditions

(3.24) $$k_s \left| A^{(s)}_{h,k,o} \right|_h \leq 2(1-\eta) \quad , \qquad s = 0, 1, \ldots, S-1 \, ,$$

(3.25) $$k_s \;\; C(h)^2 \;\leq \qquad , \quad s = 0, 1, \ldots, S-1 \quad .$$

Démonstration. Il suffit de remarquer que lorsque $A^{(s)}_{h,k,o}$ est un opérateur auto-adjoint, on a

$$A^{(s)}_{h,k,o} (\operatorname{Re} A^{(s)}_{h,k,o})^{-1/2} = (A^{(s)}_{h,k,o})^{1/2} \qquad \text{avec}$$

$$\operatorname{Re} A^{(s)}_{h,k,o} = \frac{A^{(s)}_{h,k,o} + (A^{(s)}_{h,k,o})^*}{2}$$

et

$$\left| (A^{(s)}_{h,k,o})^{1/2} \right|^2_h = \left| A^{(s)}_{h,k,o} \right|_h \, .$$

Faisons maintenant deux hypothèses .dont la première ne fait que préciser l'hypothèse (3.6)

(3.26) $$\begin{cases} \text{Pour tout } s = 0, 1, \ldots, S-1 \text{ , il existe une constante } \beta_s > 0 \\ \text{telle que} \\ \operatorname{Re} (A^{(s)}_{h,k,o} \; v_h, v_h)_h \geq \beta_s \|v_h\|^2_h \quad , \quad \forall v_h \in V_h \quad , \end{cases}$$

(3.27) $$\begin{cases} \text{pour tout } s = 0, 1, \ldots, S-1, \text{ il existe une constante } M_s > 0 \\ \text{telle que} \\ (A^{(s)}_{h,k,o} \; u_h, v_h)_h \leq M_s \|u_h\|_h \|v_h\|_h \quad , \quad \forall u_h \; v_h \quad V_h \; . \end{cases}$$

Bien entendu, on a $\beta_s \geq \beta$ lorsque (3.6) et (3.26) sont verifiés.

Corollaire 3.2.

Sous les hypothèses (3.5), (3.6), (3.7), (3.26) et (3.27), le schèma explicite (2.7) est stable si

$$(3.28) \qquad k_s\, C(h)^2 \leq \frac{2\beta_s}{M_s^2}(1-\eta) \leq \rho \quad , \quad s = 0,\ldots,S-1$$

Si de plus $A^{(s)}_{h,k,o} = (A^{(s)}_{h,k,o})^*$ pour tout s, la stabilité est assurée si

$$(3.29) \qquad k_s\, C(h)^2 \leq \frac{2}{M_s}(1-\eta) \leq \rho \quad , \quad s = 0,1,\ldots,S-1.$$

Démonstration. On part de la condition de stabilité (3.14). On déduit de l'hypothèse (3.26) que

$$|(\mathrm{Re}\, A^{(s)}_{h,k,o})^{-1/2} v_h|_h \leq \sqrt{\beta_s}\, \|v_h\|_h \quad , \quad \forall v_h \in V_h \, ,$$

et de l'hypothèse (3.27) que

$$|A^{(s)}_{h,k,o} v_h|_h \leq M_s\, C(h)\, |v_h|_h \quad , \quad \forall v_h \in V_h \, .$$

Il en résulte que

$$|A^{(s)}_{h,k,o}\,(\mathrm{Re}\, A^{(s)}_{h,k,o})^{-1/2}|_h^2 \leq \frac{M_s^2}{\beta_s} C(h)^2 \, , s=0,1,\ldots,S-1.$$

Les conditions de stabilité (3.14) et (3.15) sont donc entraînées par (3.28). Lorsque $A^{(s)}_{h,k,o} = (A^{(s)}_{h,k,o})^*$ pour tout s, la condition (3.29) entraîne (3.24) et (3.25).

Remarque 3.2.

En pratique, on aura toujours $2\beta_s/M_s^2 \leq \rho$ (resp. $2/M_s \leq \rho$) pour ρ assez grand indépendant de h, k, s.

4. Exemples

Nous allons appliquer les résultats précédents à l'étude de l'approximation par des méthodes de différences finies de problèmes évoqués aux n°. 1.2 et 1.3. Précisons d'abord quelques notations. On choisit

$$h = (h_1, \ldots, h_n) \quad , \quad h_i > 0 \; ; \qquad i = 1, \ldots, n \; , \tag{4.1}$$

et on considère le réseau $\mathcal{R}_h$ des points M de la forme

$$M = (m_1 h_1, \ldots, m_n h_n) \quad , \quad m_i \in \mathbb{Z} \, .$$

A tout point $M \in \mathcal{R}_h$ on associe le pavé ω_h^M et l'ensemble σ_h^M définis par

$$\bar{\omega}_h^M = \prod_{i=1}^{n} \left] (m_i - \tfrac{1}{2}) h_i \; , \; (m_i + \tfrac{1}{2}) h_i \right[\, ,$$

$$\sigma_h^M = \Big(\bigcup_{i=1}^{n} \bar{\omega}_h^{M+(h_i/2)} \Big) \cup \Big(\bigcup_{i=1}^{n} \bar{\omega}_h^{M-(h_i/2)} \Big) \, , \tag{4.2}$$

où

$$M \mp \frac{h_i}{2} = (m_1 h_1, \ldots, m_{i-1} h_{i-1}, (m_i \mp \tfrac{1}{2}) h_i, m_{i+1} h_{i+1}, \ldots, m_h h_n) \; .$$

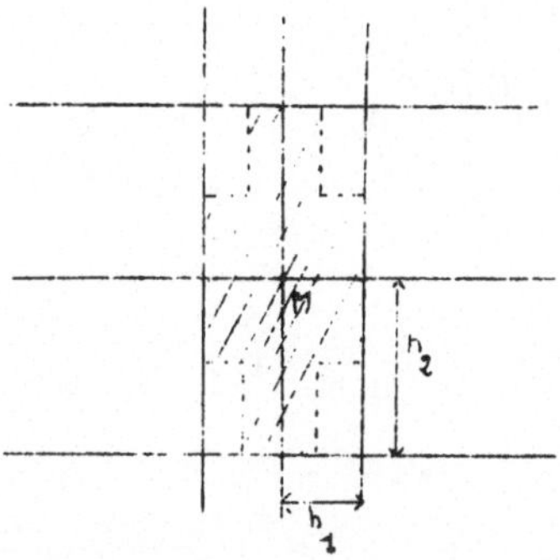

σ_h^M = partie hachurée.

On désigne par θ_h^M la fonction caracteristique du pavé $\bar{\omega}_h^M$. Enfin on introduit l'operateur aux différences finies $\hat{\nabla}_{h_i}$.

(4.3) $$\hat{\nabla}_{h_i} f(x) = \frac{1}{h_i}\left[f(x+\frac{h_i}{2}) - f(x-\frac{h_i}{2})\right]$$

avec

$$x \mp \frac{h_i}{2} = (x_1, \ldots, x_{i-1} \ , \ x_i \mp \frac{h_i}{2} \ , \ x_{i+1}, \ldots, x_n) \ .$$

4.1. Exemple I-a.

On considère l'exemple a du N° 1.2. (correspondant aux conditions aux limites de Dirichlet). On pose

(4.4) $$\Omega_h = \left\{ M \mid M \in \mathcal{R}_h \ , \ \bar{\omega}_h^M \subset \Omega \right\}$$

et on désigne par $\ell^2(\Omega_h)$ l'espace des suites $v_h = \left\{ v_h(M) \in C \ ; \ M \in \Omega_h \right\}$ de carré sommable sur Ω_h. On choisit alors

(4.5) $$V_h = \ell^2(\Omega_h)$$

muni du produit scalaire

(4.6) $$(u_h, v_h)_h = h_1 \ldots h_n \sum_{M \in \Omega_h} u_h(M) \ . \ \overline{v_h(M)} \ .$$

On définit ensuite l'opérateur $q_h \in \mathcal{L}(V_h; L^2(\Omega))$ par

(4.7) $$q_h v_h = \sum_{M \in \Omega_h} v_h(M) \ \theta_h^M \quad , \quad v_h \in V_h \quad .$$

Il est clair que

(4.8) $$(u_h, v_h)_h = \int_\Omega q_h u_h \ . \ \overline{q_h v_h} \ dx \quad , \quad u_h, v_h \in V_h .$$

On considère enfin

(4.9) $$\| v_h \|_h = (\sum_{i=1}^n \int_\Omega |\hat{\nabla}_{h_i} \ q_h \ v_h|^2 dx + \int_\Omega |q_h \ v_h|^2 dx)^{1/2} \ .$$

Calculons la constante C(h) intervenant dans (2.1). On a

$$\int_\Omega |\hat{\nabla}_{h_i} q_h v_h|^2 dx = \frac{1}{h_i^2} \int_\Omega | q_h v_h (x + \frac{h_i}{2}) - q_h v_h (x - \frac{h_i}{2})|^2 dx \leq$$

$$\leq \frac{4}{h_i^2} \int_\Omega |q_h v_h|^2 dx = \frac{4}{h_i^2} |v_h|_h^2$$

puisque les fonctions $x \longrightarrow q_h v_h (x \mp h_i/2)$ ont. leurs supports contenus dans Ω. On en déduit que

$$\|v_h\|_h^2 \leq (4 \sum_{i=1}^n \frac{1}{h_i^2} + 1) |v_h|_h^2$$

d'où

$$C(h) = (4 \sum_{i=1}^n \frac{1}{h_i^2} + 1)^{1/2} . \tag{4.10}$$

Définissons maintenant les opérateurs $A^{(s)}_{h,k,o}$ et $A^{(s)}_{h,k,1}$. Si les $_s$, s = 0, 1, ..., S-1, sont des nombres > 0 tels que

$$\text{Re} \sum_{i,j=1}^n a_{ij}(x,t) \xi_j \bar{\xi}_i \geq \beta_s \sum_{i=1}^n |\xi_i|^2 , \tag{4.11}$$

$$\forall \xi_i \in C, \text{ p.p. dans } \Omega \times (t_s, t_{s+1})$$

où

$$t_o = 0 , \quad t_s = \sum_{r=o}^{s-1} k_r , \; s = 1, \ldots, S , \tag{4.12}$$

on pose

$$a^{(s)}_{h,o}(t; u_h, v_h) = \sum_{i,j=1}^n \int_\Omega a_{ij}(x,t) . \hat{\nabla}_{h_j} q_h u_h(x) . \hat{\nabla}_{h_i} \overline{q_h v_h}(x) \, dx + \tag{4.13}$$

$$+ \beta_s \int_\Omega q_h u_h (x) . \overline{q_h v_h}(x) . dx,$$

$$a^{(s)}_{h,1}(t; u_h, v_h) = \sum_{i=1}^n \int_\Omega a_i(x,t) . \hat{\nabla}_{h_i} q_h u_h(x) . \overline{q_h v_h}(x) \, dx + \tag{4.14}$$

$$+ \int_\Omega (a_o(x,t) - \beta_s) q_h u_h(x) . \overline{q_h v_h}(x) . dx ,$$

puis

(4.15) $$a^{(s)}_{h,k,i}(u_h, v_h) = \frac{1}{k_s}\int_{t_s}^{t_{s+1}} a^{(s)}_{h,i}\,(t; u_h, v_h)\,dt\ ,\quad i = 0,1\ ,$$

$$s = 0,1,\ldots,S-1,$$

et

(4.16) $$(A^{(s)}_{h,k,i}\, u_h, v_h)_h = a^{(s)}_{h,k,i}\,(u_h, v_h)\ ,\qquad u_h, v_h \in V_h\ ,$$

$$i = 0,1,\ s = 0,1,\ldots,S-1.$$

Il est clair que toutes les hypothèses (3.5) , (3.6), (3.7) , (3.26) et (3.27) sont verifiées avec $\beta = \alpha$ et

(4.17) $$M_s = \|\mathcal{A}\|_{L^\infty(\Omega\times(t_s, t_{s+1}))}\ ,\quad s = 0,1,\ldots,S-1\ ,$$

$\mathcal{A}(x,t)$ désignant toujours la norme euclidienne de la matrice $(a_{ij}(x,t))_{1\leq i,j\leq n}$

Notons aussi que les opérateurs $A^{(s)}_{h,k,o}$ sont auto-adjoints si la matrice précédente est hermitienne p.p. dans Q_T.

Il nous reste à définir les éléments $f^{(s)}_{h,k}$ et $u^{(o)}_{h,k}$. Pour cela, considérons l'opérateur $r_h \in (L^2(\Omega)\ ;\ V_h)$:

(4.18) $$(r_h v)(M) = \frac{1}{h_1\ldots h_n}\int_{\bar{\omega}_h M} v(x)\,dx\ ,\qquad M \in \Omega_h\ .$$

Si $f \in L^2(0,T;H^{-1}(\Omega))$ est donné sous la forme $f = f_o + \sum_{i=1}^{n} \frac{f_i}{x_i}$ avec $f_i \in L^2(Q_T)$ pour $i = 0,1,\ldots,n$, on choisit

(4.19) $$f^{(s)}_{h,k} = \frac{1}{k_s}\int_{t_s}^{t_{s+1}} \Big\{(r_h\, f_o(t) + \sum_{i=1}^{n} r_h\, \hat{\nabla}_{h_i}\, f_i(t)\Big\}\ ,$$

$$s = 0,1,\ldots,S-1.$$

On prend enfin

(4.20) $$u^{(o)}_{h,k} = r_h\ u_o\ .$$

Tous les résultats obtenus au N° 3 peuvent alors s'appliquer. En particulier, en utilisant le corollaire 3.2, il y a stabilité du schéma explicite pour les k_s assez petits lorsque :

$$(4.21) \qquad k_s \left(\sum_{i=1}^{n} \frac{1}{h_i^2}\right) \leq \frac{s}{2\|\mathcal{Q}\|^2_{L^\infty(\Omega\ (t_s,t_{s+1}))}} (1-\eta)\ , \qquad s = 0,1,\ldots,S-1\ ,$$

où η' est (comme η) une constante >0 arbitrairement petite indépendante de h,k,s. Lorsque la matrice $(a_{ij}(x,t))_{1\leq i,j\leq n}$ est hermitienne p.p. dans Q_T, on obtient la condition de stabilité pour k_s assez petit

$$(4.22) \qquad k_s \sum_{i=1}^{n} \frac{1}{h_i^2} \leq \frac{1}{2\|\mathcal{Q}\|_{L^\infty(\Omega\ (t_s,t_{s+1}))}} (1-\eta')\ , \quad s = 0,1,\ldots,S-1.$$

Remarque 4.1.

Si Ω est borné, on peut prendre au lieu de (4.9)

$$(4.9') \qquad \|v_h\|_h = \sum_{i=1}^{n} \int_\Omega |\hat{\nabla}_{h_i}\ q_h\ v_h|^2\ dx)^{1/2}\ .$$

Les normes (4.9) et (4.9)' sont alors équivalentes uniformément en h. $C(h) = \left(\sum_{i=1}^{n}\frac{1}{h^2}\right)^{1/2}$ et les conditions de stabilité (4.21) et (4.22) ont lieu sans la restriction "k_s assez petit".

4.2. Exemple I.b.

Passons maintenant à l'exemple b du N° 1.2. (correspondant aux conditions aux limites de Neumann). On pose cette fois

$$(4.22) \qquad \Omega_h = \left\{ M \mid M \in \mathcal{R}_h\ ,\ \sigma_h^M \cap \Omega \neq \emptyset \right\}.$$

On choisit toujours $V_h = \ell^2(\Omega_h)$ muni du produit scalaire

$$(4.23) \qquad u_h, v_h)_h = h_1 \ldots h_n \sum_{M\in\Omega_h} u_h(M)\ .\ \overline{v_h(M)}.$$

On pose encore

$$(4.24) \qquad q_h\, v_h = \sum_{M \in \Omega_h} v_h(M)\, \theta_h^M \,, \quad v_h \in V_h \,.$$

La relation (4.8) ne subsiste plus mais on a

$$(4.25) \qquad (u_h, v_h)_h = \int_{\Omega(h)} q_h u_h \cdot \overline{q_h v_h} \cdot dx \,, \quad \Omega(h) = \bigcup_{M \in \Omega_h} \omega_h^M \,.$$

<u>Lemme 4.1.</u>

<u>L'application</u> $v_h \to \| v_h \|_h$ <u>définie par</u>

$$(4.26) \qquad \| v_h \|_h = \Big(\sum_{i=1}^{n} \int_\Omega |\hat{\nabla}_{h_i} q_h v_h(x)|^2 \, dx + \int_\Omega |q_h v_h(x)|^2 \, dx \Big)^{1/2}$$

<u>est une norme sur</u> V_h.

<u>Démonstration.</u> Le seul point non trivial est de vérifier que $\| v_h \|_h = 0$ entraîne $v_h = 0$. En effet $\int_\Omega |q_h v_h|^2 = 0$ implique $q_h v_h = 0$ sur Ω de sorte que $v_h(M) = 0$ pour tout $M \in \Omega_h^o = \{ M | M \in \mathcal{R}_h, \bar{\omega}_h^M \cap \Omega \neq \emptyset \}$. Il ne reste plus qu'à examiner le cas où $M \in \Omega_h - \Omega_h^o$. Pour un tel M, on a $\sigma_h^M \cap \Omega \neq \emptyset$ et $\bar{\omega}_h^M \cap \Omega = \emptyset$; il existe alors un i tel que $\bar{\omega}_h^{M+(h_i/2)} \cap \Omega \neq \emptyset$ ou que $\bar{\omega}_h^{M-(h_i/2)} \cap \Omega \neq \emptyset$. Choisissons par exemple la première éventualité : alors $\bar{\omega}^{M+(h_i/2)} \cap \Omega \neq \emptyset$ et $M+h_i \in \Omega_h^o$ d'où $v_h(M+h_i)=0$. Puisque

$$\hat{\nabla}_{h_i} q_h v_h = \sum_{P \in \Omega_h} v_h(P) \, (\theta_h^{P-(h_i/2)} - \theta_h^{P+(h_i/2)}) \frac{1}{h_i} = 0 \text{ sur } \Omega,$$

nous en déduisons que le coefficient de $\theta_h^{M+(h_i/2)}$, c'est-à-dire $\frac{1}{h_i} (v_h(M+h_i) - v_h(M))$, est nul d'où $v_h(M) = 0$.

Il est facile de vérifier que l'expression (4.10) de $C(h)$ reste valable. On définit ensuite les opérateurs $A_{h,k,i}^{(s)}$ exactement comme au N°4.1. Les éléments $f_{h,k}^{(s)}$ et $u_{h,k}^{(o)}$ seront donnés comme suit. On considère l'opérateur $r_h \in \mathcal{L}(L^2(\Omega), V_h)$

$$(4.27) \qquad (r_h v)(M) = \frac{1}{h_1 \ldots h_n} \int_{\bar{\omega}_h^M} \tilde{v}(x)\, dx, \quad M \in \Omega_h$$

où $\tilde{v}$ est le prolongement de v par 0 dans $\Omega(h)-\Omega$. Alors si f $L^2(Q_T)$ et $u_o \in L^2(\Omega)$, on pose

(4.28) $$f^{(s)}_{h,k} = \frac{1}{k_s} \int_{t_s}^{t_{s+1}} r_h\, f(t)\, dt\ , \quad s = 0, 1, \ldots, S-1$$

(4.29) $$u^{(o)}_{h,k} = r_h\, u_o\, .$$

Il est clair que les résultats de stabilité du N° 4.1. restent valables (inégalités (4.21) et (4.22) pour k_s assez petit), la remarque 4.1. étant évidemment mise à part.

4.3. Exemple II.

On considère l'exemple du N° 1.3. On pose

(4.30) $$\tau^M_h = \bar{\omega}^M_h \cup \left(\bigcup_{i=1}^{n} \bar{\omega}^{M+h_i}\right) \cup \left(\bigcup_{i=1}^{n} \omega_h^{M-h_i}\right)\, .$$

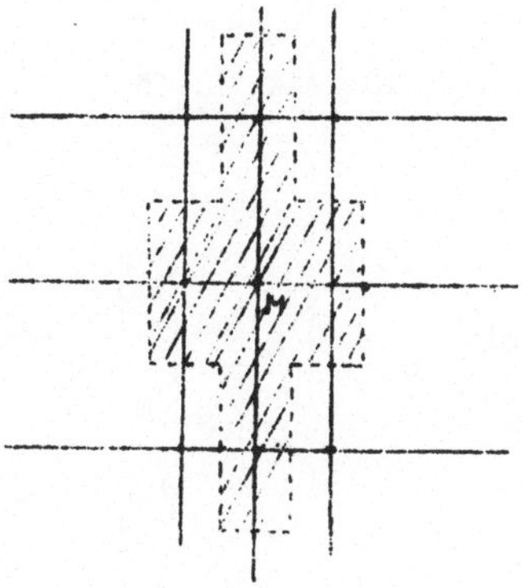

τ^M_h = partie hachurée.

On choisit

(4.31) $$\Omega_h = \left\{ M \mid M \in \mathcal{R}_h\, ,\ \tau^M_h \cap \Omega \neq \emptyset\right.\, ,$$

puis $V_h = \ell^2(\Omega_h)$ avec le produit scalaire habituel. On définit $q_h \in \mathcal{L}(V_h; L^2(\Omega))$ de la manière standard et on prend

(4.32) $$\|v_h\| = \left(\int_\Omega \left\{|\Delta_h\, q_h\, v_h(x)|^2 + |q_h\, v_h(x)|^2\right\} dx\right)^{1/2}\, ,$$

avec

(4/33) $$\Delta_h = \sum_{i=1}^{n} \hat{\nabla}_{h_i}^2 .$$

On vérifie aisément, comme pour le lemme 4.1, que (4.32) définit une norme sur V_h. D'autre part la constante $C(h)$ est donnée par

(4.34) $$C(h) = \left(16 \sum_{i=1}^{n} \frac{1}{h_i^4} + 1\right)^{1/2} .$$

On définit les opérateurs $A_{h,k,i}^{(s)}$ en posant

(4.35) $$a_{h,o}^{(s)}(t;u_h,v_h) = \int_\Omega a(x,t) \cdot \Delta_h q_h u_h(x) \cdot \overline{\Delta_h q_h v_h}(x)\, dx + $$
$$+ \beta_s \int_\Omega q_h u_h(x)\, \overline{q_h v_h}(x)\, dx,$$

(4.36) $$a_{h,1}^{(s)}(t;u_h,v_h) = -\beta_s \int_\Omega q_h u_h(x) . \overline{q_h\, v_h}(x) . dx$$

avec

(4.37) $$\beta_s = \underset{\substack{x\in\Omega \\ t\in(t_s,t_{s+1})}}{\text{inf ess}} \quad \text{Re } a(x,t) \geq \alpha ,$$

puis

(4.38) $$a_{h,k,i}^{(s)}(u_h,v_h) = (A_{h,k,i}^{(s)} u_h, v_h)_h = \frac{1}{k_s} \int_{t_s}^{t_{s+1}} a_h^{(s)}(t;u_h,v_h)\, dt ,$$
$$\forall u_h, v_h \in V_h .$$

On a

(4.39) $$M_s = \| a \|_{L^\infty(\Omega \times (t_s,t_{s+1}))} .$$

Si $f \in L^2(Q_T)$ et si $u_o \in L^2(\Omega)$, on définit $f_{h,k}^{(s)}$ et $u_{h,k}^{(o)}$ comme en (4.27), (4.28), (4.29).

Il y a stabilité du schéma explicite dans le cas général où a est complexe lorsque

$$(4.40)\qquad k_s \sum_{i=1}^{n} \frac{1}{h_i^4} \leq \frac{1}{8} - \frac{\underset{x\in\Omega\quad t\in(t_s,t_{s+1})}{\text{inf ess.}} \operatorname{Re} a(x,t)}{\|a\|^2_{L^\infty(\Omega\,(t_s,t_{s+1}))}} (1-\eta'),$$

$$s = 0, 1, \ldots, S-1,$$

pour les k_s assez petit. Dans le cas où a est réel, la stabilité a lieu lorsque

$$(4.41)\qquad k_s \sum_{i=1}^{n} \frac{1}{h_i^4} \leq \frac{1}{8\,\|a\|_{L^\infty(\Omega\; t_s,t_{s+1}))}} (1-\eta') \quad , \; s = 0, 1, \ldots\; S-1$$

toujours pour k_s assez petit.

On comparera les résultats obtenus ici avec ceux de [13] . On pourrait ici aussi multiplier les exemples. Nous renvoyons à [2] pour la discrétisation de formes a(t;u, v) plus générales et à [1] pour un point de vue un peu différent.

5. Etude de la convergence.

Nous allons donner seulement un résultat de convergence faible pour simplifier. Nous nous restreindrons au cas du schéma explicite, la méthode et le résultat se généralisant trivialement au cas du schéma implicite.

Soit X un espace de Hilbert sur C tel que H soit un sous espace fermé de X et soit π l'opérateur de projection orthogonale de X sur H. On considère ensuite une famille d'opérateurs $p_{h,k} \in \mathcal{L}(V_h^{s+1}\,;\, L^2(0,T;H))$ tels que $q_{h,k} = \pi_o\, p_{h,k}$ $(V_h^{S+1}; L^\infty(0,T;H))$ avec les propriétés

$$(5.1)\qquad \| p_{h,k}\, v_{h,k} \|_{L^2(0,T;X)} \leq C_1 \left(\sum_{s=o}^{S=1} k_s \|v^{(s)}_{h,k}\|_h^2 \right)^{1/2}$$

$$(5.2)\qquad \| q_{h,k}\, v_{h,k} \|_{L^\infty(0,T;H)} \leq C_2 \max_{s=o,\ldots,S} | v^{(s)}_{h,k} |_h \;,$$

pour tout $v_{h,k} = \left\{ v^{(s)}_{h,k} \in V_h \; ; \; s = 0, 1, \ldots, S \right\} \in V_h^{S+1}$, les constantes C_1 et C_2 étant indépendantes de h et k. Soit ω un opérateur de $\mathcal{L}(V;X)$.

Soit $\mathcal{V}$ un sous espace dense de V et soit ρ_h une application linéaire de $\mathcal{V}$ dans V_h . Nous faisons les hypothèses suivantes :

H.1 Si une famille $\{w_{h,k}\}$ est telle que

$$p_{h,k}\, w_{h,k} \rightarrow W \text{ dans } L^2(0,T;X) \text{ faible lorsque } h,k \quad 0, \tag{5.3}$$

on a

a) $$w = \pi W \in L^2(0,T;V)\,, \quad W = \omega w\,, \tag{5.4}$$

b) pour tout $v \in \mathcal{V}$ et toute fonction $\varphi \in C^1(0,T)$

$$\sum_{s=1}^{S-1} (w_{h,k}^{(s)}, \rho_h v)_h \,(\varphi(t_s) - \varphi(t_{s-1})) \rightarrow \int_0^T (w(t),v)\,\varphi'(t)\,dt \tag{5.5}$$

$$\sum_{s=0}^{S-1} k_s\,(A_{h,k}^{(s)}, \rho_h v)_h\, \varphi(t_s) \rightarrow \int_0^T a(t; w(t), v)\,\varphi(t)\,dt\,, \tag{5.6}$$

H.2. Les données $f_{h,k}^{(s)}$ et $u_{h,k}^{(o)}$ satisfont à

$$\sum_{s=0}^{S-1} k_s \| f_{h,k}^{(s)} \|_h^2 \leq C_3\,, \tag{5.7}$$

$$| u_{h,k}^{(o)} |_h \leq C_4\,, \tag{5.8}$$

où les constantes C_3 et C_4 sont indépendantes de h et k. De plus, on a pour tout $v \in \mathcal{V}$ et toute fonction $\varphi \in C^1(0,T)$

$$\sum_{s=0}^{S-1} k_s\,(f_{h,k}^{(s)}, \rho_h v)_h\, \varphi(t_s) \rightarrow \int_0^T (f(t),v)\,\varphi(t)\,dt\,, \tag{5.9}$$

$$(u_{h,k}^{(o)}, \rho_h v)_h \rightarrow (u_o, v)\,. \tag{5.10}$$

Théorème 5.1.

Sous les conditions d'application du théorème de stabilité (3.3) et sous les hypothèses de consistance H.1 et H.2 , la solution $u_{h,k}$ du schéma explicite a les propriétés de convergence suivantes lorsque h et $k \rightarrow 0$

(5.11) $\quad p_{h,k}\, u_{h,k} \rightarrow \omega u$ dans $L^2(0,T;X)$ faible,

(5.12) $\quad q_{h,k}\, u_{h,k} \rightarrow u$ dans $L^\infty(0,T;H)$ faible.

Démonstration. Soit $u_{h,k}$ la solution du schéma explicite (2.7). Lorsque les conditions de stabilité (3.11) et (3.15) sont vérifiées, on a en vertu du théorème 3.3 et des hypothèses (5.7) et (5.8)

$$\sum_{s=o}^{S-1} k_s \| u_{h,k}^{(s)} \|_h^2 \ , \ \max_{s=o,1,\dots,S} | u_{h,k}^{(s)} |_h \leq C_5$$

où C_5 est une constante indépendante de h et k. Ensuite, on déduit de (5.1) et (5.2) que $p_{h,k}\ u_{h,k}$ reste dans un borné de $L^2(0,T;X)$ tandis que $q_{h,k} u_{h,k}$ reste dans un borné de $L^\infty(0,T;H)$. On peut donc extraire de la famille $\{h,k\}$ une sous famille, notée toujours h, k telle que

(5.13)
$$\begin{aligned} & p_{h,k}\, u_{h,k} \rightarrow U_* \quad \text{dans } L^2(0,T;X) \text{ faible,} \\ & q_{h,k}\, u_{h,k} \rightarrow u_* = \pi U_* \quad \text{dans } L^\infty(0,T;H) \text{ faible.} \end{aligned}$$

D'après H.1, $u_* \in L^2(0,T;V)$ et $U_* = \omega u_*$. Il suffit maintenant de démontrer que u_* est la solution de (1.4), (1.5) et (1.6).

Soit $v \in \mathcal{V}$ et soit $\varphi \in C_o^\infty(0,T)$. Nous déduisons de (2.7) (en supprimant les indices h et k)

$$\sum_{s=o}^{S-1} (u^{(s+1)} - u^{(s)}, \rho v) \varphi(t_s) + \sum_{s=o}^{S-1} k_s (A^{(s)} u^{(s)}, \rho v) \varphi(t_s) = $$
$$= \sum_{s=o}^{S-1} k_s (f^{(s)}, \rho v) \varphi(t_s) .$$

Mais pour k assez petit

$$\sum_{s=o}^{S-1} (u^{(s+1)} - u^{(s)}, \rho v) \varphi(t_s) = - \sum_{s=1}^{S-1} (u^{(s)}, \rho v) (\varphi(t_s) - \varphi(t_{s-1}))$$

puisque $\varphi(0) = 0$ et que $\varphi(t_{S-1}) = 0$. On a donc pour k assez petit

$$(5.14)\quad \begin{cases} -\sum_{s=1}^{S-1} (u^{(s)}, \rho v)(\varphi(t_s)-\varphi(t_{s-1})) + \sum_{s=o}^{S-1} k_s(A^{(s)}u^{(s)}, \rho v)\varphi(t_s) = \\ \qquad = \sum_{s=o}^{S-1} k_s \,.(f^{(s)}, \rho v)\varphi(t_s) \,. \end{cases}$$

En passant à la limite grâce à H.1 et H.2 , on trouve

$$(5.15)\quad -\int_o^T (u_*(t), v)\varphi'(t)\, dt + \int_o^T a(t; u_*(t), v)\varphi(t)\, dt = \\ = \int_o^T (f(t), v)\,\varphi(t)\, dt$$

pour tout $v \in \mathcal{V}$ et toute fonction $\varphi \in C_o^\infty(\,0, T\,)$. Comme $\mathcal{V}$ est dense dans V , on obtient (5.15) pour tout $v \in V$. Il en resulte que, au sens des distributions sur $]0, T[$ à valeurs dans V' , on a

$$(5.16)\quad u_*'(.) + A(.)\, u_*(.) = f(.)\,.$$

Mais puisque $A(.)\, u_*(.)$ et $f(.) \in L^2(0, T; V')$, on déduit de (5.16)

$$(5.17)\quad \begin{cases} u_*' \in L^2(0, T; V')\,, \\ u_*'(t) + A(t)\; u_*(t) = f(t) \text{ dans } V'\,, \text{ p.p. en } t. \end{cases}$$

Il reste à montrer que $u_*(0) = u_o$. Soit $\varphi \in C^1(0, T)$ nulle dans un voisinage de T. Alors pour k assez petit et pour $v \in \mathcal{V}$, on a

$$\begin{cases} -\sum_{s=1}^{S-1} (u^{(s)}, \rho v)(\varphi(t_s)-\varphi(t_{s-1})) + \sum_{s=o}^{S-1} k_s\,(A^{(s)}u^{(s)}, \rho v)\varphi(t_s) = \\ \qquad = \sum_{s=o}^{S-1} k_s\,(f^{(s)}, \rho v)\varphi(t_s) - (u^{(o)}, \rho v)\varphi(0)\,. \end{cases}$$

En passant à la limite, on obtient pour tout $v \in V$

$$(5.18)\quad -\int_o^T (u_*(t), v)\varphi'(t)\, dt + \int_o^T a(t; u_*(t), v)\varphi(t)\, dt = \\ = \int_o^T (f(t), v)\varphi(t)\, dt + (u_o, v)\varphi(0)\,.$$

L'équation (5.18) et la formule de Green (1.7) entraînent

(5.19) $\quad (u_o, v) = (u_*(0), v) \quad \forall v \in V.$

De la densité de V dans H, on déduit que $u_*(0) = u_o$.

Enfin de l'unicité de la solution u du problème continu, il résulte que c'est toute la famille $u_{h,k}$ qui converge vers u au sens indiqué.

Remarque 5.1.

Sous des hypothèses raisonnables, on peut obtenir à partir du théorème précédent des résultats de convergence forte dans $L^2(0,T;X)$ et $L^\infty(0,T;H)$; nous renvoyons pour ce point à [16] et [3] . Notons que dans [10] on démontre la convergence forte dans $L^2(0,T;H)$ par un argument de compacité : ce genre de raisonnement est utile pour les problèmes non linéaires (cf. [10]) .

Remarque 5.2.

On peut utiliser différemment l'inégalité de stabilité (3.16) pour passer à la limite. Dans [1] , on construit à partir de $u_{h,k}$ une approximation $p_{h,k}\, u_{h,k}$ de la solution u dans l'espace

$$\mathcal{W} = \left\{ v \mid v \in L^2(0,T;V) , \ \frac{dv}{dt} \ L^2(0,T;V') \right\}$$

muni de la norme hilbertienne naturelle et on obtient directement la convergence de $p_{h,k}\, u_{h,k}$ vers u dans cet espace $\mathcal{W}$.

6. Retour sur les exemples.

Nous reprenons maintenant les exemples du N°4 et nous allons montrer que les hypothèses H.1 et H.2 sont vérifiées dans tous les cas.

6.1. Exemple du N° 4.1.

On choisit

(6.1) $X = (L^2(\Omega))^{n+1}$ muni de la structure hilbertienne produit.

A tout $F = (f, f_1, f_2, \ldots, f_n) \in (L^2(\Omega))^{n+1}$, on associe $\pi F = f \in L^2(\Omega)$. On définit l'opérateur $\omega \in \mathcal{L}(H^1_o(\Omega) ; (L^2(\Omega))^{n+1})$ par

(6.2) $$\omega v = (v, \frac{\partial v}{\partial x_1}, \ldots, \frac{\partial v}{\partial x_n}), \quad v \in H^1_o(\Omega).$$

On prend ensuite

(6.3) $$\mathcal{V} = C^\infty_o(\Omega),$$

(6.4) $$(\rho_h v)(M) = v(M), \quad \forall M \in \Omega_h, \ \forall v \in C^\infty_o(\Omega).$$

Définissons la famille d'opérateurs $p_{h,k}$. D'abord si $v_h \in V_h$, on pose

(6.5) $$p_h v_h = (q_h v_h, \hat{\nabla}_{h_1} q_h v_h, \ldots, \hat{\nabla}_{h_n} q_h v_h) \in (L^2(\Omega))^{n+1}$$

puis si $v_{h,k} = \left\{ v^{(s)}_{h,k} \in V_h \ ; s = 0, 1, \ldots, S \right\}$, on pose

(6.6) $$p_{h,k} \, v_{h,k}(t) = p_h \, v^{(s)}_{h,k}, \quad t_s \leq t < t_{s+1}, \quad s = 0, 1, \ldots, S-1.$$

Il est alors clair que les hypothèses (5.1) et (5.2) sont satisfaites.

Passons maintenant à la vérification de l'hypothèse H.1.a. Si $W = (w, w_1, \ldots, w_n)$ est limite faible dans $L^2(0, T; X)$ de $p_{h,k} \, w_{h,k}$; montrons d'abord que $w_i = \frac{\partial w}{\partial x_i}$, $i = 1, \ldots, n$. Soit $\varphi \in C^\infty_o(Q_T)$; on a pour h assez petit

$$\int_{Q_T} \hat{\nabla}_{h_i} q_{h,k} \, w_{h,k}(x,t) . \varphi(x,t) . dx\, dt = - \int_{Q_T} q_{h,k} \, w_{h,k}(x,t) . \hat{\nabla}_{h_i} \varphi(x,t) \, dx\, dt$$

Mais $q_{h,k} \, w_{h,k}$ (resp $\hat{\nabla}_{h_i} q_{h,k} \, w_{h,k}$) $\to w$ (resp. w_i) dans $L^2(Q_T)$ faible et il est facile de vérifier que $\hat{\nabla}_{h_i} \varphi \to \frac{\partial \varphi}{\partial x_i}$ dans $L^2(Q_T)$ fort.

En passant à la limite, on voit donc que pour tout $\varphi \in C^\infty_o(Q_T)$, on a

$$\int_{Q_T} w_i(x,t).\varphi(x,t).dx\,dt \to - \int_{Q_T} w(x,t)\frac{\partial\varphi}{\partial x_i}(x,t)\quad dx\quad dt$$

ce qui signifie que $w_i = \frac{\partial w}{\partial x_i}$ au sens des distributions sur Q_T. On en déduit que $w \in L^2(0,T;H^1(\Omega))$. Pour montrer que $w \in L^2(0,T;H^1_o(\Omega))$, il suffit de remarquer que $p_{h,k}\,\tilde{w}_{h,k}$ prolongement de $p_{h,k}\,w_{h,k}$ par 0 dans $R^n \times (0,T)$ converge vers $\tilde{W}$ prolongement de W par 0 dans $R^n \times (0,T)$ dans $(L^2(R^n \times (0,T)))^{n+1}$ faible. On voit alors comme précedemment que $\tilde{w}$ $L^2(0,T;H^1(R^n)))$. Si Ω est "assez régulier", ce que nous supposons, il en résulte que $\gamma_o w = 0$ d'où $w \in L^2(0,T;H^1_o(\Omega))$.

Consacrons nous maintenant à la verification des hypothèses H.1.b. Introduisons d'abord une notation. Si $\varphi \in C^1(0,T)$, posons

(6.7) $\qquad \varphi_k(t) = \varphi(t_s)\ ,\quad t_s \leq t < t_{s+1}\ ,\quad s = 0,1,\ldots,S-1\ ,$

(6.8) $$\hat{\nabla}_k\varphi_k(t) = \begin{cases} 0\ , & t_o \leq t < t_1\ , \\ \frac{1}{k_s}(\varphi(t_s)-\varphi(t_{s-1}))\ , & t_s \leq t < t_{s+1},\ s = 1,\ldots,S-1. \end{cases}$$

On voit aisément que $\varphi_k \to \varphi$ et $\hat{\nabla}_k\varphi_k \to \varphi'$ dans $L^2(0,T)$ fort. On remarque alors que

(6.9) $$\sum_{s=1}^{S-1}(w^{(s)}_{h,k},\rho_h v)_h\ (\varphi(t_s)-\varphi(t_{s-1})) =$$
$$= \int_o^T (q_{h,k}\,w_{h,k}(t),\ q_h\rho_h v)\hat{\nabla}_k\varphi_k(t)\,dt,$$

$$\sum_{s=o}^{S-1} k_s(A^{(s)}_{h,k}\,w^{(s)}_{h,k},\ \rho_h v)\varphi(t_s) =$$

(6.10) $$= \int_{Q_T}\Big\{\sum_{i,j=1}^{n} a_{ij}(x,t).\hat{\nabla}_{h_j} q_{h,k}\,w_{h,k}(x,t).\hat{\nabla}_{h_i} q_h\rho_h\bar{v}(x) +$$
$$+\sum_{i=1}^{n} a_i(x,t).\hat{\nabla}_{h_i} q_{h,k}\,w_{h,k}(x,t)\,q_h\rho_h\bar{v}(x) +$$
$$+ a_o(x,t)\,q_{h,k}\,w_{h,k}(x,t).\ q_h\rho_h\bar{v}(x)\Big\}\bar{\varphi}_k(t)\,dx\,dt.$$

Puisque $p_h \rho_h v \to \omega v$ dans $(L^2(\Omega))^{n+1}$ fort (vérification immédiate), la conclusion résulte de la convergence de $p_{h,k}\, w_{h,k}$ vers ωw dans $(L^2(Q_T))^{n+1}$ faible.

Il ne nous reste plus qu'à examiner les hypothèses H.2. Nous allons pour cela démontrer les deux lemmes suivants.

Lemme 6.1.

Soit $u_o \in L^2(\Omega)$; alors

(6.11) $$|r_h\, u_o|_h \leq \|u_o\|_{L^2(\Omega)} \quad ,$$

(6.12) $$(r_h\, u_o, \rho_n v)_h \to (u_o, v) \quad , \quad \forall v \in C_o^\infty(\Omega) .$$

Démonstration. On a

$$|r_h u_o|_h^2 = h_1 \ldots h_m \sum_{M \in \Omega_h} \left(\frac{1}{h_1 \ldots h_n} \int_{\bar{\omega}_h^M} u_o\, dx\right)^2 =$$

$$= \frac{1}{h_1 \ldots h_n} \sum_{M \in \Omega_h} \left(\int_{\bar{\omega}_h^M} u_o\, dx\right)^2 \leq \sum_{M \in \Omega_h} \int_{\omega_h^M} |u_o|^2\, dx \leq u_o \quad {}^2_{L^2(\Omega)}$$

D'autre part

$$(r_h\, u_o, \rho_h v)_h = (u_o, q_h \rho_h v)$$

et (6.12) résulte de la convergence de $q_h \rho_h v$ vers v dans $L^2(\Omega)$ fort.

Lemme 6.2.

Si f $L^2(0,T;H^{-1}(\Omega))$ est donné sous la forme $f = f_o + \sum_{i=1}^n \frac{\partial f_i}{\partial x_i}$ avec f_i $L^2(Q_T)$, $i = 0, 1, \ldots, n$ et s on définit

(6.13) $$f_{h,k}^{(s)} = \frac{1}{k_s} \int_{t_s}^{t_{s+1}} \left\{ r_h\, f_o(t) + \sum_{i=1}^n r_h\, \hat{\nabla}_{h_i} f(t) \right\} dt, \quad s = 0, 1, \ldots, S-1,$$

on a

(6.14) $$\sum_{s=o}^{S-1} k_s \|f_{h,k}^{(s)}\|_h^{*2} \leq \sum_{i=o}^n \int_{Q_T} |f_i(x,t)|^2\, dx\, dt,$$

$$(6.15)\qquad \sum_{s=0}^{S-1} k_s (f_{h,k}^{(s)}, \rho_h v)_h\, \varphi(t_s) \longrightarrow \int_0^T (f(t), v)\varphi(t)\, dt\ ,$$

$$\forall v \in C_0^\infty(\Omega)\ ,\ \forall \varphi \in C^1(0,T).$$

Démonstration. On a

$$(f_{h,k}^{(s)}, v_h)_h = (q_h f_{h,k}^{(s)}, q_h v_h) = \frac{1}{k_s}\int_{t_s}^{t_{s+1}} (q_h r_h f_o(t) +$$

$$+ \sum_{i=1}^{n} q_h r_h \hat{\nabla}_{h_i} f_i(t),\ q_h v_h)\, dt\ .$$

Or il est immédiat de voir que

$$\int_{\bar{\omega}_h^M} \hat{\nabla}_{h_i} f_i(x,t)\, dx = -\int_\Omega f_i(x,t)\, \hat{\nabla}_{h_i} \theta_h^M(x)\, dx$$

de sorte que

$$(q_h r_h \hat{\nabla}_{h_i} f_i(t),\ q_h v_h) = -(f_i(t),\ \hat{\nabla}_{h_i} q_h v_h)\ .$$

Il en résulte que

$$(f_{h,k}^{(s)}, v_h)_h = \frac{1}{k_s}\int_{t_s}^{t_{s+1}} \left\{\int_\Omega f_o(x,t)\, q_h \overline{v_h}(x)\, dx\ - \right.$$

$$\left. - \sum_{i=1}^{n} \int_\Omega f_i(x,t) \hat{\nabla}_{h_i} q_h \bar{v}_h(x)\, dx \right\} dt$$

d'où

$$\| f_{h,k}^{(s)} \|_h^* \leq \frac{1}{k_s}\int_{t_s}^{t_{s+1}} \left(\sum_{i=0}^{n} \int_\Omega |f_i(x,t)|^2\, dx\right)^{1/2} dt\ .$$

On en déduit que

$$\sum_{s=0}^{S-1} k_s \| f_{h,k}^{(s)} \|_h^{*2} \leq \sum_{s=0}^{S-1} \frac{1}{k_s} \left| \int_{t_s}^{t_{s+1}} \left(\sum_{i=0}^{n} \int_\Omega |f_i(x,t)|^2\, dx\right)^{1/2} dt \right|^2 \leq$$

$$\leq \sum_{i=0}^{n} \int_{Q_T} |f_i(x,t)|^2\, dx\, dt\ .$$

D'autre part

$$\sum_{s=0}^{S-1} k_s (f_{h,k}^{(s)}, \rho_h v)_h \varphi(t_s) = \int_0^T \left\{\int_\Omega f_o(x,t) . q_h \rho_h \bar{v}(x)\, dx\ - \right.$$

$$\left. - \sum_{i=1}^{n} \int_\Omega f_i(x,t) . \hat{\nabla}_{h_i} q_h \rho_h \bar{v}(x)\, dx \right\} \varphi_k(t)\, dt.$$

Puisque $p_h \rho_h v \quad \omega v$ dans $(L^2(\Omega))^{n+1}$ fort, on trouve que

$$\sum_{s=o}^{S-1} k_s (f_{h,k}^{(s)}, \rho_h v) \varphi(t_s) \to \int_o^T \Big\{ \int_\Omega f_o(x,t)\, \bar{v}(x)\, dx -$$

$$- \sum_{i=1}^{n} \int_\Omega f_i(x,t) \frac{\partial \bar{v}}{\partial x_i}(x)\, dx \Big\} \varphi(t)\, dt.$$

Mais la dernière expression n'est autre que $\int_o^T (f(t), v)\varphi(t)\, dt$. Le lemme est donc démontré.

On a ainsi vérifié les hypothèses H.2. On peut donc appliquer le théorème 5.1. Si $u_{h,k}$ est la solution d'un schéma stable, on a

$$q_{h,k}\, u_{h,k} \to u \text{ dans } L^2(Q_T) \text{ faible et dans } L^\infty(0,T;L^2(\Omega)) \text{ faible,}$$

(6.16) $$\hat{\nabla}'_{h_i}\, q_{h,k}\, u_{h,k} \to \frac{\partial u}{\partial x_i} \text{ dans } L^2(Q_T) \text{ faible, } i = 1, \ldots, n.$$

Remarque 6.1.

Conformément à la remarque 5.1, on peut démontrer que dans (6.16) toutes les convergences sont en fait des oonvergences fortes.

6.2. Exemple du N° 4.2.

Ce qui à été fait au N° 6.1. se transpose trivialement dans ce cas.

6.3. Exemple du N° 4.3.

On choisit ici

(6.17) $$X = (L^2(\Omega))^2$$

et on définit l'opérateur $\omega \in \mathcal{L}(H(\Delta;\Omega);(L^2(\Omega))^2)$ par

(6.18) $$\omega v = (v, \Delta v) \quad , \quad v \in H(\Delta;\Omega).$$

On prend ensuite

(6.19) $$\mathcal{V} = C_o^\infty(\bar{\Omega})$$

qui est bien dense dans $H(\Delta;\Omega)$ lorsque Ω est "assez régulier". La vérification des hypothèses de consistance se fait comme au N° 6.1.

7. Equations dévolution du 2ème ordre en t. Formulation abstraite.

On considère le triplet d'espaces $\{V, H, V'\}$ comme au N° I.1. Soit $u, v \to a(t;u,v)$ une famille de formes sesquilinéaires continues sur $V \times V$ dépendant du paramètre $t \in [0,T]$, $T < \infty$, avec les propriétés suivantes :

(7.1) $$a(t;u,v) = a_0(t;u,v) + a_1(t,u,v),$$

a) les formes $u, v \to a_0(t;u,v)$ sont sesquilinéaires continues sur $V \times V$ et $\forall u, v \in V$ chaque fonction $t \to a_0(t;u,v)$ est une fois continûment différentiable dans $[0,T]$,

(7.2) $$|a_0(t;u,v)| \leq K_0 \|u\| \|v\| \quad , \quad \forall u, v \in V,$$

(7.3) $$a_0(t;u,v) = \overline{a_0(t;v,u)} \quad , \quad \forall u, v \in V,$$

(7.4) $$a_0(t;v,v) \geq \alpha \|v\|^2 \quad , \quad \alpha > 0, \ \forall v \in V;$$

b) les formes $u, v \to a_1(t;u,v)$ sont sesquilinéaires continues sur $V \times H$ et $\forall u, v \in V$ chaque fonction $t \to a_1(t;u,v)$ est mesurable,

(7.5) $$a_1(t;u,v) \leq K_1 \|u\| \|v\| \quad , \quad \forall u \in V, \ \forall v \in H.$$

On désigne par $A(t) \in \mathcal{L}(V;V')$ l'opérateur défini par $a(t,u,v)$.

On a alors le théorème suivant ;

Théorème 7.1.

Sous les hypothèses précédentes et pour f donné dans $L^2(0,T;H)$, u_0 donné dans V, u_1 donné dans H, il existe une fonction u vérifiant

(7.6) $$u \ L^\infty(0,T;V), \ u' = \frac{du}{dt} \in L^\infty(0,T;H), \quad u'' = \frac{d^2u}{dt^2} \in L^2(0,T;V'),$$

(7.7) $$u''(t) + A(t)\,u(t) = f(t) \quad \text{p.p. en } t,$$

(7.8) $$u(0) = u_0 \ , \ u'(0) = u_1 \ .$$

Pour la démonstration de ce théorème, nous renvoyons à [4], [5]. On trouvera egalement dans [5] des propriétés supplémentaires de régularité de la solution u. L'unicité de la solution u n'est connue qu'avec des hypothèses supplémentaires, par exemple

(7.9) la fonction $t \rightarrow a_1\ (t;u,v)$ est une fois continûment différentiable dans $[0,T]$, $\forall u \in V$, $\forall v \in H$.

Les exemples s'obtiennent de la même façon que pour les équations du 1er ordre ; nous renvoyons à [4].

Dans la suite on supposera pour simplifier un peu que $A(t) = A$ ne dépend pas du paramètre t mais nos méthodes s'adaptent aisément au cas général (cf. [10]).

8. Formulation abstraite de l'approximation des équations d'évolution du 2ème ordre en t. Résultats de stabilité.

8.1. Les schémas d'approximation.

On introduit comme au N° 2 les espaces de Hilbert V_h. On choisit ensuite un pas de temps k constant (pour simplifier): $k = \frac{T}{S}$ avec $S \quad \infty$. Pour chaque couple $\{h, k\}$, on se donne

$$(8.1) \qquad \begin{array}{l} A_h \in \mathcal{L}(V_h;V_h) \text{ (indépendant de } k), \\ \left\{f^{(s)}_{h,k} \in V_h \;;\; s = 1,\dots,S-1\right\} \\ u^{(o)}_{h,k} \in V_h,\ u^{(1)}_{h,k} \in V_h. \end{array}$$

On définit alors un schéma implicite. Trouver $u_{h,k} = \left\{u^{(s)}_{h,k} \quad V_h \;;\; s = 0,\dots,S\right\}$ vérifiant :

$$(8.2) \qquad \frac{1}{k^2}\left(u^{(s+1)}_{h,k} - 2u^{(s)}_{h,k} + u^{(s-1)}_{h,k}\right) + A_h\, u^{(s+1)}_{h,k} = f^{(s)}_{h,k}\;,\quad s = 1,\dots,S-1\,.$$

De la même façon, on considère le schéma explicite : trouver $u_{h,k} = \left\{u^{(s)}_{h,k} \quad V_h \;;\; s = 0,1,\dots,S\right\}$ vérifiant

(8.3) $\frac{1}{k^2}(u_{h,k}^{(s+1)} - 2\,u_{h,k}^{(s)} + u_{h,k}^{(s-1)}) + A_h\, u_{h,k}^{(s)} = f_{h,k}^{(s)}$, $s = 1$, $S-1$

On va étudier maintenant la stabilité de ces deux schémas.

8.2. Stabilité du schéma implicite.

On va faire sur les opérateurs A_h les hypothèses suivantes :

(8.4) $A_h = A_{h,o} + A_{h,1}$, $A_{h,i} \in \mathcal{L}(V_h; V_h)$, $i = 0, 1$,

(8.5) $A_{h,o} = A_{h,o}^{*}$,

(8.6) $\begin{cases} \text{il existe une constante } \beta > 0 \text{ indépendante de } h \text{ telle que} \\ (A_{h,o}\, v_h, v_h)_h \geq \beta \| v_h \|_h^2 \ , \ \forall v_h \in V_h \ , \end{cases}$

il existe une constante $P > 0$ indépendante de h telle que

(8.7) $| A_{h,1}\, v_h |_h \leq P \, \| v_h \|_h$, $\forall v_h \in V_h$.

Théorème 8.1.

Sous les hypothèse précédentes, le schéma implicite (8.2) admet une solution unique $u_{h,k} \in V_h^{s+1}$ lorsque $k < \frac{\sqrt{\beta}}{2P}$.

Démonstration analogue à celle du théorème 3.1.

Théorème 8.2.

Sous les hypothèses précédentes et lorsque k est assez petit, il existe une constante $K > 0$ indépendante de h.k.s. telle que la solution $u_{h,k}$ du schéma implicite (8.2) vérifie l'inégalité de l'énergie discrète pour $s = 1, \dots, S$

(8.8)
$$\begin{cases} \left| \frac{1}{k}(u_{h,k}^{(s)} - u_{h,k}^{(s-1)} \right|_h^2 + \| u_{h,k}^{(s)} \|_h^2 \leq \\ \leq K \left\{ \left| \frac{1}{k}(u_{h,k}^{(1)} - u_{h,k}^{(o)}) \right|_h^2 + (A_{h,o}\, u_{h,k}^{(1)}, u_{h,k}^{(1)})_h + \right. \\ \qquad + k \sum_{r=1}^{s-1} | f_{h,k}^{(r)} |_h^2 \end{cases}$$

<u>Le schéma implicite (8.2) est donc inconditionnellement stable au sens de (8.8).</u>

<u>Démonstration.</u> Introduisons d'abord une notation : on pose

$$(8.9)\qquad \begin{aligned} \bar{\nabla}_k u^{(s)}_{h,k} &= \frac{1}{k}\left(u^{(s)}_{h,k} - u^{(s-1)}_{h,k}\right), \\ \nabla_k u^{(s)}_{h,k} &= \frac{1}{k}\left(u^{(s+1)}_{h,k} - u^{(s)}_{h,k}\right). \end{aligned}$$

Le schéma implicite (8.2) peut alors s'écrire (en supprimant les indices h et k) pour $r = 1, \ldots, S-1$

$$(8.10)\qquad \bar{\nabla}^2 u^{(r+1)} + A\, u^{(r+1)} = f^{(r)} .$$

Multiplions scalairement (8.10) par $\bar{\nabla} u^{(r+1)}$ et prenons deux fois la partie réelle de l'équation obtenue ; nous obtenons

$$2\mathrm{Re}\,(\bar{\nabla}^2 u^{(r+1)}, \bar{\nabla} u^{(r+1)}) + 2\mathrm{Re}\,(A\, u^{(r+1)}, \bar{\nabla} u^{(r+1)}) = (f^{(r)}, \bar{\nabla} u^{(r+1)})$$

On remarque que

$$2\mathrm{Re}\,(\bar{\nabla}^2 u^{(r+1)}, \bar{\nabla} u^{(r+1)}) = \bar{\nabla} |\bar{\nabla} u^{(r+1)}|^2 + k |\bar{\nabla}^2 u^{(r+1)}|^2 ,$$

$$2\mathrm{Re}(A_0 u^{(r+1)}, \bar{\nabla} u^{(r+1)}) = \bar{\nabla}(A_0 u^{(r+1)}, u^{(r+1)}) + k(A_0 \bar{\nabla} u^{(r+1)}, \bar{\nabla} u^{(r+1)}),$$

ce qui entraîne

$$\begin{aligned} \bar{\nabla} |\bar{\nabla} u^{(r+1)}|^2 + \bar{\nabla}(A_0 u^{(r+1)}, u^{(r+1)}) &\leq \\ &\leq 2\mathrm{Re}\,(f^{(r)}, \bar{\nabla} u^{(r+1)}) - 2\mathrm{Re}\,(A_1 u^{(r+1)}, \bar{\nabla} u^{(r+1)}) . \end{aligned}$$

En majorant le second membre de cette inégalite, on trouve

$$\begin{aligned} \bar{\nabla} \left\{ |\bar{\nabla} u^{(r+1)}|^2 + (A_0 u^{(r+1)}, u^{(r+1)}) \right\} &\leq 2(P \| u^{(r+1)} \| + |f^{(r)}|)\, |\bar{\nabla} u^{(r+1)}| \leq \\ &\leq P^2 \| u^{(r+1)} \|^2 + |f^{(r)}|^2 + 2 |\bar{\nabla} u^{(r+1)}|^2 . \end{aligned}$$

En multipiant par k et en sommant de $r = 1$ à $r = s-1$ ($s \leq S$), on en déduit compte tenu de l'inégalité de coercivité (8.6)

$$\begin{cases} |\bar{\nabla} u^{(s)}|^2 + \beta \|u^{(s)}\|^2 \leq |\bar{\nabla} u^{(1)}|^2 + (A_o u^{(1)}, u^{(1)}) + k \sum_{r=1}^{s-1} |f^{(r)}|^2 + \\ \qquad + 2k \sum_{r=2}^{s} |\bar{\nabla} u^{(r)}|^2 + P^2 k \sum_{r=2}^{s} \|u^{(r)}\|^2 , \end{cases}$$

d'où pour s = 1, ..., S

$$(8.11) \quad \begin{cases} (1-2k) |\bar{\nabla} u^{(s)}|^2 + (\beta - P^2 k) \|u^{(s)}\|^2 \leq |\bar{\nabla} u^{(1)}|^2 + (A_o u^{(1)}, u^{(1)}) + \\ \qquad + k \sum_{r=1}^{s-1} |f^{(r)}|^2 + 2k \sum_{r=2}^{s-1} |\bar{\nabla} u^{(r)}|^2 + P^2 k \sum_{r=2}^{s-1} \|u^{(r)}\|^2 \end{cases}$$

Il suffit maintenant d'appliquer le lemme de Gronwall discret à (8.11) (ce qui est loisible si k assez petit) pour obtenir l'inégalité de stabilité (8.8).

8.3. Stabilité du schéma explicite.

Passons maintenant à l'étude de la stabilité du schéma explicite sous les mêmes hypothèses que précédemment.

Théorème 8.3.

On fait les hypothèses (8.4), (8.5), (8.6) et (8.7). Sous le condition de stabilité

$$(8.12) \qquad k \, | A_{h,o} |_h^{1/2} \leq 2(1-\eta)$$

où $\eta > 0$ est une constante arbitrairement petite et independante de h, k, la solution $u_{h,k}$ du schéma explicite (8.3) vérifie l'inégalité de l'énergie discrète pour $s = 0, 1, \ldots, S-1$ et pour k assez petit

$$(8.13) \quad \begin{cases} \eta \left| \frac{1}{k} (u_{h,k}^{(s+1)} - u_{h,k}^{(s)}) \right|_h^2 + \beta \eta \| u_{h,k}^{(s)} \|_h^2 \leq \\ \leq K \left\{ \left| \frac{1}{k} (u_{h,k}^{(1)} - u_{h,k}^{(o)}) \right|_h^2 + (A_{h,o} u_{h,k}^{(o)}, u_{h,k}^{(o)})_h + k \sum_{r=1}^{s} | f_{h,k}^{(r)} |_h^2 \right\} \end{cases}$$

où K est une constante > 0 indépendante de h, k, s et η.

Démonstration. Le schéma explicite (8.3) s'écrit avec les notations (8.9) et en supprimant les indices h et k

$$\bar{\nabla}\nabla u^{(r)} + A\,u^{(r)} = f^{(r)}\,, \qquad r = 1,\dots,S-1\,. \tag{8.14}$$

Multiplions scalairement (8.14) par $\bar{\nabla} u^{(r)} + \nabla u^{(r)}$; on obtient, en prenant deux fois la partie réelle de l'équation obtenue et en remarquant que

$$2\mathrm{Re}\,(\bar{\nabla}\nabla u^{(r)},\ \bar{\nabla}u^{(r)} + \nabla u^{(r)}) = 2\,\bar{\nabla}|\nabla u^{(r)}|^2\,,$$

$$(8.15)\quad \begin{cases} 2\,\bar{\nabla}|\nabla u^{(r)}|^2 + 2\mathrm{Re}\,(A\,u^{(r)},\ \bar{\nabla}u^{(r)}) + 2\mathrm{Re}\,(A\,u^{(r)},\ \nabla u^{(r)}) = \\ \qquad = 2\mathrm{Re}\,(f^{(r)},\ \bar{\nabla}u^{(r)} + \nabla u^{(r)})\,. \end{cases}$$

Mais

$$2\mathrm{Re}\,(A\,u^{(r)},\ \bar{\nabla}u^{(r)}) = (A_0 u^{(r)},\ u^{(r)}) + k\,(A_0\,\bar{\nabla}u^{(r)},\ \nabla u^{(r)}) + $$
$$+ 2\mathrm{Re}\,(A_1 u^{(r)},\ \nabla u^{(r)})\,,$$

$$2\mathrm{Re}\,(A\,u^{(r)},\ \nabla u^{(r)}) = \nabla(A_0 u^{(r)},\ u^{(r)}) - k(A_0\,\nabla u^{(r)},\ \nabla u^{(r)}) + $$
$$+ 2\,\mathrm{Re}\,(A_1 u^{(r)},\ \nabla u^{(r)})\,.$$

On en déduit que (8.15) peut s'écrire

$$2\,\bar{\nabla}|\nabla u^{(r)}|^2 + \bar{\nabla}(A_0 u^{(r)},\ u^{(r)}) + \nabla(A_0 u^{(r)},\ u^{(r)}) - k^2\,\bar{\nabla}(A_0\,\nabla u^{(r)},\ \nabla u^{(r)}) =$$
$$= 2\mathrm{Re}\,(f^{(r)},\ \bar{\nabla}u^{(r)} + \nabla u^{(r)}) - 2\mathrm{Re}\,(A_1 u^{(r)},\ \bar{\nabla}u^{(r)} + \nabla u^{(r)})$$

ce qui donne en majorant le second membre de cette équation de la façon habituelle

$$(8.16)\quad \begin{cases} 2\,\bar{\nabla}|\nabla u^{(r)}|^2 + \bar{\nabla}(A_0 u^{(r)},\ u^{(r)}) + \nabla(A_0 u^{(r)},\ u^{(r)}) - k^2\,\bar{\nabla}(A_0\,\nabla u^{(r)},\ \nabla u^{(r)}) \leq \\ \leq 2(|f^{(r)}|^2 + P^2\,|u^{(r)}|^2) + (|\bar{\nabla}u^{(r)}|^2 + |\nabla u^{(r)}|^2) \end{cases}$$

Multiplions l'inégalité (8.16) par k et sommons de r = 1 à s ; on trouve pour s = 1,..., S-1

$$
(8.17)\quad \begin{cases} 2|\nabla u^{(s)}|^2 + (A_0 u^{(s)}, u^{(s)}) + (A_0 u^{(s+1)}, u^{(s+1)}) - k^2(A_0 \nabla u^{(s)}, \nabla u^{(s)}) \leq \\ \leq 2 |\nabla u^{(o)}|^2 + (A_0 u^{(o)}, u^{(o)}) + (A_0 u^{(1)}, u^{(1)}) - k^2(A_0 \nabla u^{(o)}, u^{(o)}) + \\ + 2k \sum_{r=1}^{s} |f^{(r)}|^2 + 2P^2 k \sum_{r=1}^{s} \|u^{(r)}\|^2 + 2k \sum_{r=1}^{s} (|\bar{\nabla} u^{(r)}|^2 + |\nabla u^{(r)}|^2). \end{cases}
$$

Mais

$$
(A_0 u^{(r+1)}, u^{(r+1)}) = (A_0 u^{(r)}, u^{(r)}) + 2k \operatorname{Re} (A_0 u^{(r)}, \nabla u^{(r)}) + k^2(A_0 \nabla u^{(r)}, \nabla u^{(r)})
$$

si bien que (8.17) peut s'écrire

$$
(8.18)\quad \begin{cases} 2|\nabla u^{(s)}|^2 + 2(A_0 u^{(s)}, u^{(s)}) + 2k \operatorname{Re}(A_0 u^{(s)}, \nabla u^{(s)}) \leq \\ \leq (2+2k)|\nabla u^{(o)}|^2 + 2(A_0 u^{(o)}, u^{(o)}) + 2k \operatorname{Re}(A_0 u^{(o)}, \nabla u^{(o)}) + \\ + 2k \sum_{r=1}^{s} |f^{(r)}|^2 + 2P^2 k \sum_{r=1}^{s-1} \|u^{(r)}\|^2 + 4k \sum_{r=1}^{s-1} |\nabla u^{(r)}|^2 + \\ + 2P^2 k \|u^{(s)}\|^2 + 2k |\nabla u^{(s)}|^2 . \end{cases}
$$

D'après le lemme 3.1., on a

$$
\begin{aligned} 2\operatorname{Re} (A_0 u^{(r)}, \nabla u^{(r)}) &\leq 2 |A_0|^{1/2} (A_0 u^{(r)}, u^{(r)})^{1/2} |\nabla u^{(r)}| \leq \\ &\leq |A_0|^{1/2} \left[(A_0 u^{(r)}, u^{(r)}) + |\nabla u^{(r)}|^2\right] \end{aligned}
$$

Dans ces conditions l'inégalité (8.18) devient

$$(8.19)\quad \begin{cases} (2-k|A_o|^{1/2}-2k)\,|\nabla u^{(s)}|^2 + ((2-k|A_o|^{1/2})\beta - 2P^2k)\,\|u^{(s)}\|^2 \leq \\ \leq (2+k\,|A_o|^{1/2}+2k)\,|\nabla u^{(o)}|^2 + (2+k\,|A_o|^{1/2})\,(A_o u^{(o)}, u^{(o)}) + \\ + 2k \sum_{r=1}^{s} |f^{(r)}|^2 + 2P^2k \sum_{r=1}^{s-1} \|u^{(r)}\|^2 + 4k \sum_{r=1}^{s-1} |\nabla u^{(r)}|^2 . \end{cases}$$

Sous la condition de stabilité (8.12) l'inégalité (8.19) entraine pour k assez petit de façon que $2k \leq \eta$, $2P^2k \leq \beta\eta$,

$$(8.20)\quad \begin{cases} \eta|\nabla u^{(s)}|^2 + \beta\eta\|u^{(s)}\|^2 \leq K' \Big\{ |\nabla u^{(o)}|^2 + (A_o u^{(o)}, u^{(o)}) + \\ \qquad + k \sum_{r=1}^{s} |f^{(r)}|^2 \Big\} + 2P^2k \sum_{r=1}^{s-1} \|u^{(r)}\|^2 + 2k \sum_{r=1}^{s-1} |\nabla u^{(r)}|^2 , \end{cases}$$

où K' est une constante indépendante de h, k, s. Il suffit maintenant d'appliquer le lemme 3.2. (lemme de Gronwall discret) pour obtenir le résultat.

Remarque 8.1.

La condition de stabilité (8.12) ne dépend que de $A_{h,o}$ partie principale de l'opérateur A_h.

Faisons maintenant l'hypothèse

$$(8.21)\quad \begin{cases} \text{il existe une constante } M > 0 \text{ telle que} \\ (A_{h,o} u_h : v_h)_h \leq M \|u_h\|_h \, \|v_h\|_h \quad , \; \forall u_h, v_h \in V_h . \end{cases}$$

Corollaire 8.1.

Sous les hypothèses (8.4), (8.5), (8.6), (8.7) et (8.21), le schéma explicite est stable (au sens de 8.13) si

$$(8.22)\qquad k\;C(h) \leq \frac{2}{M}\,(1-\eta) .$$

La démonstration est immédiate.

On pourrait, comme pour les équations du ler ordre, illustrer la théorie faite par de nombreux exemples. Ce point n'offre aucune nouvelle difficulté et nous renvoyons, comme au N° 4, à [2] , [10] , [1] pour la discrétisation de l'opérateur A(t) . L'étude de la convergence se fait comme au N° 5 . On obtient avec des hypothèses convenables des résultats de convergence faible dans $L^{\infty}(0,T;V)$ (cf. [10]) . Pour des résultats de convergence forte nous renvoyons à [10] , [16], [3].

Bibliographie

1 J.P. AUBIN - Approximation des espaces de distributions et des opérateurs differentiels, Mémoires Soc. Math. France, 12, 1967.

2 J. CEA - Approximation variationnelle des problèmes aux limites, Ann. Inst. Fourier (1964), 14, p. 345-444.

3 J. LIEUTAUD - Thèse (à paraître).

4 J.L. LIONS - Equations différentielles opérationnelles et problèmes aux limites, Springer Verlag, Berlin (1961).

5 J.L. LIONS - Equations différentielles operationnelles dans les espaces de Hilbert, Cours CIME (1963), Editions Cremonese, Rome.

6 J.L. LIONS - E. MAGENES - Problèmes aux limites non homogènes (II), Ann. Inst. Fourier (1961), 11, p. 137-178.

7 J.L. LIONS - P.A. RAVIART - Remarques sur la résolution et l'approximation d'équations d'évolution couplées, I.C.C. Bull. (1966), 5, p. 1-20.

8 J.L. LIONS - P.A. RAVIART - Remarques sur la résolution, exacte et approchée, d'équations d'évolution paraboliques à coefficients non bornés, Calcolo (1967), 4, 2, p. 221-234.

9 A. MIGNOT - Méthodes d'approximation des solutions de certains problems aux limites linéaires, Thèse, Paris (1967).

10 P.A. RAVIART - Sur l'approximation de certaines equation d'évolution linéaires et non linéaires, Jour. Math. pur. et appl. (1967), 46, p. 11-183.

11 P.A. RAVIART - On the approximation of weak solutions of linear parabolic equations by a class of multistep difference methods Technical report CS 31, Stanford University (1965).

12 P.A. RAVIART - Sur la résolution et l'approximation de certaines équations paraboliques non linéaires dégénérées, Arch. Rat. Mech. Anal. (1967), 25, p. 64-80.

13 R.D. RICHTMYER - Difference methods for initial value problems, Interscience New-York (1957).

14 L. SCHWARTZ - Théorie des distributions, Hermann, Pariz, t.I (1951), t.II (1957).

15 L. SCHWARTZ - Théorie des distributions à valeurs vectorielles, lère partie Ann. Inst. Fourier, (1957), 7, p. 1-139.

16 R. TEMAN - Sur la stabilité et la convergence de la méthode des pas fractionnaires, Thèse Paris (1967).

CENTRO INTERNAZIONALE MATEMATICO ESTIVO

(C.I.M.E.)

H. BREZIS et M. SIBONY

"METHODES D'APPROXIMATION ET D'ITERATION POUR LES OPERATEURS MONOTONES"

Corso tenuto ad Ispra dal 3-11 Luglio 1967

INTRODUCTION

Soit V un Banach sur $\mathbb{R}$; V' son dual. On se propose de résoudre numériquement certaines équations de la forme

(1) $Au = f$ pour f donné dans V'

où A est un opérateur monotone non nécessairement linéaire de V dans V'. Nous donnons un théorème d'existence et d'unicité.

On considère ensuite l'équation :

(2) $Au_\lambda + \lambda Bu_\lambda = f$

avec B monotone borné de V dans V'. On montre, sous certaines hypothèses que $u_\lambda \xrightarrow[\lambda \to 0]{} u$ dans V fort où u est la solution de (1).

A l'espace V, nous associons un espace V_h de dimension finie. Dans V_h nous avons alors l'équation discrétisée :

(3) $A_h u_h = f_h$.

Sous certaines hypothèses la solution u_h de (3) converge fortement dans V vers la solution u de (1).

Enfin nous donnons une méthode itérative permettant de résoudre explicitement (3).

Le plan est le suivant :

1 - un théorème d'existence et d'unicité.

2 - Propriétés de l'opérateur A^{-1}

3 - Propriété de la solution de l'équation $Au_\lambda + \lambda Bu_\lambda = f$

4 - Applications

5 - Méthodes d'approximations numériques

6 - Application aux familles d'approximation $\{w_i^{1,p}, V_h, p_h^i, r_h\}$

7 - Résolution du problème discrétisé : méthode itérative

8 - Applications à la résolution numérique de certaines équations aux dérivées partielles

I - Un théorème d'existence et d'unicité

Définition 1.1

On dit qu'un espace normé V est uniformément convexe si, $\forall\ \varepsilon$ tel que $0 < \varepsilon < 2$, $\exists\ \delta(\varepsilon) > 0$ tel que les relations :

$$\|u\| \leq 1 \ , \ \|v\| \leq 1 \quad \text{et} \quad \|u-v\| \geq \varepsilon$$

$$\Longrightarrow \|u+v\| \leq 2-\delta \quad , \quad \forall\ u,v \in V$$

Soit V un Banach sur $\mathbb{R}$ uniformément convexe de norme $\|\ \|$. Soit V' son dual. Soit A un opérateur non nécessairement linéaire de V dans V'. On cherche $u \in V$ vérifiant :

(1.1) $\qquad Au = f$ pour f donné dans V'.

Théorème 1.1.

Si l'opérateur A vérifie les conditions

(1.2) $\qquad (Au-Av,\ u-v) \geq (\varphi(\|u\|) - \varphi(\|v\|))\ (\|u\| - \|v\|)$

pour tout u et $v \in V$, où φ est une application strictement croissante de $\mathbb{R}_+ \longrightarrow \mathbb{R}$ telle que

$$\lim_{r \to +\infty} \varphi(r) = +\infty$$

(1.3) $\qquad$ Les restrictions de A aux segments de V sont continues dans V' faible.

Alors $\forall\ f \in V'$, $\exists\ u \in V$ unique tel que $Au = f$.

Existence

La condition φ strictement croissante entraîne :

(1.4) $(Au - Av\ ,\ u-v) \geq 0$

D'autre part, si on fait $v = 0$ dans (1.2) il vient

$$(Au,u) \geq (A(0),u) + (\varphi(\|u\|) - \varphi(0))\ \|u\|$$

$$\frac{(Au,u)}{\|u\|} \geq -\|A(0)\|_{V'} + (\varphi(\|u\|) - \varphi(0))$$

d'où

(1.5) $$\lim_{\|u\|\to+\infty} \frac{(Au,u)}{\|u\|} = +\infty$$

et l'on sait qu'avec les hypothèses (1.3) (1.4) et (1.5) le problème (1.1) admet une solution. cf. [3] et [4].

Avant de démontrer l'unicité, nous avons besoin de deux lemmes

Lemme 1.1.

On suppose (1.2) et (1.3). Alors l'ensemble S des solutions de l'équation $Au = f$ est un convexe fermé de V.

Démonstration : L'équation $Au = f$ est équivalente à la relation

(1.6) $(Av - f\ ,\ v - u) \geq 0 \qquad \forall\ v \in V$

En effet : $Au = f \Rightarrow (Au - f\ ,\ v-u) \geq 0$ et

$$(Av-f,v-u) = (Au-f,v-u) + (Av-Au,v-u) \geq 0$$

d'après la monotonie de A.

Réciproquement posons $v = u + t\varphi$, $t > 0$; (1.6) donne $\forall \varphi \in V$ $(A(u+t\varphi)-f,\varphi) \geq 0$ et l'on fait $t \longrightarrow 0$, D'où

$$Au = f.$$

$S_{v_0} = \{u \mid (Av_0-f,v_0-u) \geq 0\}$ est un demi espace fermé de V.

Donc S = ensemble des solutions de Au = f est un convexe fermé de V, car

$$S = \bigcap_{v_o \in V} S_{v_o} .$$

Lemme 1.2.

L'ensemble des solutions de Au = f est situé sur une sphère de V.

Démonstration : Soient u_1 et u_2 deux solutions de l'équation Au = f.

En reportant dans (1.2) il vient :

$$(\varphi(\|u_1\|) - \varphi(\|u_2\|)) (\|u_1\| - \|u_2\|) = 0$$

D'où : $\|u_1\| = \|u_2\|$

Unicité

L'ensemble S des solutions est un convexe fermé de V situé sur une sphère. Comme V est uniformément convexe, S est réduit à un point, d'où l'unicité.

II- Propriétés de l'opérateur A^{-1}

Lemme 2.1.

Soit $r \in \mathbb{R}_+$ et $r_\lambda \in \mathbb{R}_+$, $\forall \lambda \geq 0$. Soit φ une application strictement croissante de $\mathbb{R}_+ \longrightarrow \mathbb{R}$ telle que

$$(\varphi(r_\lambda) - \varphi(r)) (r_\lambda - r) \xrightarrow[\lambda \to 0]{} 0$$

Alors $r_\lambda \longrightarrow r$ quand $\lambda \longrightarrow 0$.

Démonstration : On montre facilement que r_λ est borné. Supposons maintenant que $r_\lambda \not\to r$. Alors on pourrait extraire $r_\mu \longrightarrow \rho \neq r$ et $\varphi(r_\mu) \to \varphi(r)$

Deux cas :

1) Si $\rho < r$

Soit r_1 tel que $\rho < r_1 < r$ à partir d'un certain rang $r_\mu \leq r_1$ ce qui donne :

$$\varphi(r_\mu) \leq \varphi(r_1) < \varphi(r)$$

et à la limite :

$$\varphi(r) \leq \varphi(r_1) < \varphi(r)$$

ce qui est absurde.

2) Si $\rho > r$

Soit r_1 tel que $r < r_1 < \rho$ à partir d'un certain rang $r_\mu \geq r_1$ ce qui entraîne :

$$\varphi(r_\mu) \geq \varphi(r_1) > \varphi(r)$$

et à la limite

$$\varphi(r) \geq \varphi(r_1) > \varphi(r)$$

ce qui est absurde.

C.Q.F.D.

Théorème 2.1.

On suppose (1.2) et (1.3). Alors A^{-1} est monotone, borné (c'est à dire transforme les ensembles bornés en des ensembles bornés) et continu de V' fort dans V fort.

Démonstration : Il est évident que A^{-1} est monotone. Soit $Au = f$ avec $\|f\| \leq M$. On a

$$(Au - A(0),u) = (f - A(0),u) \geq (\varphi(\|u\|) - \varphi(0)) \|u\|$$

D'où : $\varphi(\|u\|) \leq M + \|A(0)\| + \varphi(0)$

Ceci montre que A^{-1} est borné.

Montrons que A^{-1} est continu de V' fort dans V fort. Il suffit de montrer que si $f_n = Au_n$ et si $f_n \longrightarrow f = Au$ dans V' fort, alors $u_n \longrightarrow u$ dans V fort.

La suite f_n est bornée, donc u_n aussi. Suivant un ultrafiltre plus fin $u_n \longrightarrow \xi$ dans V faible, $Au_n \longrightarrow f$ dans V' fort. Par suite $A\xi = f = Au$. D'où $u = \xi$ d'après l'unicité. Il en résulte que $u_n \longrightarrow u$ dans V faible.

D'autre part

$$(Au_n - Au\ ,\ u_n - u) \geq (\varphi(\|u_n\|) - \varphi(\|u\|))(\|u_n\| - \|u\|) \geq 0$$

et

$$(Au_n - Au\ ,\ u_n - u) = (Au_n - f\ ,\ u_n - u) \longrightarrow 0$$

Donc

$$(\varphi(\|u_n\|) - \varphi(\|u\|)\ (\|u_n\| - \|u\|) \longrightarrow 0$$

D'après le lemme 2.1., $\|u_n\| \longrightarrow \|u\|$. D'où $u_n \longrightarrow u$ dans V fort.

III - Propriété de la solution de l'équation $Au_\lambda + \lambda Bu_\lambda = f$

Théorème 3.1.

Soit A un opérateur de V dans V' vérifiant (1.2) et (1.3) et soit B un opérateur monotone borné de $V \longrightarrow V'$ tel que les restrictions aux segments de V soient continues dans V' faible. Alors

1) $\forall f \in V'$, $\forall \lambda \geq 0$ l'équation

(3.1) $$Au_\lambda + \lambda\, Bu_\lambda = f$$

admet une solution unique u_λ.

2) $u_\lambda \longrightarrow u$ dans V fort quand $\lambda \longrightarrow 0$ où u est la solution de l'équation $Au = f$.

Démonstration : 1) Pour l'existence et l'unicité on applique le théorème 1.1. à l'équation (3.1).

2) Montrons que $u_\lambda \longrightarrow u$ dans V fort, quand $\lambda \to 0$

a) $\exists\, M > 0$ tel que $\|u_\lambda\| \leq M$. Pour tout λ assez petit. Nous avons en effet:

$$(Au_\lambda - A(0),u_\lambda) \geq (\varphi(\|u_\lambda\| - \varphi(0))\ \|u_\lambda\|$$

or

$$Au_\lambda = f - \lambda\, Bu_\lambda$$

$$(f - \lambda\, Bu_\lambda - A(0),u_\lambda) \geq (\varphi(\|u_\lambda\|) - \varphi(0))\ \|u_\lambda\|$$

D'autre part

$$(Bu_\lambda - B(0),u_\lambda) \geq 0$$

$$\Longrightarrow (f - A(0),u_\lambda) \geq (\varphi(\|u_\lambda\|) - \varphi(0))\|u_\lambda\| + \lambda(B(0),u_\lambda)$$

$$|\varphi(\|u_\lambda\|) - \varphi(0)|\ \|u_\lambda\| \leq \|f - A(0) - \lambda\ B(0)\|_{V'}\ \|u_\lambda\|_V$$

D'où

$$\varphi(\|u_\lambda\|) \leq \varphi(0) + \|f - A(0) - \lambda\ B(0)\|_{V'} \tag{3.2}$$

ce qui entraîne $\|u_\lambda\|$ borné, car sinon il existerait u_μ telle que $\|u_\mu\| \longrightarrow +\infty$ et on aurait alors $\varphi(\|u_\mu\|) \longrightarrow +\infty$ ce qui est contraire à (3.2).

b) $u_\lambda \xrightarrow[\lambda \to 0]{} u$ faiblement.

En effet comme B est borné, $\|u_\lambda\| \leq M$ entraîne $\|Bu_\lambda\| \leq M'$. Donc quand $\lambda \longrightarrow 0$, $Au_\lambda \longrightarrow f$ fortement.

Soit $\mathcal{U}$ un ultrafiltre plus fin. Suivant $\mathcal{U}$, $u_\lambda \longrightarrow v$ faiblement et $Au_\lambda \longrightarrow f$ fortement. Il en résulte que Av = f = Au. Donc v = u. En effet, on va montrer que si $u_\lambda \longrightarrow v$ faiblement $(\|u_\lambda\| \leq M)$

et $Au_\lambda \longrightarrow f$ fortement

$\Longrightarrow Av = f.$

Nous avons $(Au_\lambda - Aw, u_\lambda - v) \geq 0 \qquad \forall v \in V$

et à la limite $(f - Av , v - v) \geq 0 \qquad \forall v \in V,$

or ceci est équivalent à $Av = f$. Donc $u_\lambda \xrightarrow[\lambda \to 0]{} u$ faiblement.

c) Montrons que $u_\lambda \xrightarrow[\lambda \to 0]{} u$ fortement. On a

$$(Au_\lambda - Au , u_\lambda - u) \geq (\varphi(\|u_\lambda\|) - \varphi(\|u\|))(\|u_\lambda\| - \|u\|)$$

$$(f - \lambda Bu_\lambda - Au, u_\lambda - u) \geq (\varphi(\|u_\lambda\| - \varphi(\|u\|))(\|u_\lambda\| - \|u\|).$$

ce qui donne

$$-\lambda(Bu_\lambda, u_\lambda - u) \geq (\varphi(\|u_\lambda\|) - \varphi(\|u\|))(\|u_\lambda\| - \|u\|)$$

Comme $u_\lambda - u$ et Bu_λ sont bornés ; quand $\lambda \longrightarrow 0$ le 1er membre tend vers 0.

Donc :

$$(\varphi(\|u_\lambda\|) - \varphi(\|u\|) (\|u_\lambda\| - \|u\|) \xrightarrow[\lambda \to 0]{} 0$$

Il résulte du lemme 2.1. que $\|u_\lambda\| \longrightarrow \|u\|$ et comme V est uniformément convexe on en déduit que $u_\lambda \longrightarrow u$ dans V fort.

IV - Applications

Exemple 4.1.

Ω étant un ouvert borné de $\mathbb{R}^n$. Soit $V = W_o^{1,p}(\Omega)$ (1) muni de la norme

$$\|u\|_V = \left(\sum_{i=1}^{n} \|D_i u\|_{L_p}^p\right)^{1/p} \qquad p \geq 2$$

V est un Banach uniformément convexe pour cette norme. L'opérateur

$Au = -\sum_{i=1}^{n} D_i\,(|D_i u|^{p-2} D_i u)$ applique $W_o^{1,p}(\Omega)$ dans $W^{-1,q}(\Omega)$ avec $\frac{1}{p} + \frac{1}{q} = 1$.

Montrons que A vérifie les hypothèses du théorème 1.1. avec $\varphi(r) = r^{p-1}$, $r \geq 0$. En effet :

$$(Au,u) = \sum_{i=1}^{n} \|D_i u\|_{L_p}^p = \|u\|_V^p$$

$$(Au,v) = \sum_{i=1}^{n} (|D_i u|^{p-2} D_i u, D_i v) \leq \sum_{i=1}^{n} \|D_i u\|_{L_p}^{p-1} \|D_i v\|_{L_p}$$

$$(Au,v) \leq \left(\sum_{i=1}^{n} \|D_i u\|_{L_p}^p\right)^{1/q} \left(\sum_{i=1}^{n} \|D_i v\|_{L_p}^p\right)^{1/p}$$

$$(Au,v) \leq \|u\|_V^{p-1} \|v\|_V$$

Donc

$$(Au-Av,u-v) \geq \|u\|_V^p - \|u\|_V^{p-1} \|v\|_V - \|v\|_V^{p-1} \|u\|_V + \|v\|_V^p$$

D'où

$$(Au-Av,u-v) \geq (\|u\|_V^{p-1} - \|v\|_V^{p-1})(\|u\|_V - \|v\|_V) \quad \forall u,v \in V$$

---

(1) On désigne par $W^{1,p}(\Omega)$, $1<p<\infty$ l'espace des (classes des) fonctions u $L^p(\Omega)$ telles que $D_i u = \frac{\partial u}{\partial x_i} \in L^p(\Omega)$, les dérivées étant prises au sens des distributions sur Ω. $W_o^{1,p}(\Omega)$ est l'adhérence de $\mathcal{D}(\Omega)$ dans $W^{1,p}(\Omega)$. L'espace $W^{-1,q}(\Omega)$, $1/p + 1/q = 1$ désigne le dual fort de $W_o^{1,p}(\Omega)$.

Donc la condition (1.2) est vérifiée. Il est évident que l'application $t \longrightarrow (A(u+tv),w)$ est continue $\forall\, u,v,w \in V$, d'où (1.3).

Par conséquent pour f donné dans $W^{-1,q}(\Omega)$, il existe $u \in W_0^{1,p}(\Omega)$ unique tel que

$$Au = f.$$

Exemple 4.2.

Plus généralement les problèmes

(4.1) $\qquad Au + su = f$

(4.2) $\qquad Au - s\Delta u = f$

avec $Au = -\sum_{i=1}^{n} D_i\,(|D_i u|^{p-2}\, D_i u) \qquad p \geq 2$

$\Delta u = \sum_{i=1}^{n} D_i^2 u \quad , \qquad s \geq 0 \quad s$ réel.

admettent une solution unique $u \in W_0^{1,p}(\Omega)$ pour f donné dans $W^{-1,q}(\Omega)$.

V - Méthodes d'approximation numérique.

Soit V un espace de Banach de norme $\|\ \|$, et soit $h = (h_1,\dots,h_n) \in \mathbb{R}_+^n$ $h \neq 0$ un paramètre destiné à tendre vers 0.

Définition 5.1. (cf. [1])

Nous appellerons h-approximation de l'espace V, le quadruplet $\{V, V_h, p_h, r_h\}$ défini par la donnée de

1) Un espace V_h de dimension finie et de norme $\|\ \|_h$.

2) Une application $p_h \in \mathcal{L}(V_h, V)$ appelée prolongement de V_h dans V.

3) Une application $r_h \in \mathcal{L}(V, V_h)$ appelée restriction de V dans V_h.

Définition 5.2.

La famille $\{V, V_h, p_h, r_h\}$ est dite consistante si dans V fort

$$\lim_{h \to o} p_h\, r_h\, v = v \qquad \forall\, v \in V$$

Soit V_h' le dual de V_h. Nous notons par $(\ ,\ )_h$ la dualité entre V_h et V_h'. V_h' est muni de la norme duale

$$\|f_h\|_h^* = \sup_{v_h \in V_h} \frac{|(f_h, v_h)_h|}{\|v_h\|_h} \qquad \forall\, f_h \in V_h'$$

Désignons par r_h^* l'adjoint de p_h et par p_h^* l'adjoint de r_h. Nous avons

$$(r_h^* f, u_h)_h = (f, p_h\, u_h) \qquad \forall f \in V' \ ,\ u_h \in V_h$$

$$(p_h^* f_h, u) = (f_h, r_h u)_h \qquad \forall f_h \in V_h' \ ,\ u \in V$$

Définition 5.3.

On dit que $f_h \in V_h'$ converge discrètement vers $f \in V'$ si

$$\lim_{h \to o} \|r_h^* f - f_h\|_h^* = 0.$$

Soit V un Banach sur $\mathbb{R}$, uniformément convexe, de norme $\|\ \|$. Soit V' son dual. Soit A un opérateur non nécessairement linéaire de V dans V'. On se donne une famille d'approximation $\{V, V_h, p_h, r_h\}$ de l'espace V et l'on fait les hypothèses suivantes :

(H 1) A est un opérateur borné vérifiant (1.2) et (1.3).

H 2) La famille $\{V, V_h, p_h, r_h\}$ est consistante et p_h est injective.

On pose $A_h = r_h^* \, A \, p_h : V_h \longrightarrow V_h'$.

Nous avons alors le schéma suivant :

$$\begin{array}{ccc} V & \xrightarrow{\quad A \quad} & V' \\ r_h \downarrow \uparrow p_h & & p_h^* \uparrow \downarrow r_h^* \\ V_h & \xrightarrow[\quad A_h \quad]{} & V_h' \end{array}$$

On munit V_h de la norme $\|u_h\|_h = \|p_h u_h\|_V$. Il en résulte que

$$\|r_h^* \, f\|_h^* \leq \|f\|_{V'} \qquad \forall f \in V'$$

D'autre part si f_h converge discrètement vers f alors $\|f_h\|_h^*$ est borné.

En effet :

$$\|f_h\|_h^* = \|r_h^* f - f_h\|_h^* + \|r_h^* f\|_h \leq c^{te}.$$

Théorème 5.1.

Pour tout $f_h \in V_h'$, il existe $u_h \in V_h$ unique tel que

(5.1) $\qquad A_h \, u_h = f_h$

De plus si f_h converge discrètement vers $f \in V'$, alors $p_h u_h$ converge dans V fort vers la solution u de (1.1).

Lemme 5.1.

L'équation (5.1) admet une solution unique.

Démonstration : Appliquons le théorème 1.1. à l'opérateur A_h.

1) Montrons que V_h muni de la norme $\|u_h\|_h = \|p_h u_h\|$ est uniformé-

memnt convexe. En effet :

$$\forall \varepsilon > 0 \quad , \quad 0 < \varepsilon < 2 \qquad \exists\ \delta(\varepsilon) \text{ tel que si}$$

$$\|p_h\, u_h\| = 1 \quad , \quad \|p_h\, v_h\| = 1 \quad \text{et} \quad \|p_h(u_h - v_h)\| > \varepsilon$$

on a

$$\left\|\frac{p_h\, u_h + p_h\, v_h}{2}\right\| \leq 1 - \delta$$

d'après l'uniforme convexité de V, et ceci entraîne

$$\left\|\frac{u_h + v_h}{2}\right\|_h \leq 1 - \delta$$

2) Formons

$$(A_h u_h - A_h v_h, u_h - v_h)_h = (r_h^* A p_h u_h - r_h^* A p_h v_h \ , \ u_h - v_h)_h = X$$

$$X = (A\, p_h\, u_h - A\, p_h\, v_h \ , \ p_h\, u_h - p_h\, v_h)$$

$$X \geq (\varphi(\|p_h u_h\|) - \varphi(\|p_h v_h\|))\ (\|p_h u_h\| - \|p_h v_h\|)$$

$$X \geq (\varphi(\|u_h\|_h) - \varphi(\|v_h\|_h))\ (\|u_h\|_h - \|v_h\|_h)$$

3) L'application $t \longrightarrow (A_h(tu_h + v_h), w_h)_h = (Ap_h(tu_h + v_h), p_h w_h)$ est continue d'après (1.3). D'où l'existence et l'unicité de u_h solution de (5.1).

<u>Lemme 5.2.</u>

Si f_h converge discrètement vers f alors

$$\|p_h\, u_h\| = \|u_h\|_h \quad \text{est borné.}$$

<u>Démonstration</u> : $(A_h u_h - A_h(0), u_h)_h \geq (\varphi(\|u_h\|_h) - \varphi(0))\ \|u_h\|_h$

ce qui donne :

$$(f_h - A_h(0), u_h)_h \geq (\varphi(\|u_h\|_h) - \varphi(0)) \|u_h\|_h$$

D'où $$\varphi(\|u_h\|_h) \leq \varphi(0) + \|f_h - A_h(0)\|_h^*$$

La convergence discrète de f_h vers $f \Longrightarrow \|f_h\|_h^*$ borné et

$$\|A_h(0)\|_h^* = \|r_h^* A(0)\|_h^* \leq \|A(0)\|_{V'}$$

On a donc $\varphi(\|u_h\|_h) \leq C^{te}$. Ceci entraîne que $\|u_h\|_h$ est borné car sinon, il existerait une suite extraite u_k telle que $\|u_k\|_k \longrightarrow +\infty$ et on aurait $\varphi(\|u_k\|_k) \longrightarrow +\infty$ ce qui est impossible.

Lemme 5.3.

Si f_h converge discrètement vers f, alors $p_h\, u_h$ converge vers u dans V faible et $Ap_h\, u_h$ converge vers f dans V' faible.

Démonstration : D'après le lemme précédent $\|p_h\, u_h\| \leq C^{te}$. Donc suivant un ultrafiltre $\mathcal{U}$, $p_h\, u_h \longrightarrow \xi$ dans V faible et comme $Ap_h\, u_h$ est borné, $Ap_h\, u_h \longrightarrow \eta$ dans V' faible.

Montrons que $\eta = Au = f$. Soit $v \in V$

$$(A_h\, u_h\, , r_h\, v)_h = (f_h\, , r_h\, v)_h = (Ap_h\, u_h\, , p_h\, r_h\, v)$$

or

$$(Ap_h u_h\, , p_h r_h v) \longrightarrow (\eta, v)$$

(d'après l'hypothèse de consistance).

D'autre part $(f_h, r_h v)_h \longrightarrow (f,v)$. En effet :

$$(f_h, r_h v)_h = (f_h - r_h^* f, r_h v)_h + (r_h^* f, r_h v)_h$$

Mais :

$$|(f_h - r_h^* f, r_h v)_h| \leq \|f_h - r_h^* f\|_h^* \|r_h v\|_h$$

$$|(f_h - r_h^* f, r_h v)_h| \leq \|f_h - r_h^* f\|_h^* \|p_h r_h v\| \longrightarrow 0$$

(d'après la convergence discrète de f_h vers f).

et $\quad (r_h^* f, r_h v)_h = (f, p_h r_h v) \longrightarrow (f,v)$

d'où $\quad (\eta, v) = (f,v) \qquad \forall\, v \in V$

et par conséquent : $\eta = f = Au$.

Par ailleurs : $(Ap_h u_h, p_h u_h) = (f_h, u_h)_h = (f_h - r_h^* f, u_h)_h + (r_h^* f, u_h)_h$

et

$$(f_h - r_h^* f, u_h)_h \longrightarrow 0$$

$$(r_h^* f, u_h)_h = (f, p_h u_h) \longrightarrow (f, \xi)$$

D'où

$$\limsup_{h} (Ap_h u_h, p_h u_h) \leq (f, \xi)$$

Rappelons cf. [2] que si un filtre x_i converge vers x dans V faible, $Ax_i \longrightarrow y$ dans V' faible et

$$\limsup (Ax_i, x_i) \leq (y,x)$$

Alors Ax = y.

Il en résulte que $A\xi = f = Au$ et donc d'après l'unicité de la solution $u = \xi$. D'où le lemme.

<u>Démonstration du théorème 5.1</u> : On a

$$(Ap_h u_h - Au, p_h u_h - u) = (f_h, u_h)_h - (Ap_h u_h, u) - (f, p_h u_h - u)$$

Mais $(f_h, u_h)_h \longrightarrow (f,u)$ puisque

$$(f_h,u_h)_h = (f_h - r_h^* f, u_h)_h + (f, p_h u_h)$$

on en déduit que

$$\lim_{h\to o} (Ap_h u_h - Au, p_h u_h - u) = 0$$

D'où

$$\lim_{h\to o} (\varphi(\|p_h u_h\|) - \varphi(\|u\|))(\|p_h u_h\| - \|u\|) = 0$$

et par conséquent $\|p_h u_h\| \longrightarrow \|u\|$.

On en déduit que $p_h u_h \longrightarrow u$ dans V fort.

VI - Application aux familles d'approximation $\{W_i^{1,p}, V_h, p_h^i, r_h\}$

Soit $W_i^{1,p} = \{\varphi \mid \varphi \in L^p(\Omega)\ ,\ D_i\varphi \in L^p(\Omega)\}$. Ω étant un ouvert borné de $\mathbb{R}^n$ et $D_i\varphi = \dfrac{\partial\varphi}{\partial x_i}$ prise au sens des distributions.

Choix de Ω_h

Nous désignons par R_h le réseau des points M de la forme

$M = (m_1h_1,\dots,m_nh_n)$, $m_i \in \mathbb{Z}$, $i = 1,n$, $h = (h_1,\dots,h_n)$, $h \in \mathbb{R}_+^n$, $h \neq 0$.

Nous posons :

$$\omega_{h,q}^M = \prod_{i=1}^n \left[(m_i - \frac{q_i+1}{2})\, h_i\ ,\ (m_i + \frac{q_i+1}{2})\, h_i\right]$$

$q = (q_i)_{1\leq i\leq n}$, q_i entier positif $|q| = \sum_{i=1}^n q_i$

$$\rho_h^M = \bigcup_{|q|\leq 1} \omega_{h,q}^M$$

$$\Omega_h' = \{M \mid M \in R_h\ ,\ \rho_h^M \subset \Omega\}$$

Enfin :

$$\Omega_h = \bigcup_{M\in\Omega_h'} \omega_{h,o}^M$$

Choix de V_h

V_h = espace des suites de nombres réels

$u_h = (u_h^M)_{M \in \Omega'_h}$.

Soit $x \in \Omega_h$. On pose $u_h(x) = u_h^M$ si $x \in \omega_{h,o}^M$. V_h est un espace de dimension finie, que nous mettons en dualité avec V'_h par le produit scalaire :

$$(u_h, v_h)_h = h_1 \ldots h_n \sum_{M \in \Omega'_h} u_h^M v_h^M$$

Posons

$$\nabla_i u_h(x) = \frac{u_h(x + \frac{h_i}{2}) - u_h(x - \frac{h_i}{2})}{h_i}$$

$\nabla_i u_h$ étant défini , $\forall u_h \in V_h \quad x \in \Omega_h$

$\nabla_i^2 u_h = \nabla_i (\nabla_i u_h)$

Choix de p_h^i

Notons

$\theta_{h_i}^{m_i}$ = fonction caractéristique de $[(m_i - \frac{1}{2})h_i \, , \, (m_i + \frac{1}{2})h_i]$

$\chi_{h_i} = \frac{1}{h_i} \theta_{h_i}^o$

θ_h^M = fonction caractéristique de $\omega_{h,o}^M = \overset{n}{\underset{i=1}{\otimes}} \theta_{h_i}^{m_i}$

$\chi_h = \frac{1}{h_1 \ldots h_n} \theta_h^o = \overset{n}{\underset{i=1}{\otimes}} \chi_{h_i}$

Alors $p_h^i u_h = \sum_{M \in \Omega_h} u_h^M \chi_{h_i} * \theta_{h_i}^M$

(($*$) étant l'opérateur de convolution), est un prolongement de V_h dans $W_i^{1,p}$

On a identifié u_h à la fonction $u_h = \sum_{M \in \Omega_h} u_h^M \theta_h^M$. V_h est muni de la norme

$$\|u_h\|_h = (\sum_{i=1}^n \|\partial_i p_h^i u_h\|^p_{W_i^{1,p}})^{1/p} = (\sum_{i=1}^n \|\nabla_i u_h\|^p_{L_p})^{1/p}$$

Choix de r_h

Soit une suite de fonctions $\gamma_h \in W^{1,\infty}$ à support dans $\Gamma_h = \sum_{M \in \Omega_h'} \Theta_h^M$ tendant vers 1 dans $W^{1,\infty}$. Posons alors

$$r_h\, \varphi(M) = \langle \gamma_h \varphi , \chi_h \rangle \qquad \forall\ M \in \Omega_h' \quad , \quad \varphi \in W_i^{1,p}$$

On sait (cf. [1]) que les approximations $\{W_i^{1,p} , V_h(\Omega_h) , p_h^i , r_h\}$ sont consistantes.

Application 6.1.

Soit $A : W_o^{1,p}(\Omega) \longrightarrow W^{-1,q}(\Omega)$ défini par

$$Au = - \sum_{i=1}^{u} D_i \left(|D_i u|^{p-2} D_i u\right) + su \qquad s \geq 0 \quad p \geq 2$$

Ω étant un ouvert borné de $\mathbb{R}^n$ de frontière Γ "très régulière . Au problème de Dirichlet :

$$(6.1) \qquad Au = f \qquad u|_\Gamma = 0 \quad \text{pour } f \in L^q(\Omega)$$

nous associons la formulation variationnelle

$$(6.2) \qquad a(u,v) = (f,v) \qquad \forall\ v \in \dot{W}_o^{1,p}(\Omega)$$

avec

$$a(u,v) = \sum_{i=1}^{n} \int_\Omega |D_i u|^{p-2} D_i u\, D_i v\, dx + s \int_\Omega uv\, dx$$

comme

$$W_o^{1,p}(\Omega) \subset \bigcap_{|i| \leq 1} W_i^{1,p}(\Omega)$$

la famille $\{W_i^{1,p} , V_h , p_h^i , r_h\}$ précédemment considérée permet de discrétiser (6.2). Nous avons alors :

$$(6.3) \qquad a_h(u_h,v_h) = (Ap_h\, u_h , p_h\, v_h) = (f, p_h^o\, v_h)$$
$$\forall\ v_h \in V_h \text{ pour } f \text{ donné dans } L^q(\Omega)$$

avec

$$(Ap_h u_h, p_h v_h) = \sum_{i=1}^{n} \int_\Omega |D_i p_h^i u_h|^{p-2} D_i p_h^i u_h \, D_i p_h^i v_h \, dx + s \int_\Omega p_h^o u_h \, p_h^o v_h \, dx$$

$$(Ap_h u_h, p_h v_h) = \sum_{i=1}^{n} \int_\Omega |\nabla_i u_h|^{p-2} \nabla_i u_h \, \nabla_i v_h \, dx + s \int_\Omega u_h \, v_h \, dx$$

D'après la définition de A_h, nous avons :

$$A_h u_h = - \sum_{i=1}^{n} \nabla_i \, (|\nabla_i u_h|^{p-2} \nabla_i u_h) + s \, u_h$$

En posant $f_h = \frac{1}{h_1 \ldots h_n} \int_\Omega f \, \theta_h^M \, dx$

on se ramène à une équation discrétisée de la forme :

(6.4) $$A_h \, u_h = f_h$$

et f_h converge discrètement vers f.

D'après le théorème 5.1, il existe $u_h \in V_h$ unique vérifiant (6.4) et l'on a, $p_h^i u_h \longrightarrow u$ dans $W_i^{1,p}(\Omega)$ fort ; u étant la solution du problème (6.1).

Application 6.2.

On cherche $u \in W_o^{1,p}(\Omega)$ vérifiant

(6.5) $$Au - s\Delta u = f \qquad s \geq 0 \qquad f \text{ donné dans } L^q(\Omega)$$

avec

$$Au = - \sum_{i=1}^{n} D_i \, (|D_i u|^{p-2} D_i u)$$

On vérifie aisément que le théorème 1.1. s'applique et l'on a donc l'existence et l'unicité de la solution u de (6.5).

Comme dans l'exemple précédent (6.5) discrétisée se met sous la forme

(6.6) $$- \sum_{i=1}^{n} \nabla_i \, (|\nabla_i u_h|^{p-2} \nabla_i u_h) - s \sum_{i=1}^{n} \nabla_i^2 \, u_h = f_h \quad , \quad f_h \in L^q(\Omega)$$

Si f_h converge discrètement vers f on peut appliquer le théorème 5.1, qui assure l'existence et l'unicité de la solution u_h de (6.6) et de plus nous avons :

$$p_h^i u_h \longrightarrow u \quad \text{dans} \quad W_i^{1,p}(\Omega) \text{ fort.}$$

VII - Résolution du problème discrétisé par une méthode itérative.

On se propose dans ce paragraphe de donner une méthode itérative permettant de résoudre dans V_h (de dimension finie) l'équation

$$A_h u_h = f_h \quad ; \quad f_h \text{ donné dans } V_h'$$

Plus généralement soit $\mathcal{H}$ un espace de Hilbert sur $\mathbb{R}$ muni du produit scalaire (,) et de la norme $\|\ \|$. On se propose de résoudre dans $\mathcal{H}$ l'équation $Au = f$; pour f donné dans $\mathcal{H}'$, où A est un opérateur non-linéaire de $\mathcal{H}$ dans $\mathcal{H}'$. On se donne une application $S \in \mathcal{L}(\mathcal{H},\mathcal{H}')$ verifiant :

(7.1) $\qquad (Su,v) = (u,Sv) \qquad \forall u,v \in \mathcal{H}$

(7.2) $\qquad$ Il existe une constante $\sigma > 0$ telle que $(Su,u) \geq \sigma \|u\|^2$

On pose $[u]^2 = (Su,u)$. $[u]$ définit sur $\mathcal{H}$ une norme équivalente à $\|\ \|$.

On suppose que l'opérateur A vérifie :

(7.3) $\qquad (Au-Av,u-v) \geq k\,[u-v]^2 \qquad$ avec $k > 0$, $\forall u,v \in \mathcal{H}$

(7.4) $\qquad$ Quelque soit $N > 0$, $\exists$ une constante $C(N)$ telle que

$\forall u,v \in \mathcal{H}$, avec $[u] \leq N$, $[v] \leq N$ alors

$$(Au-Av, w) \leq C(N)\,[u-v]\,[w] \qquad \forall w \in \mathcal{H}$$

Nous avons avons le

Théorème 7.1

Avec les hypothèses (7.1), (7.2), (7.3), (7.4) l'équation $Au = f$ admet une solution unique $u \in \mathcal{H}$ et la suite u_n définie par l'itération

$$Su_{n+1} = Su_n - \rho(Au_n - f)$$

converge fortement vers u pour $\rho > 0$ et $u_o \in \mathcal{H}$ convenablement choisis.

Démonstration : L'existence et l'unicité résultent du théorème 1.1. D'autre part, l'itération a un sens puisque S est bijective.

$$Su_{n+1} = Su_n - \rho(Au_n - f)$$

$$Su = Su - \rho(Au - f)$$

En retranchant membre à membre :

$$Su_{n+1} - Su = Su_n - Su - \rho(Au_n - Au)$$

Posons $\varepsilon_n = u_n - u$ il vient :

$$S\varepsilon_{n+1} = S\varepsilon_n - \rho(Au_n - Au)$$

$$\varepsilon_{n+1} = \varepsilon_n - \rho S^{-1}(Au_n - Au)$$

$$[\varepsilon_{n+1}]^2 = (S\varepsilon_{n+1}, \varepsilon_{n+1}) = (S\varepsilon_n - \rho(Au_n - Au), \varepsilon_n - \rho S^{-1}(Au_n - Au))$$

$$[\varepsilon_{n+1}]^2 = [\varepsilon_n]^2 - 2\rho(Au_n - Au, \varepsilon_n) + \rho^2 (Au_n - Au, S^{-1}(Au_n - Au))$$

En utilisant (7.3) :

$$[\varepsilon_{n+1}]^2 \leq [\varepsilon_n]^2 - 2\rho k[\varepsilon_n] + \rho^2 (Au_n - Au, S^{-1}(Au_n - Au))$$

Soit N_o tel que $[u] \leq N_o/2$; on fait l'hypothèse de récurrence $[u_n - u] = [\varepsilon_n] \leq N_o/2$. Ceci entraîne que $[u_n] \leq N_o$. Or

$$(Au_n - Au, S^{-1}(Au_n - Au)) = [S^{-1}(Au_n - Au)]^2$$

et d'après (7.4) et l'hypothèse de récurrence on a pour $C(N_o) = C$

$$(Au_n - Au, S^{-1}(Au_n - Au)) \leq C\ [\varepsilon_n]\ [S^{-1}(Au_n - Au)]$$

D'où

$$[S^{-1}(Au_n - Au)] \leq C\ [\varepsilon_n]$$

ce qui entraîne

$$[\varepsilon_{n+1}]^2 \leq (1 - 2\rho k + \rho^2 C^2)\ [\varepsilon_n]^2 = \Theta[\varepsilon_n]^2$$

On choisit ρ tel que $\Theta < 1$ et l'hypothèse de récurrence est vérifiée. Par conséquent $\varepsilon_n \longrightarrow 0$. Ce qui achève la démonstration.

Remarque 7.1.

En particulier la suite u_n converge vers u pour $u_o = 0$, $N_o = \frac{2\|f-A(0)\|}{k\sqrt{\sigma}}$, $C = C(N_o)$ et $\rho_{opt} = k/C^2$ ce qui donne $\Theta = 1 - \frac{k^2}{C^2} < 1$.

VIII - Applications à la résolution numérique de certaines équations aux dérivées partielles.

Dans ce qui suit nous nous proposons d'appliquer à quelques exemples les méthodes précédentes.

Exemple 8.1.

On cherche $u \in W_o^{1,p}(\Omega)$ vérifiant l'équation

(8.1) $\qquad Au = f$

avec

$$Au = -\sum_{i=1}^{n} D_i\ (|D_i u|^{p-2}\ D_i u) + su \qquad s > 0$$

pour f donné dans $L^q(\Omega)$

Considérons le problème discrétisé associé à (8.1) : On cherche $u_h \in V_h$ tel que

(8.2) $$A_h u_h = f_h$$

avec

$$A_h u_h = - \sum_{i=1}^{n} \nabla_i (|\nabla_i u_h|^{p-2} \nabla_i u_h) + s u_h \qquad s > 0$$

$$f_h = \frac{1}{h_1 \dots h_n} \int_\Omega f \, \theta_h^M \, dx$$

Théorème 8.1

i) l'équation (8.1) admet une solution unique $u \in W_o^{1,p}(\Omega)$

ii) Le problème (8.2) admet une solution unique $u_h \in V_h$. De plus nous avons

$$p_h u_h \longrightarrow u \text{ dans } W_o^{1,p}(\Omega) \text{ fort}$$

iii) L'équation (8.2) peut se résoudre par l'itération suivante :

$$u_h^{n+1} = u_h^n - \rho(A_h u_h^n - f_n)$$

avec ρ convenablement choisi et $u_h^o = 0$.

Démonstration : Les points i) et ii) ont déjà été démontrés aux paragraphes 4 et 6. Pour démontrer iii), appliquons les résultats du théorème 7.1. en posant $\mathcal{H} = V_h$ muni du produit scalaire

$$(u_h, v_h)_h = (u_h, v_h)_{L^2}$$

Posons

$$S u_h = u_h \qquad [u_h] = \|u_h\|_{L^2}$$

Comme : $(A_h u_h - A_h v_h, u_h - v_h)_h \geq s \|u_h - v_h\|^2_{L^2}$

on a avec les notations du théorème 7.1. $k = s$

Calcul de C(N)

Supposons que $[u_h] \leq N_o$, $[v_h] \leq N_o$

on a :

$$[u_h]^2 = h_1 \, h_2 \ldots h_n \sum_{M \in \Omega'_h} |u_h(M)|^2$$

ce qui entraîne

$$[u_h]^2 \geq h_1 \ldots h_n \sup |u_h(M)|^2$$

De même nous avons $\forall x$

$$|\nabla_i \, u_h(x)| \leq \frac{2}{h_i} \sup_x |u_h(x)|$$

D'où

$$(8.3) \qquad |\nabla_i \, u_h(x)| \leq \frac{2}{h_i \sqrt{h_1 \ldots h_n}} [u_h] \leq \frac{2N_o}{h_i \sqrt{h_1 \ldots h_n}} = N'_i$$

Par ailleurs

$$(A_h u_h - A_h v_h, w_h)_h = \sum_{i=1}^{n} (|\nabla_i u_h|^{p-2} \nabla_i u_h - |\nabla_i v_h|^{p-2} \nabla_i v_h, \nabla_i w_h)_h + $$
$$+ s(u_h - v_h, w_h)_h$$

ce qui donne

$$|(A_h u_h - A_h v_h, w_h)_h| \leq (\sum_{i=1}^{n} \| |\nabla_i u_h|^{p-2} \nabla_i u_h - |\nabla_i v_h|^{p-2} \nabla_i v_h \|^2)^{1/2} (\sum_{i=1}^{n} \|\nabla_i w_h\|^2)^{\frac{1}{2}}$$
$$+ s \, \|u_h - v_h\| \, \|w_h\|$$

or pour $|\alpha| \leq N'$, $|\beta| \leq N'$, α et β réels on a :

$$||\alpha|^{p-2}\alpha - |\beta|^{p-2}\beta| \leq (p-1) \, N'^{p-2} \, |\alpha - \beta|$$

En posant $\alpha = \nabla_i u_h(x)$, $\beta = \nabla_i v_h(x)$, il vient compte tenu de (8.3)

$$\sum_{i=1}^{n} \| |\nabla_i u_h|^{p-2} \nabla_i u_h - |\nabla_i v_h|^{p-2} \nabla_i v_h \| \leq (p-1) \sum_{i=1}^{n} N'^{p-2}_i \, \|\nabla_i (u_h - v_h)\|$$

or : $\|\nabla_i u_h\|_{L^2} \leq \frac{2}{h_i} \|u\|_{L^2}$

Donc :

$$|(A_h u_h - A_h v_h, w_h)_h| \leq (p-1) \left(\sum_{i=1}^{n} \frac{4N_i'^{2(p-2)}}{h_i^2} \|u_h - v_h\|^2\right)^{1/2} \left(\sum_{i=1}^{n} \frac{4}{h_i^2} \|w_h\|^2\right)^{1/2}$$

$$+ s \|u_h - v_h\| \|w_h\|$$

D'où

$$|(A_h u_h - A_h v_h)_h| \leq \{(p-1)\left(\sum_{i=1}^{n} \frac{4N_i'^{2(p-2)}}{h_i^2}\right)^{1/2} \left(\sum_{i=1}^{n} \frac{4}{h_i^2}\right)^{1/2} + s\} [u_h - v_h] [w_h]$$

Or

$$N_i' = \frac{2N_o}{h_i \sqrt{h_1 \ldots h_n}}$$

Donc on a avec les notations du théorème 7.2.

$$C(N_o) = (p-1) N_o^{p-2} \frac{2^p}{(h_1 \ldots h_n)^{(p-2)/2}} \left(\sum_{i=1}^{n} \frac{1}{h_i^{2p-2}}\right)^{1/2} \left(\sum_{i=1}^{n} \frac{1}{h_i^2}\right)^{1/2} + s$$

Calcul de N_o

$$N_o = 2[u_h] = 2 \|u_h\|_{L^2}$$

Or $(A_h u_h, u_h)_h = (f_h, u_h)_h \geq \sum_{i=1}^{n} \|\nabla_i u_h\|_{L^p}^p$

ce qui donne

$$\sum_{i=1}^{n} \|\nabla_i u_h\|_{L^p}^p \leq \|f_h\|_{L^q} \|u_h\|_{L^p}$$

or

$$\|u_h\|_{L^p}^p \leq C_1(\Omega_h) \sum_{i=1}^{n} \|\nabla_i u_h\|_{L^p}^p \qquad \textbf{(Poincaré)}$$

D'où

$$\|u_h\|_{L_p}^{p-1} \leq C_1 \|f_h\|_{L_q}$$

Donc $N_o = 2 \|u_h\|_{L^2}$ est une constante indépendante de h. En posant :

$\rho = \dfrac{s}{C^2(N_o)}$ l'itération $u_h^{n+1} = u_h^n - \rho(A_h u_h^n - f_h)$ est convergente d'après le théorème 7.1.

Remarque 8.1.

Si $f \in L^2(\Omega)$ on pourrait prendre aussi $N_o = \dfrac{2\,\|f_h\|_{L^2}}{s}$

Remarque 8.2.

Posons $\Theta = 1 - \dfrac{s^2}{C^2}$, $u_h^o = 0$, nous avons alors la formule de majoration de l'erreur suivante :

$$[u_h^n - u_h]^2 = \|u_h^n - u_h\|^2_{L^2} \leq \Theta^{n+1}\,\|u_h\|^2_{L^2}$$

Le nombre d'itérations nécessaires pour avoir $[u_h^n - u_h] < \varepsilon$ peut être estimé à l'aide de la formule

$$n \sim \frac{2\,\mathrm{Log}\,\|u\| - 2\,\mathrm{Log}\,\varepsilon}{\mathrm{Log}\,\Theta}$$

Exemple 8.2.

On cherche $u \in W_o^{1,p}(\Omega)$ vérifiant l'équation

(8.6) $\quad Au = f \quad$ pour f donnée dans $L_p(\Omega)$

avec

$$Au = -\sum_{i=1}^{n} D_i\,(|D_i u|^{p-2}\,D_i u) - s\Delta u \qquad s > 0$$

Considérons le problème discrétisé associé : on cherche $u_h \in V_h$ vérifiant :

(8.7) $\quad A_h\,u_h = f_h$

avec

$$A_h u_h = -\sum_{i=1}^{n} \nabla_i\,(|\nabla_i u_h|^{p-2}\,\nabla_i u_h) - s\sum_{i=1}^{n} \nabla_i^2\,u_h \qquad s > 0$$

$$f_h = \frac{1}{h_1 \dots h_n} \int_\Omega f\,\Theta_h^M\,dx$$

Théorème 8.2.

i) L'équation (8.6) admet une solution unique $u \in W_o^{1,p}(\Omega)$

ii) Le problème (8.7) admet une solution unique $u_h \in V_h$. De plus on a

$$p_h u_h \xrightarrow[h\to o]{} u \text{ dans } W_o^{1,p}(\Omega) \text{ fort}$$

iii) L'équation discrétisé (8.7) peut se résoudre à l'aide de la formule itérative suivante :

$$S_h u_h^{n+1} = S_h u_h^n - \rho(A_h u_h^n - f_h)$$

avec

$$S_h u_h = -\sum_{i=1}^{n} \nabla_i^2 u_h \quad , \ \rho \text{ convenablement choisi et } u_h^o = 0$$

Démonstration : Les points i) et ii) ont été démontré dans les paragraphes 4 et 6. Pour iii), nous appliquons les résultats du théorème 7.1. en posant : $\mathcal{H} = V_h(\Omega)$ muni du produit sclaire $(u_h,v_h)_h = (u_h,v_h)_{L^2}$

$$S_h u_h = -\sum_{i=1}^{n} \nabla_i^2 u_h$$

$$[u_h]^2 = \sum_{i=1}^{n} \|\nabla_i u_h\|_{L^2}^2$$

Comme

$$(A_h u_h - A_h v_h , u_h - v_h)_h \geq s \sum_{i=1}^{n} \|\nabla_i (u_h - v_h)\|_{L^2}^2$$

il en résulte que $k = s$ avec les notations du théorème 7.1.

Calcul de $C(N_o)$

Supposons que $[u_h] \leq N$; $[v_h] \leq N$, on a

$$[u]_h^2 = \sum_{i=1}^{n} \|\nabla_i u_h\|_{L^2}^2 = \sum_{i=1}^{n} h_1 \ldots h_n \sum_{M \in \Omega_h'} |\nabla_i u_h(M)|^2$$

Donc

$$|\nabla_i u_h(x)| \leq \frac{[u_h]}{\sqrt{h_1 \ldots h_n}} = N' \qquad \forall x \text{ et } \forall i$$

Il en résulte comme dans l'exemple précédent :

$$|(A_h u_h - A_h v_h, w_h)| \leq ((p-1)\, N'^{p-2} + s)\, [u_h - v_h]\, [w_h]$$

Donc

$$C(N) = (p-1)\, N'^{p-2} + s$$

D'où

$$(8.8) \qquad C(N_o) = (p-1) \frac{N_o^{p-2}}{(h_1 \ldots h_n)^{(p-2)/2}} + s$$

Calcul de N_o

On a

$$(A_h u_h, u_h) \geq \sum_{i=1}^{n} \|\nabla_i u_h\|^p_{L_p}$$

$$\sum_{i=1}^{n} \|\nabla_i u_h\|^p_{L_p} \leq \|f_h\|_{L_q} \|u_h\|_{L_p} \leq C_1(\Omega) \|f_h\|_{L_q} \left(\sum_{i=1}^{n} \|\nabla_i u_h\|^p_{L_p}\right)^{1/p}$$

$$\Longrightarrow \left(\sum_{i=1}^{n} \|\nabla_i u_h\|^p_{L_p}\right)^{1/p} \leq C_1^{q/p} \|f_h\|^{q/p}_{L_q}$$

or $L_p \subset L_2$ avec injection continue pour $p \geq 2$. Donc

$$\left(\sum_{i=1}^{n} \|\nabla_i u_h\|^2_{L^2}\right)^{1/2} = [u_h] \leq C^{te}$$

Donc $N_o = C^{te}$ indépendante de h. En posant $\rho = \frac{s}{C^2(N)}$ l'itération

$$S_h u_h^{n+1} = S_h u_h^n - \rho(A_h u_h^n - f_h)$$

est convergente. D'où le théorème.

Remarque 8.3.

1) si $f \in L^2(\Omega)$ on pourrait prendre aussi

$$(8.9) \qquad N_o = \frac{2\, \|f_h\|_{L_2}}{s}$$

2) Si on pose $u_h^o = 0$, $\theta = 1 - \frac{s^2}{C^2(N)}$

nous avons alors la formule de majoration de l'erreur suivante :

$$C_1(\Omega)\ \|u_h^n-u_h\|^2_{L^2} \le \sum_{i=1}^{n} \|\nabla_i u_h^n-\nabla_i u_h\|^2_{L^2} \le \Theta^{n+1}\ \|u_h\|^2_{L^2}$$

où $C_1(\Omega)$ est une constante ne dépendant que de l'ouvert Ω.

3) Si $n \sim \dfrac{2\ \mathrm{Log}\ \|u\| - 2\ \mathrm{Log}\ \varepsilon}{\mathrm{Log}\ \Theta}$ alors on a :

$$\|u_h^n-u_h\|^2_{L^2} \le \frac{\varepsilon}{C_1(\Omega)} \quad \text{et} \quad \sum_{i=1}^{n} \|\nabla_i u_h^n-\nabla_i u_h\|^2_{L^2} < \varepsilon$$

Remarque 8.4.

On suppose que $f \in L^2(\Omega)$. Si l'ouvert $\Omega = (0,1)\times(0,1)$ et $h_1 = h_2 = h$ alors on a :

$$(8.10) \qquad C(N) = (p-1)\ \frac{N^{p-2}}{h^{p-2}} + s \quad \text{avec} \quad N = \frac{\sqrt{2}}{s}\ \|f_h\|_{L_2}$$

En effet : $\|u_h\|_{L^2} \le \|\nabla_i u_h\|_{L^2}$ pour $i = 1,2$

$$\Longrightarrow 2\ \|u_h\|^2_{L^2} \le \|\nabla_1 u_h\|^2_{L^2} + \|\nabla_2 u_h\|^2_{L^2} = [u_h]^2$$

D'où $\sqrt{2}\ \|u_h\| \le [u]$ comme $A(0) = 0$, on a $N = \frac{\sqrt{2}}{s}\|f_h\|_{L_2}$ D'où (8.10).

Cas particulier

Si $\Omega = (0,1)$, nous avons alors

$$(8.11) \qquad C(N) = (p-1)\ \frac{N^{p-2}}{h^{p-2}} + s \quad \text{avec} \quad N = \frac{2\ \|f_h\|_{L_2}}{s}$$

Remarque 8.5

Pour ceux qui s'interessent aux resultats numériques cf l'article de Brezis_Sibony à paraître dans Arch.Rat. Math. and Mech.

BIBLIOGRAPHIE

[1] AUBIN Approximation des espaces de distributions et des opérateurs différentiels (Thèse 1966)

[2] H. BREZIS Opérateurs monctones C.R. Ac. Sc. de Paris (Avril 1967)

[3] F. BROWDER Existence ans uniqueness theorems for solutions of non linear boundary value problems.
Proceding of symposia in apllied mathematics, vol. 17, p. 24-49, Amer. Math. Soc. 1965.

[4] J. LERAY et J.L. LIONS Quelques résultats de Visik sur les problemes elliptiques non linéaires par les méthodes de Minty-Browder.
Bull. Soc. Math. France 93, 1965, p. 97-107.

[6] R.S. VARGA Matrix iterative analysis, Prentice Hall.

[5] PETRYSHYN 1) On the extension and the solution of non linear equations - Illinois Journal of Mathematics vol. 10 N° 2, June 1966

2) Cours C.I.M.E. 1967

CENTRO INTERNAZIONALE MATEMATICO ESTIVO

(C.I.M.E.)

V. THOMEE

"SOME TOPICS IN STABILITY THEORY FOR PARTIAL DIFFERENCE OPERATORS"

Corso tenuto ad Ispra dal 3-11 Luglio 1967

"SOME TOPICS IN STABILITY THEORY FOR PARTIAL DIFFERENCE OPERATORS"

by

Vidar Thomée

(University of Göteborg)

1. Stability in L_p

Let W_p denote $L_p(R^d)$ for $1 \leq p < \infty$ and let $W_\infty = \mathcal{C}$ be the set of bounded, uniformly continuous complex-valued functions on R^d. Then W_p; $1 \leq p \leq \infty$, is a Banach-space with norm $(x = (x_1, \ldots, x_d))$,

$$(1.1) \qquad \|u\|_{W_p} = \begin{cases} \left(\int_{R^d} |u(x)|^p dx\right)^{1/p}, & 1 \leq p < \infty, \\ \sup_{R^d} |u(x)|, & p = \infty. \end{cases}$$

For later use we shall also introduce the Sobolev space W_p^m of distributions such that $D^\alpha u = i^{-|\alpha|} \partial^{|\alpha|} / \partial x_1^{\alpha_1} \ldots \partial x_d^{\alpha_d} \in W_p$ for $|\alpha| = \sum_j \alpha_j \leq m$ $(\alpha = (\alpha_1, \ldots, \alpha_d))$. This is also a Banach space with norm

$$\|u\|_{W_p^m} = \sum_{\alpha \leq m} \|D u\|_{W_p}.$$

Let $\mathcal{C}^m = W_\infty^m$, $W_p^\infty = \bigcap_{m>0} W_p^m$, $\mathcal{C}^\infty = W_\infty^\infty$. By Sobolev's imbedding theorem, W_p^∞ consists of all infinitely differentiable functions with $D^\alpha u \in W_p$ for all α. For $1 \leq p \leq \infty$, W_p^∞ is dense in W_p. When u is a N-vector $u = (u_1, \ldots, u_N)$, we use again the above definitions where in (1.1) $|u| = (\sum_j |u_j|^2)^{1/2}$.

Consider the initial-value problem

$$(1.2) \qquad \frac{\partial u}{\partial t} = P(x, D)u \equiv \sum_{|\alpha| \leq M} P_\alpha(x) D^\alpha u, \quad t \geq 0,$$

$$(1.3) \qquad u(x, 0) = v(x),$$

where $x \in R^d$, $u = u(x,t)$, and $v = v(x)$ are complex N-vectors and $P_\alpha(x)$ are NxN matrices with elements which we assume for simplicity to be in $\mathcal{C}^\infty$.

The initial-value problem (1.2), (1.3) is said to be correctly posed in W_p if $P = P(x,D)$ (considered as a densely defined closed operator in W_p) is the infinitesimal generator of a C_0 semi-group of operators $E(t)$ on W_p for $t \geq 0$, that is the family of bounded operators $E(t)$, $t > 0$, satisfies

$$E(0) = I = \text{the identity operator},$$

$$E(t_1 + t_2) = E(t_1)E(t_2), \qquad t_1, t_2 \geq 0,$$

$$\|E(t)v\|_{W_p} \leq C_T \|v\|_{W_p}, \qquad 0 \leq t \leq T, \qquad v \in W_p,$$

$$\left\|\left[k^{-1}(E(k) - I) - P\right] v\right\|_{W_p} \to 0, \quad k \to 0, \qquad v \in W_p^\infty.$$

The operator $E(t)$ is referred to as the solution operator connected with the initial-value problem.

The definition of correctness thus depends on p; the case most discussed in the literature is the case of correctness in L_2. In the case of constant coefficients, using Fourier transforms one easily finds that the initial-value problem is correctly posed in L_2 if and only if for any $T \geq 0$,

$$\text{(1.4)} \qquad \sup\left\{\left|\exp(tP(\xi))\right| ; \quad 0 \leq t \leq T ; \quad \xi \in R^d\right\} < \infty .$$

For a NxN matrix A with eigenvalues λ_j, $j = 1, \ldots, N$, we define

$$\Lambda(A) = \max_j \operatorname{Re} \lambda_j .$$

We also set

$$\operatorname{Re} A = \frac{1}{2}(A + A^*) .$$

Clearly Re A is a hermitian matrix. For hermitian NxN matrices,

$A \leq B$ means $(Au, u) \leq (Bu, u)$ for all N-vectors u, where

$(u, v) = \sum_{j=1}^{N} u_j \bar{v}_j$.

A necessary condition for (1.4) to hold is that (1.2), (1.3) is correctly posed in Petrowsky's sense, namely

$$(1.5) \qquad \sup \left\{ \Lambda(P(\xi)) ; \quad \xi \in R^d \right\} < \infty .$$

Necessary and sufficient conditions have been given by Kreiss [14] . It is not our aim here to give a complete account of his work, but we state one such result which we shall need in Section 2 :

Theorem 1.1. The condition (1.4) holds if there are positive constants C_1, C_2 and for each $\xi \in R^d$ a hermitian matrix $H(\xi)$ such that

$$(1.6) \qquad C_1^{-1} I \leq H(\xi) \leq C_1 I$$

and

$$\mathrm{Re}(H(\xi)P(\xi)) \leq C_2 I .$$

On the other hand, if (1.4) folds we have (1.5) and there are positive constants C_1, C_2, C_3 and for each $\xi \in R^d$ a positive definite hermitian matrix H( ) satisfhying (1.6) and

$$\mathrm{Re}(H(\xi) P(\xi)) \leq \left[C_2 + C_3 \Lambda(P(\xi)) \right] I .$$

We shall consider the case of a hermitian (hyperbolic) system

$$(1.7) \qquad \frac{\partial u}{\partial t} = \sum_{j=1}^{d} A_j(x) \frac{\partial u}{\partial x_j} , \qquad A_j = A_j^{*} .$$

In the constant coefficient case we obtain the correctness in L_2 at once from the fact that $\exp(tP(\xi)) = \exp(ti \sum A_j \xi_j)$ is then a unitary matrix. The correctness in L_2 in the case of variable coefficients was proved by Friedrichs [7] . As for the case of W_p, $p \neq 2$, we have the following theorem by Brenner [4] :

Theorem 1.2. Let $1 \leq p \leq \infty$, $p \neq 2$. Assume that the system (1.7) has constant coefficients. Then the corresponding initial-value problem is correctly posed in W_p if and only if

$$A_j A_k = A_k A_j , \qquad j, k = 1, \ldots, d .$$

This condition is the necessary and sufficient condition for the simultaneous diagonalizability of the matrices A_j, $j = 1, \ldots, d$, and so the result means that apart from the case when the system can be brought into the form

$$(1.8) \qquad \frac{\partial v}{\partial t} = \sum_{j=1}^{d} D_j \frac{\partial v}{\partial x_j} ,$$

with real diagonal matrices D_j, the initial -value problem for (1.7) is not correctly posed in W_p, $p \neq 2$.

For the approximate solution of the initial-value problem (1.2), (1.3) we consider operators of the form

$$E_h v(x) = A_{1,h}^{-1} A_{2,h} v(x) ,$$

where h is a small positive number, and $A_{i,h}$ are explicit difference operators

$$A_{i,h} v(x) = \sum_{\beta} a_{i\beta} (x, h) v(x+\beta h) ,$$

where the summation is over a finite set of $\beta = (\beta_1, \ldots, \beta_d)$, β_j integer, and $a_{i\beta}(x, h)$ are $N \times N$ matrices which are polynomials in h with coefficients in $\mathcal{C}^\infty$. If $A_{1,h} = I$, the operator is explicit, in other cases implicit.

Assume that (1.2), (1.3) is correctly posed in W_p. Let k be a small positive number tied to h by the relation $k/h^M = \lambda$ = constant. The idea is then to choose the operator E_h so as to be able to approximate $E(nk)v$ by $E_h^n v$: E_h is said to be consistent with $E(k)$ if there is a dense set of initial-values v in W_p such that $u(x, t) = E(t) v$ W_p^∞ for $t \geq 0$ and

$$(1.9) \qquad u(x,t+k) = E_h u(x,t) + o(k) \qquad \text{in } W_p \text{ when } h \to 0 .$$

The operator E_h is said to approximate $E(k)$ with order of accuracy μ if $o(k)$ in (1.9) can be replaced by $k0(h^{\mu})$. In applications, condition (1.9) turns out to be a purely algebraic condition on the coefficients.

The operator E_h is said to be stable in W_p if for any $T > 0$,

$$(1.10) \qquad \sup\left\{ \|E_h^n\|_{W_p} \; ; \; nk \leq T \right\} < \infty .$$

We have the well-known Lax' equivalence theorem [16] :

<u>Theorem 1.3.</u> Assume that the initial-value problem (1.2), (1.3) is correctly posed in W_p and let E_h be consistent with $E(k)$. Then the stability condition (1.10) is equivalent to convergence ; for any $v \in W_p$, $t \geq 0$, and nay pair of sequences $\{h_j\}_1^{\infty}$, $\{n_j\}_1^{\infty}$ with $h_j \to 0$, $n_j k_j \to t$ when $j \to \infty$, one has

$$(1.11) \qquad \left\| E_{h_j}^{n_j} v - E(t)v \right\|_{W_p} \to 0, \qquad j \to \infty .$$

It should be noticed that stability is necessary for convergence only if one demands (1.11) for all $v \in W_p$; for individual $v \in W_p$ one can have convergence even without stability. Generally this depends on the regularity of v; for an interesting example with analytic initial data and highly unstable difference operator, see Dahlquist [5] . Although in principle one may thus have convergence without stability, in practices round-off errors then cause problems. We will return to these questions in Section 3 .

Again the case most discussed in the literature is the case of stability in L_2. For constant coefficients, using Fourier transforms one finds that E_h is stable in L_2 if and only if for any $T > 0$,

$$(1.12) \qquad \sup\left\{ \left| E_h(\xi)^n \right. \; ; \; 0 \leq nk \leq T , \quad \xi \in R^d \right\} < \infty ,$$

where $E_h(\xi)$ is the symbol (or amplification matrix) of E_h,

(1.13) $$E_n(\xi) = \Big(\sum_\beta a_{1\beta}(h)\exp(i<\beta, h\xi>)\Big)^{-1}\Big(\sum_\beta a_{2\beta}(h)\exp(i<\beta, h\xi>)\Big).$$

A necessary condition for stability is the von Neumann condition, ($\rho(A)$ is the spectral radius of A),

(1.14) $$\rho(E_h(\ )) \le 1 + C_k, \quad \xi \in R^d, \quad h \le h_0 .$$

Setting

$$|u|_H = (Hu, u)^{1/2}, \qquad |A|_H = \sup |Au|_H / |u|_H ,$$

one has the following analogue of Theorem 1.1 for difference operators:

Theorem 1.4. The condition (1.12) holds if there are positive constants h_0, C_1, C_2, and for each $\xi \in R^d$ and $h \le h_0$ a hermitian matrix $H_h(\ )$ such that

(1.15) $$C_1^{-1}\, I \le H_h(\xi) \le C_1\, I$$

and

$$|E_h(\xi)|_{H_h(\xi)} \le 1 + C_2 k .$$

On the other hand, if (1.12) holds we have (1.14) and if $0 \le \nu < 1$ there are positive constants h_0, C_1, C_2, and for each $\xi \in R^d$ and $h \le h_0$ a hermitian matrix $H_h(\xi)$ satisfying (1.15) and

$$|E_h(\xi)|_{H_h(\xi)} \le \nu\rho(E_h(\xi)) + (1 - \nu) + C_2 k$$

Proof. For $\nu = 0$, see Kreiss [13]. The variant with $\nu > 0$ is due to Widlund [26].

Consider the case of a hermitian system (1.7). For operators consistent with such systems, Kreiss [15] defined the difference operator E_h to be dissipative of order ν (ν even) if (with the natural generalization of (1.13) to variable coefficients),

(1.16) $$\rho(E_h(x,\xi)) \leq 1 - C_1 |h\xi|^{\nu} + C_2 k, \quad \xi \quad R^d, h \quad \xi_j \leq \pi ,$$

and has been able to prove :

Theorem 1.5. Let ν be an even natural number. Assume that E_h is consistent with the hermitian system (1.7), has order of accuracy $\nu - 1$, and is dissipative of order ν. Then it is stable in L_2.

Certain results have been proved also in the case of accuracy of order $\nu - 2$.

Consider now a hermitian system (1.7) with constant coefficients. It follows from the proof of Theorem 1.2. that if $p \neq 2$, except for the case that the system can be brought into the form (1.8), there do not exist difference operators consistent with (1.7) which are stable in W_p. We therefore do not essentially restrict the generality by considering the scalar equation

(1.17) $$\frac{\partial u}{\partial t} = \rho \frac{\partial u}{\partial x} , \qquad \rho \text{ real} .$$

We then want to discuss the stability in W_p, $p \neq 2$, of consistent operators

(1.18) $$E_h v(x) = \sum_j a_j \, v(x + jh) ,$$

with constant a_j. In the case of an implicit operator, the sum may be infinite. Introducing the function

$$a(\xi) = E_h(h^{-1}\xi) = \sum_j a_j e^{ij\xi} ,$$

we have stability in L_2 if and only if $|a(\xi)| \leq 1$ for $\xi \in R^d$. For $p \neq 2$ we need :

Theorem 1.6. Let $1 \leq p \leq \infty$, $p \neq 2$. The operator E_h in (1.18) is stable in W_p if and only if one of the following two conditions is satisfied, namely

a) $a(\xi) = ce^{ij\xi}$, $|c| = 1$;

b) $|a(\xi)| < 1$ except for at most a finite number of points ξ_q, $q = 1,\ldots, Q$, in $|\xi| \leq \pi$ where $a(\xi) = 1$. For $q = 1,\ldots,Q$, there are constants α_q, β_q, ν_q where α_q is real, $\operatorname{Re}\beta_q > 0$, and ν_q is an even natural number such that

$$a(\xi_q + \xi) = a(\xi_q)\exp(i\alpha_q\xi - \beta_q\xi^{\nu_q}(1 + o(1))), \quad \xi \to 0 .$$

<u>Proof.</u> See [22], [23] . Here we only want to sketch the proof in the case $p = \infty$ and

(1.19) $$|a(\xi)| < 1 \quad \text{for} \quad 0 < |\xi| \leq \pi , \qquad a(0) = 1 .$$

When $p = \infty$ we have

$$\| E_h^n \|_{\ell} = \sum_j |a_{nj}| ,$$

where

(1.20) $$a_{nj} = \frac{1}{2\pi}\int_{-\pi}^{\pi} a(\xi)^n e^{-ij\xi}\, d\xi .$$

To prove the sufficiency of the conditions in the theorem, let

(1.21) $$a(\xi) = \exp(i\alpha\xi - \beta\xi^{\nu}(1 + o(1))), \qquad \xi \to 0 ,$$

$$\alpha \text{ real}, \quad \operatorname{Re}\beta > 0, \quad \nu \text{ even} .$$

(Notice that (1.19) and (1.21) are exactly the assumptions of Theorem 1.5 in this particular case.) By (1.20) and (1.21) we get by simple estimates, the second of which after two integrations by parts ,

$$|a_{nj}| \leq C \min\left\{ n^{-1/\nu} , n^{1/\nu}(j + \alpha n)^{-2} \right\} ,$$

and hence we obtain easily

$$\|E_h^n\|_{\ell} = \sum_{|j+\alpha n| \leq n^{1/\nu}} |a_{nj}| + \sum_{|j+\alpha n| > n^{1/\nu}} |a_{nj}| \leq C.$$

To prove the necessity, assume that $a(\xi)$ does not satisfy (1.21). Because of (1.19) we then have

$$a(\xi) = \exp(i\alpha\xi + i\xi^{\mu} q(\xi) - \gamma\xi^{\nu}(1 + o(1))), \quad \xi \to 0, \tag{1.22}$$

α real, $q(\xi)$ polynomial, $q(0) \neq 0$, $1 < \mu < \nu$, even,

$\mathrm{Re}\ \gamma > 0$.

It is then easy to prove

$$\sum_j |a_{nj}|^2 = \frac{1}{2\pi}\int_{-\pi}^{\pi} |a(\xi)|^{2n} d\xi > C_1 n^{-1/\nu},$$

and using a lemma by van der Corput,

$$\max_j |a_{nj}| \leq C_2 n^{-1/\mu},$$

and so

$$\|E_h^n\|_{\mathscr{C}} = \sum_j |a_{nj}| \geq \frac{\sum_j |a_{nj}|^2}{\max_j |a_{nj}|} \geqslant \frac{C_1}{C_2} n^{\frac{1}{\mu} - \frac{1}{\nu}},$$

which contradicts the stability since $\mu < \nu$.

As an application, consider for the solution of the equation (1.17) the Lax-Wendroff [17] operator,

$$E_h v(x) = \frac{1}{2}(\rho^2\lambda^2 + \rho\lambda)v(x+h) + (1 - \rho^2\lambda^2)v(x) +$$
$$+ \frac{1}{2}(\rho^2\lambda^2 - \rho\lambda)v(x-h)$$

with

$$a(\xi) = \rho^2\lambda^2 \cos\xi + i\rho\lambda\sin\xi + 1 - \rho^2\lambda^2.$$

This can be described as the most accurate explicity operator for (1.17) based on the three points $x \pm h$ and x. Here

$$|a(\xi)|^2 = 1 - 4\rho^2\lambda^2(1 - \rho^2\lambda^2)\sin^4\frac{\xi}{2},$$

and so E_h is stable in L_2 if and only if $|\rho|\lambda \leq 1$. On the other hand, if $0 < |\rho|\lambda < 1$ we have $|a(\xi)| < 1$ for $0 < |\xi| \leq \pi$ and

$$a(\xi) = \exp(\rho\lambda i\xi - \frac{1}{6}\rho\lambda(1 - \rho^2\lambda^2)i\xi^3 + 0(\xi^4)), \quad \xi \to 0.$$

It follows from Theorem 1.6 that E_h is unstable in W_p for $p \neq 2$. By the above proof we have

$$\|E_h^n\|_{\mathcal{b}} \geq Cn^{1/12} .$$

Serdjukova [19], [20] and Hedstrom [8], [9] have, by using more refined techniques of estimating the a_{nj} above, been able to give more precise estimates of the growth of $\|E_h^n\|_{\mathcal{b}}$ for the case when E_h is stable in L_2 but unstable in the other W_p. In the particular case of the Lax-Wendroff scheme, the exact result is

$$C_1 n^{1/8} \leq \|E_h^n\|_{\mathcal{b}} \leq C_2 n^{1/8} ;$$

more generally, when $a(\xi)$ has the form (1.22) one has

$$C_1 n^{\frac{1}{2}(1-\frac{\mu}{\nu})} \leq \|E_h^n\|_{\mathcal{b}} \leq C_2 n^{\frac{1}{2}(1-\frac{\mu}{\nu})} . \tag{1.23}$$

The instability present here is of course quite weak; we shall return to its influence on convergence in Section 3.

The proof of the sufficiency part of Theorem 1.6. for $p = \infty$ is due to John [11] and Strang [21]. The proof of the necessity part in [23] for $p < \infty$ uses results about Fourier multipliers on L_p; the sufficiency part is a trivial consequence of the result for $p = \infty$. Theorem 1.6 has also been generalized to variable coefficients in [24]

2. Parabolic difference operators.

Consider as in Section 1 an initial-value problem

(2.1) $$\frac{\partial u}{\partial t} = P(x,D)u \equiv \sum_{|\alpha|\leq M} P_\alpha(x)D^\alpha u,$$

(2.2) $$u(x,0) = v(x) ,$$

with coefficients in $\mathcal{C}^\infty$. With the notation of Section 1, we say that the system (2.1) is parabolic in Petrowsky's sense if

(2.3) $$\Lambda(P(x,\xi)) \leq - C_1 |\xi|^M + C_2 , \quad \xi \in R^d .$$

In this case we have correctness in $\mathcal{C}$; one can actually even estimate the derivatives of the solution :

Theorem 2.1. Let (2.1) be parabolic in Petrowsky's sense. Then the initial value problem (2.1), (2.2) is correctly posed, and for any differential operator Q of order q and any $T > 0$ there is a positive constant C such that

(2.4) $$\| Q\ E(t)v \|_{\mathcal{C}} \leq t^{-q/M} \|v\|_{\mathcal{C}} , \qquad 0 < t \leq T.$$

Proof. See e.g. Friedman [6] .

For difference operators E_h of the form discussed in Section 1, Widlund [26] has defined a concept of parabolicity which can be considered as a discrete counterpart of (2.3) , namely ,

(2.5) $$\rho(E_h(x,\xi)) \leq 1 - C_1 k|\xi|^M + C_2 k, \quad \xi \quad R^d , \quad h|\xi_j| \leq \pi , \ h \leq h_0.$$

(Compare the definition (1.16) of a dissipative operator E_h). It can then be proved :

Theorem 2.2. Assume that E_h is consistent with (2.1) and satisfies (2.5), and that Q_h is a difference operator consistent with a differential operator of order q. Then for any $T > 0$ there is a positive constant C such that

(2.6) $$\| Q_h E_h^n v \|_{\mathcal{C}} \leq C((n+1)k)^{-q/M} \| v \|_{\mathcal{C}}, \quad (n+1)k \leq T .$$

<u>Proof.</u> See [27]. The proof goes back to John [11] and Aronson [2], [3], and depends on estimates of a discrete fundamental solution. Actually the result is valid in much more general situations, and permits multi-step schemes, variations in the coefficients also in t, and low regularity.

It might be considered natural to make inequalities like (2.4) and (2.6) definitions of parabolicity. In order to investigate where such definitions lead we shall restrict ourselves in the rest of this section to the case of constant coefficients in P and E_h, and we shall consider only L_2 so that Fourier-transforms can be conveniently applied. We then say that the system (2.1) with constant $P_\alpha(x) = P_\alpha$ is parabolic if the initial value problem (2.1), (2.2) is correctly posed in L_2 and if for any differential operator Q and any positive τ, T,

(2.7) $$\sup \left\{ \| Q \; E(t) \|_{L_2} \; ; \quad 0 < \tau \leq t \leq T \right\} < \infty .$$

Similarly, we say that the operator E_h with coefficients independent of x is parabolic in L_2 if it is stable in L_2, and if for any difference operator Q_h consistent with a differential operator Q,

(2.8) $$\sup \left\{ \| Q_h \; E_h^n \|_{L_2} \; ; \; 0 < \tau \leq nk \leq T \right\} < \infty .$$

We have the following analogue of Theorem 1.3:

<u>Theorem 2.3.</u> Assume that E_h is consistent with the parabolic equation (2.1) (with constant coefficients). Then E_h is parabolic if and only if for any difference operator Q_h consistent with a differential operator Q

of order q, any $v \in L_2$, any $t \geq 0$ which is > 0 if $q > 0$, and any pair of sequences $\{h_j\}_1^\infty$, $\{n_j\}_1^\infty$, with $h_j \to 0$, $n_j k_j \to t$ when $j \to \infty$, we have

$$\left\| Q_{h_j} E_{h_j}^{n_j} v - Q\, E(t) v \right\| \to 0, \text{ when } j \to \infty .$$

Proof . See [25] .

We now want to characterize algebraically systems (2.1) and operators E_h which are parabolic in the present sense. By Parseval's relation conditions (2.7) and (2.8) are equivalent to

(2.9) $$\sup \left\{ \left| Q(\xi) \exp (t\, P(\xi)) \right| \; ; \; 0 < \tau \leq t \leq T, \xi \in R^d \right\} < \infty ,$$

and

(2.10) $$\sup \left\{ \left| Q_h(\xi) E_h(\xi)^n \right| \; ; \; 0 < \tau \leq nk \leq T , \xi \in R^d \right\} < \infty ,$$

respectively. Here $E_h(\xi)$ is the symbol of the difference operator E_h and similarly for $Q_h(\xi)$.

Consider first the case of a system. We then have :

Theorem 2.4. Assume that the initial-value problem (2.1) , (2.2) is correctly posed in L_2. Then it is parabolic in L_2 if and only if there are positive constants C_1, C_2, such that

(2.11) $$\Lambda (P(\xi)) \leq - C_1 |\xi|^\mu + C_2 , \xi \in R^d .$$

Proof. For details see 25 . We give a short sketch . The sufficiency of the conditions follows easily from the inequality

(2.12) $$\exp (t A) \leq \exp(t \Lambda (A)) \sum_{j=0}^{N-1} (2t \quad A \quad)^j ,$$

which holds for any NxN matrix A. To prove the necessity of condition (2.11) , we notice that by the parabolicity condition (2.9) , we have

$$(1 + |\xi| \quad) \exp (\Lambda P(\xi))) \leq (1 + |\xi|) \quad \exp(P(\xi)) \leq C ,$$

and so with $\widetilde{\Lambda}(r) = \max\limits_{\xi = r} \Lambda\,(P(\xi))$,

$$\widetilde{\Lambda}(r) \leq \log C - \log(1 + r) \to -\infty \text{ when } r \to \infty \; . \tag{2.13}$$

Using the Seidenberg-Tarski elimination theorem one can prove that $\widetilde{\Lambda}(r)$ is algebraic in r for large r and thus by (2.13), there is a (rational) positive number μ and a positive C_1 such that

$$\widetilde{\Lambda}(r) = -2C_1 r^{\mu}\,(1 + o(1)) \quad \text{when} \quad r \to \infty \, .$$

This implies (2.11).

In general it is necessary in the formulation of Theorem 2.4 to explicitly assume the correctness of the initial-value problem. There are, however, some cases when the correctness follows from (2.11). One such case is when $P(\xi)$ is a normal matrix, in particular if $P(\xi)$ is scalar. An example of this is offered by the equation

$$\frac{\partial u}{\partial t} = \frac{\partial^2 u}{\partial x^2} + \frac{\partial^3 u}{\partial x^3} \, ,$$

where

$$\operatorname{Re} P(\xi) = \operatorname{Re}\,\left((i\xi)^2 + (i\xi)^3\right) = -\xi^2 \, .$$

One other case when correctness is automatic is when (2.8) holds with $= M$; this is the case of parabolicity in Petrowsky's sense. An example where (2.11) is satisfied without correctness is given by $(N=2)$

$$\frac{\partial u}{\partial t} = \frac{\partial^2 u}{\partial x^2} + \begin{pmatrix} 0 & 1 \\ 0 & 0 \end{pmatrix} \frac{\partial^4 u}{\partial x^4} \, ,$$

where

$$P(\xi) = \begin{pmatrix} -\xi^2 & \xi^4 \\ 0 & -\xi^2 \end{pmatrix}$$

Systems which satisfy (2.11) are called parabolic in Silov's sense; the present parabolicity concept is thus more restrictive than Silov's.

Using Theorem 1.1 we can give a characterization of parabolic equations which contain at the same time the correctness and the condition (2.11):

Theorem 2.5. The initial-value problem (2.1), (2.2) is parabolic if and only if there are positive constants C_1, C_2, C_3, μ and for each real a hermitian matrix $H(\xi)$ such that

$$C_1^{-1} I \le H(\xi) \le C_1 I$$

and

$$\text{Re}\,(H(\xi)P(\xi)) \le (-C_2|\xi|^{\mu} + C_3) I . \tag{2.14}$$

Our aim is now to similarly characterize parabolic difference operators. We have :

Theorem 2.6. Assume that the operator E_h is consistent with the equation (2.1) and stable in L_2. Then E_h is parabolic in L_2 if and only if there are positive constants C_1, C_2, h_0, ν such that

$$\rho(E_h(\xi)) \le 1 - C_1 k\,|\xi|^{\nu} + C_2 k,\ \xi \in R^d,\ h|\xi|_j \le \pi,\quad h \le h_0 . \tag{2.15}$$

Proof. See [25]. The proof is similar to that of Theorem 2.4 but more complicated. The inequality

$$|A^n| \le C_N \rho^{n-N+1}\left[\rho^{N-1} + (n|A - I|)^{N-1}\right],\ n \ge N,$$

$(\rho = \rho(A))$, which is analogous to (2.12), is used to prove the sufficiency of (2.15) for (2.10). To prove the necessity one uses again the Seidenberg-Tarski theorem, but this time trigonometric polynomials take the place of ordinary polynomials.

Also in this case the stability has to be explicitly assumed; it is easy to give examples where (2.15) is satisfied by an unstable operator E_h. Again, the case of a normal matrix $E_h(\xi)$, in particular a scalar $E_h(\xi)$, and the case $\nu = M$ in (2.15) are exceptions; in this latter case E_h

is parabolic in the sense of Widlund.

Using Theorem 1.4 one can prove the following counterpart of Theorem 2.5 :

Theorem 2.7. The operator E_h is parabolic in L_2 if and only if there are positive constants C_1, C_2, C_3, h_0, ν and for each $h \leq h_o$ and $\xi \in R^d$ a positive definite matrix $H_h(\xi)$ with

$$C_1^{-1} I \leq H_h(\xi) \leq C_1 I$$

and

$$\left| E_h(\xi) \right|_{H_h(\xi)} \leq 1 - C_2 k |\xi|^{\nu} + C_3 k, \qquad h|\xi_j| \leq \pi. \tag{2.16}$$

The largest possible μ and ν in (2.11) (or (2.14)) and (2.15) (or (2.16)), respectively, are referred to as the orders of parabolicity. It can be proved that

i) if (2.1) is parabolic of order μ, then there exists an operator E_h consistent with (2.1) and parabolic of order μ ;

ii) if E_h is consistent with (2.1), and (2.11) and (2.16) hold, then $\nu \leq \mu$;

iii) if (2.1) is parabolic of order μ, then there exist operators E_h consistent with (2.1) which are unstable, others which are stable but not parabolic, and still others which are parabolic, but of order $\nu < \mu$.

For details, see [25] .

3. The rate of convergence.

Consider again an initial-value problem

$$\frac{\partial u}{\partial t} = P(x, D) u, \qquad t \geq 0, \tag{3.1}$$

(3.2) $$u(x, 0) = v(x) \quad .$$

In the sequel we shall demand not only that the initial-value problem be correctly posed in W_p, but that it satisfies the stronger requirement of the following definition. We say that the initial-value problem is strongly correctly posed in W_p if for any $m > 0$, $v \in W_p^m$ implies $E(t)\, v \in W_p^m$ and there is a constant $C_{m,T}$ such that for all $v \in W_p^m$,

$$\| E(t)\, v \|_{W_p^m} \leq C_{m,T} \| v \|_{W_p^m} , \quad 0 \leq t \leq T.$$

In particular, this definition implies that $E(t) \cdot W_p^\infty \subseteq W_p^\infty$.

It can be proved that if $P = P(x, D)$ has constant coefficients, or if it is of first order, then strong correctness in W_p is an automatic consequence of correctness. Further, systems which are parabolic in Petrowsky's sense are strongly correctly posed in W_p for any p with $1 \leq p \leq \infty$.

We shall also introduce a more general parabolicity concept than in Section 2. We shall say that the system (3.1) is strongly parabolic of order b in W_p if the initial-value problem (3.1), (3.2) is correctly posed in W_p, if $v \in W_p$ implies $D^\alpha E(t)v \in W_p$ for all α when $t > 0$, and if

(3.3) $$\| E(t)\, v \|_{W_p^m} \leq C_{m,T}\, t^{-\frac{m-j}{b}} \| v \|_{W_p^j} , \quad 0 < t \leq T ; \quad j \leq m .$$

One can prove that the order b of strong parabolicity is at most equal to the order M of the system. Systems which are parabolic in Petrowsky's sense are strongly parabolic of order M in W_p for any p with $1 \leq p \leq \infty$, and systems which are parabolic of order b in L_2 in the sense of Section 2 are also strongly parabolic in L_2 of order b.

Consider difference operators of the same form as previously, namely

$$E_h v(x) = \sum_\beta a_\beta(x, h) v(x + \beta h) .$$

For the sake of simplicity we shall assume here that E_h is explicit, so that the summation is over a finite number of terms only. We have previously defined consistency of E_h with $E(k)$ to mean that for any sufficiently smooth solution $u(x,t)$ of (3.1),

$$u(x, t+k) = E_h u(x,t) + 0(k) , \qquad h \to 0 \quad ;$$

more precisely, E_h is said to approximate $E(k)$ with order of accuracy μ if for any such u,

(3.4) $$u(x, t + k) = E_h u(x,t) + k0(h^\mu), \qquad h \to 0 .$$

When (3.1), (3.2) is strongly correctly posed in W_p, it is sufficient to assume this condition locally to obtain the following global estimate :

<u>Theorem 3.1.</u> Assume that the initial-value problem (3.1), (3.2) is strongly correctly posed in W_p and that E_h approximates $E(k)$ with order of accuracy . Then there exists a constant C such that for any $v \in W_p^{M+\mu}$,

$$\| (E_h - E(k)) v \|_{W_p} \leq C h^{M+\mu} \| v \|_{W_p^{M+\mu}} .$$

<u>Proof.</u> See [18]. The proof consists in expanding $E(k)v = u(x,k)$ and $F_v v$ in Taylor series around the point $(x, 0)$, using (3.4), and estimating the remainder terms in integral form. In doing so, it is sufficient to consider v in the dense subset W_p^∞ of W_p.

We now easily obtain the following estimate for the rate of convergence :

<u>Theorem 3.2.</u> : Assume that the initial-value problem (3.1), (3.2) is

strongly correctly posed in W_p and that E_h in stable in W_p and approximates $E(k)$ with order of accuracy μ . Then there is a constant $C = C_T$ such that for $v \; W_p^{M+}$, $nk \leq T$,

$$\| (E_n^n - E(nk))v \|_{W_p} \leq Ch^{\mu} \|v\|_{W_p^{M+\mu}} .$$

Proof. We have

$$(3.5) \qquad (E_h^n - E(nk))\, v = \sum_{j=0}^{n-1} E_h^{n-1-j} (E_h - E(k))\; E\,(jk)\, v \;,$$

and so by the stability of E_h , Theorem 3.1, and the strong correctness,

$$\| (E_h^n - E(nk))v \|_{W_p} \leq C \sum_{j=0}^{n-1} kh^{\mu} \| E(jk)v \|_{W_p^{M+\mu}} \leq Cnk \; h^{\mu} \| v \|_{W_p^{M+\mu}} \;,$$

which proves the theorem. In special cases, this theorem appears in many places.

Thus, the situation is that for initial-values in W_p we have (by Lax' equivalence theorem) convergence without any added information on its rate, and if the initial-values are known to be in $W_p^{M+\mu}$ we can conclude that the rate of convergence is $0(h^{\mu}\;)$ when $h \to 0$. It is natural to ask what one can say if the initial data belong to a space "intermediate" to W_p and $W_p^{M+\mu}$. To answer this question we shall introduce some spaces of functions which are interpolation spaces between W_p and W_p^m in the sense of the theory of interpolation of Banach spaces (cf. [18] and references).

Let s be a positive real number and write $s = S + \sigma$, S integer, $0 < \sigma \leq 1$. Set $T_{\tau}\, u(x) = u(x+ \tau\,)$. We then denote by B_p^s the space of $u \in W_p$ such that the following norm is finite, namely

$$\|u\|_{B_p^s} = \|u\|_{W_p} + \sum_{|\alpha|=S} \sup_{\tau\neq 0} |\tau|^{-\sigma} \|T_\tau D^\alpha u - D^\alpha u\|_{W_p} .$$

Thus B_p^s is defined by Lipschitz type condition for the derivatives of order S; these spaces are sometimes called Lipschitz spaces. For the Heavyside function

$$\chi(x) = \begin{cases} 0, & x < 0 \\ 1, & x \geq 0, \end{cases} \qquad (d = 1), \tag{3.6}$$

we have for $1 \leq p < \infty$,

$$\|T_\tau \chi - \chi\|_{W_p} = |\tau|^{1/p} ,$$

and it follows that if $\varphi \in C_0^\infty$, then $\varphi\chi \in B_p^{1/p}$.

One can prove that

$$B_p^{s_1} \subset B_p^{s_2} \quad \text{if} \quad s_1 > s_2 , \tag{3.7}$$

and that for integer s and $\epsilon > 0$ arbitrary,

$$B_p^{s+\epsilon} \subset W_p^s \subset B_p^s .$$

The main property of these spaces that we will need is then the following interpolation property; assume that $1 \leq p \leq \infty$, m is a natural number, and s is a real number with $0 < s < m$. Then there is a constant C such that any bounded linear operator A in W_p with

$$\|Au\|_{W_p} \leq \begin{cases} C_1 \|u\|_{W_p} , \\ C_2 \|u\|_{W_p} , \end{cases}$$

whe have

$$\|Au\|_{W_p} \leq C\, C_1^{1-\theta} C_2^{\theta} \|u\|_{B_p^s} , \quad \theta = \frac{s}{m} . \tag{3.8}$$

Theorem 3.2 and (3.8) with $A = E_h^n - E(nk)$ proves immediately the following result :

Theorem 3.3. Assume that the initial-value problem (3.1), (3.2) is strongly correctly posed in W_p and that E_h is stable in W_p and approximates $E(k)$ with order of accuracy μ. Then for $0 < s < M + \mu$ there is a constant $C = C_{s,T}$ such that for any $v \in B_p^s$, $nk \leq T$,

$$(3.9) \qquad \left\| (E_h^n - E(nk))\, v \right\|_{W_p} \leq C h^{s\nu} \|v\|_{B_p^s}, \qquad \nu = \frac{\mu}{M+\mu} .$$

Notice that $\nu = \mu (M+\mu)^{-1}$ grows with μ and $\lim_{\infty} \nu = 1$. This means that the estimate (3.9) becomes increasingly better for fixed s when μ grows. In other words, if for a given strongly correctly posed initial-value problem one can construct stable difference schemes of arbitrarily high order of accuracy, then given any $s > 0$ one can obtain rates of convergence arbitrarily close to $0(h^s)$ when $h \to 0$ for all initial-values in B_p^s.

As an application, consider an L_2-stable operator E_h with order of accuracy μ for the hyperbolic equation

$$\frac{\partial u}{\partial t} = \rho(x) \frac{\partial u}{\partial x}, \qquad \rho(x) \text{ real},$$

and let $v = \varphi \chi$ where $\varphi \in C_0^\infty$ and χ is the Heavyside function (3.6). By above we have in this case

$$\left\| (E_h^n - E(nk) v \right\|_{L_2} = 0(h^{\frac{1}{2}\mu/(1+\mu)}), \qquad h \to 0 .$$

For dissipative operators E_h, stronger results have been obtained in Apelkrans [1], where also the spreading of discoundinuities is discussed.

It is natural to ask if for a parabolic system the smoothing property of the solution operator can be used to reduce the regularity demends on the initial data in Theorems 3.2 and 3.3. This is indeed the case; the result on the rate of convergence is then the following.

Theorem 3.4. Assume that the system (3.1) is strongly parabolic of order b in W_p and that E_h is stable in W_p and approximates E(k) with order of accuracy μ. Then for any $s > 0$, $T > 0$, there is a constant $C = C_{s,T}$ such that for $v \in B_p^s$, $nk \leq T$,

(3.10)
$$\| E_h^n - E(nk))v\|_{W_p} \leq 0\|v\|_{B_p^s} \cdot \begin{cases} h^{\mu}, & s > M + \mu - b, \\ h^{\mu}\log h^{-1}, & s = M + \mu - b, \\ \dfrac{s}{h^{M+\mu-b}}, & 0 < s < M + \mu - b. \end{cases}$$

Proof. For details, see [18]. Here we will only sketch a proof for the case $v \in B_p^s$ where $M + \mu - b < s < M + \mu$. When $b = M$ the other cases can be treated similarly; if $b < M$ the proof in [18] uses slightly more sofisticated interpolation theory.

We shall use (3.5). For $j = 0$ we have by the stability and Theorem 3.1,

$$\|E_h^{n-1}(E_h - E(k))v\|_{W_p} \leq \begin{cases} C h^{M+\mu}\|v\|_{W_p^{M+\mu}}, \\ C\|v\|_{W_p}, \end{cases}$$

and so by (3.7) and (3.8), since $s > \mu$,

$$\|E_h^{n-1}(E_h - E(k))v\|_{W_p} \leq Ch \|v\|_{B_p^s}.$$

For $j > 0$ we have by Theorem 3.1 and the strong parabolicity,

$$\| E_h^{n-1-j}(E_h - E(k))E(jk)v\|_{W_p} \leq C h^{M+\mu} E(jk)v_{W_p^{M+\mu}} \leq$$

$$\leq \begin{cases} C\, h^{M+\mu} \| v \|_{W_p^{M+\mu'}} \\ C\, h^{M+\mu} (jk)^{-\frac{M+\mu}{b}} \| v \|_{W_p} \end{cases} ,$$

and hence by (3.8) ,

$$\left\| E^{n-1-j}(E_h - E(k))E(jk))E(jkv \right\|_{W_p} \leq C\ h^{\mu} k(jk)^{-\nu} \|v\|_{B_p^s} ,$$

where

$$\nu = \frac{M+\mu}{b} \left(1 - \frac{s}{M+\mu}\right) = \frac{M+\mu-s}{b} < 1 .$$

Adding over j we get

$$\| (E_h^n - E(nk))v \|_{W_p} \leq C\ h^{\mu} \| v \|_{B_p^s} ,$$

which proves (3.10) in the case considered.

For an earlier particular result in the same direction, see Juncosa and Young [12] .

We shall complete this section with a simple case of a recent result by Hedstrom [10] . Consider as in Section 1 the initial-value problem for the equation

$$\frac{\partial u}{\partial t} = \rho \frac{\partial u}{\partial x} , \qquad \rho \text{ real}$$

and a consistent difference operator of the form

$$E_h v(x) = \sum_j a_j v(x + jh) .$$

Set again

$$a(\xi) = E_h(h^{-1}\xi) = \sum_j a_j e^{ij\xi} ,$$

and assume that we have the case that E_h is stable in L_2 but unstable in W_p , $p \neq 2$; in particular assume that

$$|a(\xi)| < 1 \quad \text{for} \quad 0 < |\xi| \leq \pi, \qquad a(0) = 1,$$

$$a(\xi) = \exp(i\alpha\xi + i\xi^{\mu+1} q(\xi) - \gamma\xi^{\nu}(1 + o(1))), \quad \xi \to 0,$$

$$\alpha \text{ real}, \quad q(\xi) \text{ real polynomial}, \quad q(0) \neq 0, \; 1 < \mu + 1 < \nu,$$

$$\nu \text{ even}, \quad \operatorname{Re} \gamma > 0 .$$

(In Section 1 we used μ instead of $\mu + 1$; here μ is as above the order of accuracy.) We then have :

<u>Theorem 3.5.</u> Under the above assumptions, for any $t > 0$ and s with $0 < s \leq \mu + 1$ there is a constant C such that for $v \in B_\infty^s$, $nk = t$,

$$\|(E_h^n - E(t))v\|_{\mathcal{C}} \leq C \|v\|_{B_\infty^s} \begin{cases} h^{s\frac{\mu}{1+\mu}}, & \frac{\mu+1}{2} < s \leq \mu + 1, \\ h^{\frac{\mu}{2}} \log^{-1}, & s = \frac{\mu+1}{2}, \\ h^{\frac{2s(\nu-1)-(\nu-\mu-1)}{2}}, & 0 < s < \frac{\mu+1}{2} . \end{cases}$$

In particular this means that for $v \in B_\infty^s$ where $\frac{\mu+1}{2} < s \leq \mu + 1$, the estimate for the rate of convergence is the same as the one we would have obtained by Theorem 3.3. in the stable case . For $0 < s \leq \frac{\mu+1}{2}$, however, the rate of convergence becomes slower. In the extreme case $s = 0$, we recognize the growth in (1.23) .

REFERENCES

1. Apelkrans , On difference schemes for hyperbolic equations with discountinuous initial values. To appear.

2. D.G. Aronson, The stability of finite difference approximations to second order linear parabolic differential equations. Duke Math. J. 30 (1963), 117-128 .

3. D.G. Aronsons, On the stability of certain finite difference approximations to parabolic systems of differential equatinns. Numer.Math. 5(1963), 118-137.

4. Ph. Brenner, The Cauchy problem for symmetric hyperbolic systems in L_p. Math. Scand. 19 (1966), 27-37.

5. G. Dahlquist, Convergence and stability for a hyperbolic difference equation with analytic initial-values. Math. Scand. 2(1954), 91-102.

6. A. Friedman, Generalized functions and partial differential equations. Prentice-Hall, Englewood Cliffs, New Jersey, 1963.

7. K.O. Friedrichs, Symmetric hyperbolic linear differential equations. Comm. Pure Appl. Math. 7(1954), 345-392.

8. G.W. Hedstrom, The near-stability of the Lax-Wendroff method. Numer. Math. 7(1965), 73-77 .

9. G.W. Hedstrom, Norms of powers of absolutely convergent Fourier series. Michigan Math. J. 13(1966), 393-416.

10. G.W.Hedstrom, The rate of convergence of some difference schemes. To appear.

11. F.John, On integration of parabolic equations by difference methods. Comm.Pure Appl. Math. 5(1962), 155-211.

12. M.L. Juncosa and D.M. Young, On the order of convergence of solutions of a difference equation to a solution of the diffusinn equation. J. Soc. Indust. Appl. Math. 1(1953), 111-135.

13. H.O. Kreiss, Über die Stabilitatsdefinition für Differenzengleichungen die partielle Differentialgleichungen approximieren. Nordisk Tidskr. Information-Behandling 2(1962), 153-181.

14. H.O. Kreiss, Über sachgemässe Cauchyprobleme. Math. Scand. 13 (1963), 109-128.

15. H.O. Kreiss, On difference approximations of dissipative type for hyperbolic differential equations. Comm. Pure Appl. Math. 17(1964), 335-353.

16. P.D. Lax and R.D. Richtmyer, Survey of the stability of linear finite difference equations. Comm. Pure Appl. Math. 9(1956) , 267-293.

17. P.D. Lax and B. Wendroff, Systems of conservation laws. Comm. Pure Appl. Math. 13(1960), 217-237.

18. J. Peetre and V. Thomée, On the rate of convergence for discrete initial-value problems. To appear.

19. S.I. Serdjukova, A study of stability of explicit schemes with constant real coefficients. Z. Vycisl. Mat. i Mat. Fiz. 3(1963), 365-370.

20. S.I. Serdjukova, On the stability in C of linear difference schemes with constant real coefficients. Z. Vycisl. Mat. i Mat. Fiz. 6 (1966); 477-486.

21. W.G. Strang, Polynomial approximation of Bernstein type. Trans Amer. Math. Soc. 105(1962), 525-535.

22. V. Thomée, Stability if difference schemes in the maximum-norm. J. Differential Equations 1(1965), 273-292.

23. V. Thomée, Stability of difference schemes in L^p. XIV Congr. Math. Scand. 1964. To appear.

24. V. Thomée, On maximum-norm stable difference operators, Numerical Solution of partial differential Equations (Proc. Sympos. Univ. Maryland, 1965), pp. 125-151. Academic Press. New York.

25. V. Thomée, Parabolic difference operators. Math. Scand. 19(1966), 77-107.

26. O.B. Widlund, On the stability of parabolic difference schemes, Math. Compt. 19(1965), 1-13.

27. O.B. Widlund, Stability of parabolic difference schemes in the maximum-norm. Numer. Math. 8 (1966), 186-202.

Finito di stampare il 30 dicembre 1968
dall'Editoriale Grafica - Roma